D1458429

The Role of Intellectual Property Rights in Biotechnology Innovation

The Role of Intellectual Property Rights in Biotechnology Innovation

Edited by
David Castle
University of Ottawa, Canada

Edward Elgar
Cheltenham, UK • Northampton, MA, USA

Published by
Edward Elgar Publishing Limited
The Lypiatts
15 Lansdown Road
Cheltenham
Glos GL50 2JA
UK

Edward Elgar Publishing, Inc.
William Pratt House
9 Dewey Court
Northampton
Massachusetts 01060
USA

A catalogue record for this book
is available from the British Library

Library of Congress Control Number: 2009936223

ISBN 978 1 84720 980 1 (cased)

Printed and bound by MPG Books Group, UK

Contents

Figures

Tables

Contributors

Bjørn Asheim
Department of Social and Economic Geography
Lund University
Lund, Sweden

Cecile Ayerbe
Departement Techniques de Commercialisation
Université de Nice-Sophia Antipolis
Nice, France

Richard Y. Boadi
Legal Counsel
African Agricultural Technology Foundation
Nairobi, Kenya

Jasper A. Bovenberg
Amsterdam Medical Centre
Department of Health Law
Amsterdam, The Netherlands

Tania Bubela
School of Business
University of Alberta
Edmonton, Canada

David Castle
Faculty of Arts and Faculty of Law
University of Ottawa
Ottawa, Canada

Timothy Caulfield
Health Law Institute
University of Alberta
Edmonton, Canada

Sachin Chaturvedi
Research and Information System for Developing Countries
New Delhi, India

L. Martin Cloutier
Department of Management and Technology
L'Université de Quebec á Montréal
Montreal, Canada

Abdallah S. Daar
Public Health Science
University of Toronto
Toronto, Canada

Karen L. Durell
Faculty of Law
McGill University
Montreal, Canada

Clinton W. Francis
School of Law
Northwestern University
Evanston, United States

Amy J. Glass
Department of Economics
Texas A&M University
College Station, United States

E. Richard Gold
Faculty of Law
McGill University
Montreal, Canada

J. Adam Holbrook
Centre for Policy Research on Science and Technology
Simon Fraser University
Vancouver, Canada

Marc Ingham
International Management and Marketing
International University of Monaco
Monaco

Ian Inkster
Department of History
Nottingham Trent University
Nottingham, UK

Alan G. Isaac
Department of Economics
American University
Washington, District of Columbia

F. Scott Kieff
Washington University Law
Washington University in St. Louis
St. Louis, United States

Christopher May
Department of Political Economy
Lancaster University
Lancaster, United Kingdom

Emmanuel Métais
People, Markets and Humanities
EDHEC Business School
Nice, France

Liliana Mitkova
Université de Marne-la-Vallée – Paris XII
Meaux, France

Fabricio X. Nunez
Department of Applied Economics
University of Minnesota
Minneapolis, United States

Sharon Oriel
Licensing Executives Society
Australia and New Zealand Inc.
Midland, United States

Walter G. Park
Department of Economics
American University
Washington, District of Columbia

Tina Piper
Faculty of Law
McGill University
Montreal, Canada

Susanne Sirois
Department of Chemistry
L'Université de Quebec á Montréal
Montreal, Canada

Patrick H. Sullivan
Intellectual Capital Management Group, Inc.
Palo Alto, United States

Koichi Sumikura
National Graduate Institute for Policy Studies
Tokyo, Japan

Finn Valentin
Department of Industrial Economics and Strategy
Copenhagen School of Business
Copenhagen, Denmark

Christian Zeller
Institute of Geography
University of Berne
Bern, Switzerland

Acknowledgements

This book arose from a workshop that was hosted by the Intellectual Property Modelling Group which is housed at the Centre for Intellectual Property Policy at the Faculty of Law, McGill University.

The IPMG is supported through by the Social Sciences and Humanities Research Council of Canada.

The book's editor, David Castle, is supported by Genome Canada through the Ontario Genomics Institute.

Introduction

David Castle

WHY INTELLECTUAL PROPERTY RIGHTS IN BIOTECHNOLOGY INNOVATION?

Intellectual property rights feature prominently in all innovation systems, yet characterizing their role is a difficult task, one that always fosters debate. Intellectual property rights (IPRs), particularly in the form of patent rights, are widely viewed as catalysts for innovation in high-value, knowledge-intensive sectors like biotechnology because they reward risk-taking innovative behaviour while providing public access to invention disclosures. Some challenge this incentive–access paradigm, claiming that IPRs' principal function is to coordinate actors in innovation systems. Others are sceptical about IPRs' capacity to stimulate innovation, and point to cases in which IPRs act as impediments to innovation by setting a high entry barrier to an innovation system, generating patent thickets, creating anti-commons or leading to defensive or blocking behaviour. Moreover, differences of opinion about the correct description of the role of IPRs in innovation systems are typically aggravated when the discussion migrates from descriptive to normative issues. Heated disagreement dominates discussions regarding the design and reform of intellectual property systems, the rules, institutions and practices they support, and the conditions under which regional and national innovation systems thrive.

At least four problems stand in the way of having a complete understanding of the role of IPRs in innovation systems. The first is that 'innovation system' is a term of art the meaning of which is often derived from the context in which the term is being used. Freeman, who coined the term, described innovation systems as 'the network of institutions in the public and private sectors whose activities and interactions initiate, import, modify and diffuse new technologies' (Freeman 1987). While this definition encompasses the structures and dynamics of the social framework in which innovation can occur, its openness to interpretation means that it can be substantiated with many different kinds of real-life cases. Consequently, as will be amply demonstrated throughout this book, 'innovation system' can mean: all innovation, in some abstract or global sense; regional innovation systems spanning country borders; national innovation systems; sector-specific innovation, or clusters of innovation. This divesity is important to remember, for while no particular stand on what 'really' constitutes an innovation system will be advanced in this book, authors of chapters work with different conceptions of the term. Their focus is on IPRs; differing conceptions of 'innovation system' lurk in the background.

The second problem arises because the relationship between IPRs and innovation is difficult to describe. The link between IP and innovation is difficult to delineate in a way that allows one to evaluate the effects of IPRs, particularly patents, on innovation. IPRs are among a number of elements in an innovation system. These elements include licence arrangements, trade secrets, contracts, business models, institutional culture, risk taking behaviour and other factors such as lead time generated by business practices, complementary asset management, financing, marketing and firm-to-firm collaboration. Innovation cannot be attributed to any one of these elements in isolation from the others, and yet IPRs, particularly patents, are often discussed as if their role can be wholly disaggregated from the other elements and treated in isolation. Given that innovation is a complex phenomemon arising in dynamic and quickly changing scientific, technological, business and governance contexts, attempts to understand the role of IPRs separate from other elements is not likely to be a fruitful approach. The implication is clear: to understand how IPRs contribute to innovation requires consideration of the entire system of innovation.

The third problem is methodological. Studying IPRs is challenging because there are different kinds of intellectual property protection, each of which occupy a slightly different niche in innovation systems. These niches are not well-described, which means that evaluating the effects of patents on innovation, for example, is not a straightforward undertaking. Part of the problem is that governments, industry and universities frequently assume patents have a positive impact on the rate of innovation. The traditional argument takes the form of a positive feedback loop in which patents reward risk takers by providing limited-term rights to exclude others. Since the role of IPRs in innovation is not fully understood, the reliability of this feedback loop is questionable. What if growth in innovation and more patents are coincidental, not causal? Or what if there are better ways to manage intangible assets than patents in knowledge-intensive industries where patenting has not had a significant historical role? If either of these turns out to be true, then it is more than misguided to make policy based on the traditional view that IPRs provide an incentive structure that increase rates of innovation. In particular, such an analysis ignores not only potential negative feedback loops – for example, slowing the next generation of innovation – but other positive roles that IPRs may have. These may include, for example, coordinating the flow of information between people and institutions, creating distribution channels or acting as markers of technological capacity in order to attract foreign investment.

In addition to the challenges raised about description and methodology, a gap in evidence dogs all discussions of IPRs in innovation systems. Empirical economic studies of IPRs are few in number, and have different methodologies and outcomes. The fourth problem, then, is that one cannot simply point to 'the data' and settle arguments about how patents are linked to innovation, and whether that linkage on balance stimulates or slows innovation. Innovation systems can be roughly divided into different sectors of activity, which further confounds the potential for having accurate and meaningful data from which to draw comparisons and generalizations. The IP protection the pharmaceutical industry believes necessary to protect its products in a costly and lengthy product development pipeline is not mirrored by the agricultural biotechnology sector, and the data available for each sector are rarely an apples-to-apples match of indicators. In addition, even the most thoughtful contributions of industry participants are inherently limited. They can report on how they believe that IPRs relate to their current business strategies and market structures, but can rarely say much about how different business models that rely less (or more) on IPRs would actually function. Experiential data will not settle the discussion about the role of IPRs in innovation in general. Furthermore, industrialized countries are likely to collect data, or have the collection done for them by international organizations, but developing countries do not tend to collect data. Consequently whatever data exist cannot lead to conclusions for the vast majority of countries.

Motivated by ongoing disputes about IPRs, and in light of the problems of description, methodology and verification, the purpose of this book is to bring together into one volume a collection of original contributions on the role of intellectual property rights (IPRs) in near- and medium-term innovation. This book pursues two related objectives. The first is to bring clarity, rigour and fresh perspectives to the analysis of IPRs in innovation systems. Without good descriptive work that delves deeply into the main issues at the crossroads of innovation and IPRs, it will always be difficult to evaluate claims of the positive and negative impact of IPRs on innovation. The second objective is to evaluate how IPRs actually operate in innovation systems, not just from the perspective of theory but grounded in the contexts in which they appear – global, regional, national, present and historical. To meet these objectives, this book draws on perspectives from several regions of the world and from a variety of disciplinary and professional perspectives. The aim is to uncover deeply held assumptions about the role intellectual property rights have in innovation systems and move debate beyond these assumptions by encouraging the authors of these original chapters to engage in frank exchange across regional and disciplinary perspectives.

Each of the original contributions made in this book seeks to resolve aspects of the four challenges just mentioned, and they have come together in an effort to develop a more complete understanding of the role of IPRs in innovation systems. This approach will have immediate appeal to a diverse audience of scholars of science and technology innovation, academic and public or private sector specialists in IP, and to students of innovation theory, regulation and intellectual property law. Furthermore, there are five 'unique selling propositions' that are worth pointing out: first, this book situates IPRs within the broader context of innovation systems theory, an uncommon approach that facilitates the study of IPRs in their social and historical context unrestricted by economic and legal considerations alone. Second, the book provides new insights on IPRs' capacity to foster or retard innovation in different jurisdictions and historical periods. Third, several chapters in the book provide useful methods for measuring the contribution of IPRs to innovation systems, and provide concrete examples demonstrating the consequences of making changes to IPR policies and practices. Fourth, taken as a whole, the book offers the beginning of a consensus view that the role of IPRs in innovation systems is less about strictly legal rights and duties and more about the economic and social impact of the ways in which researchers, industry and governments deploy IPRs to bring new technology forward. Finally, this book makes an important contribution by being among the first directly to investigate the role of IPRs in biotechnology innovation specifically, an area of intensive scientific and technological change, with correspondingly little research on the role of IPRs in biotechnology innovation. As life science and associated biotechnology continues to grow and diversify, knowing how to evaluate the contribution of IPRs to biotechnology innovation will become increasingly important.

THE STRUCTURE OF THIS BOOK

This book is comprised of six parts, each of which begins with a short introduction, followed by three chapters in which different facets of the role of IPRs in innovation are critically examined. The first part, 'Intellectual Property Rights in Innovation Systems', takes the view that intellectual property is one among many components that constitute an innovation system. Starting with innovation systems as the overall context for IPRs, it is then possible to make decisions about how IPRs are defined, measured, and how they are related to other factors comprising innovation systems. Furthermore, one can then study how IPRs are used by governments, industry and academe to achieve specific goals. For example, changes to university invention ownership policies are almost always deliberate, but do not always attain the goal of fostering more innovation. In this way, this part of the book provides the perspective necessary for thinking about IPRs within the broader context of innovation systems, and is the crucial first step toward evaluating the relative contribution of IPRs to near- and medium-term innovation.

The second part, 'Intellectual Property Management in Biotechnology', begins by drawing the distinction between tangible and intangible assets in the portfolios of firms and other institutions. The idea of the knowledge-based economy may be relatively new, but in merely four or so decades since innovation and the knowledge economy became talking points it has become commonplace for a firm's intangible assets to account for 80 per cent or more of the firm's worth. This inverts the conventional ratio of tangible to intangible assets observed in the 1970s. IPRs, the most conspicuous form of intangible assets, need to be actively managed. A firm wishing to add value, or capture it from innovations, faces complex management decisions. It must evaluate options such as an intellectual asset focus versus a competitive intelligence focus, and it needs to make decisions about how it can make the most effective use of intellectual property within its organization and in competition or partnership with other organizations. This part provides an overview of management goals in the private and public sectors, an identification of the forms of intellectual property that are routinely overlooked, as well as those the importance of which are over-emphasized. The study of strategies for intellectual property management leads to a more subtle understanding of how firms believe that IPRs stimulate innovation, and the tactical steps they take to stay competitive. In addition intellectual property management highlights how factors such as market performance, capacity building and strategies for technology transfer subtly structure the relationship between firms managing IP portfolios and the rest of society.

A measurement problem is one of the greatest obstacles to a full understanding of the role of IPRs in innovation systems. IPRs, particularly patents, are often promoted as measurable outputs of innovative activity. The third part, 'Intellectual Property Rights in Relation to Other Measures of Innovation', addresses the value and accuracy of using IPRs as measures of innovation, typically by counting the number of patents filed and granted. Of course, measurement implies that an innovation, whether a product, process, organization or market, can be quantified with standardized units of measurement. The temptation to use patents as direct measures of innovative activity tends to arise in situations where the dominant discourse interprets patents as cash substitutes magically caught at the bottom line. IPRs may be poor substitutes for cash, with no impact on GDP, and yet they may be useful indirect measures of what is valued in an innovation system, by whom, and what social relations will be maintained to produce valuable goods and services. When the simple story about creating value by filing patents is discarded, the role of IPRs in measuring and valuing the output of innovation systems becomes much more complicated.

Patent length is one of the most discussed aspects of IPRs since it is foundational to the nature of the limited monopoly granted to patent holders. Those who uphold the incentive–access paradigm will argue that longer patent length induces innovative activity because limited monopoly rewards accrue for longer. Innovators will see longer term patents as a way to off-set past losses and to balance taking greater risks in the future. Yet patent length is not the only method by which one can manipulate an innovation system to generate different kinds of outcomes, possibly even to trigger greater levels of innovation. The fourth part, 'Beyond Patent Length', considers different forms of incentives apart from patent length, and apart from patents themselves. It addresses the impact of various forms of intellectual property protection (patents, copyrights, trademarks, trade secrets, and plant breeders' rights), and examines the effects on consumers, producers and governments of alternative strategics for stimulating innovation. These include identification and manipulation of underlying political-economic conditions, open source strategies and treating IPRs as mechanisms for coordinating knowledge flows between people and institutions. While often starting with generic issues in regional and national innovation systems, the chapters in this part anchor the discussion in biotechnology innovation. Additionally, since it could be conjectured that industrialized countries suffer innovation gaps only with respect to small differences in comparative advantage, this part also examines the pressing issue of how innovation can be stimulated in developing countries.

Governments actively manage innovation. The fifth part, 'Innovation Governance', considers some of the mechanisms, pitfalls and issues that goverments need to consider as they adopt policies for controlling IPRs in innovation systems. The issues in this part are broad and diverse, covering the problems that arise in biotechnology innovation at institutional, national and international levels. Part of the issue for governance of innovation is to be clear about the economic, social and ethical goals of IPRs as a component of innovation governance and consider whether the stated goals are being met. Do people have access to valuable data for downstream innovation? Another problem to address is the quality and quantity of empirical evidence that can be gathered to make evaluations of innovation governance. Governments must also consider the social milieu in which there are ethically charged media portrayals of IPRs, such as patents that allegedly restrict access to crops, diagnostics and medicines. Similar issues arise in thinking about the difference beween the approaches to governance which are appropriate in industrialized versus developing countries. Governments have strong incentives to govern innovation systems optimally. Can they plan for the future?

In the last part of this book, 'National, International and Historical Comparisons', lessons are drawn for our contemporary situation from other times and other places. Many studies of IPRs often seek general truths about the nature, benefits, challenges and remedies associated with IPRs, but fail to acknowledge the historical contingency of individual IPR systems and their associated innovation systems. Do IPRs have the same role in innovation in a G8 country with elaborate innovation cultures and systems, with access to financing and with legal and scientific infrastructures, as they would in developing countries lacking these significant elements? Equally, does it matter that an IPR system comes into force at a time when professional cultures of science and medicine develop in a similarly evolving legal system? One might ask what modern dynamics arise when developing countries lacking tens or hundreds of years of evolution of their intellectual property and innovation systems find themselves competing with industrialized countries in an environment of globalized world trade.

THE ORIGINS OF THIS BOOK

The Role of Intellectual Property Rights in Biotechnology Innovation consolidates papers originally presented at a workshop held in the fall of 2005 at the Robert Schuman Centre for International Studies at the European University Institute in Florence, Italy. The chapters were at first workshop presentations, and then became chapters. They are not all uniform in length, a function of the scope of the topic, length of the workshop presentation, ensuing discussion at the workshop the author wished to incorporate, and revisions in response to referees' comments. In addition to the variation in length, author style is preserved as much as possible to let the diversity in the authors' voices prevail.

This workshop was convened by the Intellectual Property Modelling Group (IPMG), a research group based at McGill University in Montreal, Canada. The following discussion of the IPMG's goals and methodology provides background about the generation of this book and the program of research that led to its development.

The Intellectual Property Modelling Group

The Intellectual Property Modelling Group (IPMG) is a transdisciplinary group of researchers in the fields of law, economics, ethics, philosophy, management, political science and biomedical science coordinated through McGill University's Centre for Intellectual Property Policy (www.cipp.mcgill.ca). The IPMG seeks new models of intellectual property better to describe, evaluate, and change intellectual property systems (Gold et al. 2002). In light of the complexity of the field of biotechnology intellectual property and the diversity of the IPMG membership, it was necessary for the IPMG to develop a common conceptual framework, vocabulary, and a set of complementary research methods to investigate biotechnology intellectual property. The result is a conceptual and empirical methodology that is designed to provide empirically grounded, systems-level alternative policy strategies to governments, universities and industry.

Conceptual Methodology

The basis of the IPMG methodology was the development of a set of critical questions that can be used to probe aspects of IP systems (Gold et al. 2004). The IPMG used a Delphi method modified to take into account the group structure and composition. The 'probes' were developed by enumerating the common assumptions made by specialized academic disciplines about the structure and function of IP systems. Cross-discipline comparison generated a shortlist of widely held assumptions about IP systems which were then problematized by reversing the assumption into a probitive question. With respect to this book's subject, the role of IPRs in innovation systems, one common assumption made in several disciplines is that IPRs provide a necessary incentive to innovate, and that patents in particular are an optimal policy tool for stimulating research and development. As noted earlier, neither assumption is beyond dispute and neither has compelling empirical support.

The development of the analytical probes allowed the IPMG to begin a process of mapping out IP systems, including all of the variables, and the relationships between them, that are involved in biotechnology IP systems. The IPMG was able to record the knowledge it was accumulating and share analyses among team members and between IPMG and other interested parties, including its research partners and advisory board. Over the course of a few years, the IPMG has developed two principal research tools that take the form of an ontology and the formulation of the dynamic hypotheses. The dynamic hypothesis is represented by an influence diagram, a two-dimensional diagram of empirically-grounded variables and their inter-relationships. Each variable is nested in a common set of definitions that cross traditional disciplinary boundaries. The ontology represents the relationship between the variables that the IPMG has identified (for example, providing funding to develop an invention prototype) and particular policies and strategies (for example, income tax credits for research and development expenditures). The ontology is used in the development of policy alternatives from the results of modelling the interaction of variables in the dynamic hypotheses. These research tools are further described in the following descriptions of the empirical methodology and the IPMG's impact through the development of policy recommendations.

Empirical Methodology

The conceptual methodology developed by the IPMG provides a framework for analysing intellectual property. It goes further by providing justification for which data collection methods are appropriate, and for providing guidance in the collection and interpretation of data. Three inter-related data collection tools have been incorporated into the IPMG methodology: Dynamic simulation modelling, case studies, and workshops.

1. *Dynamic Simulation Modelling*
   The IPMG provides such a framework by modelling IP systems as a dynamic system rather than as a static set of legal rules. The methodological basis is adopted from systems theory used by the IPMG to understand the interactions of various factors, or variables, in leading to the creation, dissemination and use of biotechnological innovation over time. Causal relationships between the variables, for example a relationship between patent terms and the availability of generic products, consist of probabilistic inferences in the form of either positive or negative feedback loops. The IPMG has identified approximately 130 variables and thousands of inferential relationships between them. It has also gathered empirical data on those variables necessary to perform a dynamic simulation of the process of biotechnological innovation under various real-world conditions. Where possible,

existing or researched data are collected. Where not possible, inferences from incomplete data sources are used.

The IPMG's simulation model begins with a dynamic hypothesis about relationships between variables in a qualitative representation of the IP system. This hypothesis is then translated into a computer-based simulation model built with dedicated simulation software in which variables are represented as stock and flow interactions on a diagram. The simulation model permits testing of the hypothesis, and enables predictions. This simulation model is generic, in the sense that it aims to show the behaviour of specific variables over time, based on the underlying theory that pertains to these variables, or other historical data and information available about that structure. Since the model is generic, the selection of some, and elimination of other variables allows for problem- or topic-specific models to be spawned from the generic model. For example, as discussed below, a model containing mainly variables relevant to the issue of IPRs in innovation has been developed, which allows for specific kinds of policy questions to be answered. In another respect, models of specific case studies can also be developed, discussed next.

2. *Case Studies*

   Case studies have many purposes and are used extensively by the IPMG. Retrospective case studies are an effective way of validating the dynamic model. Prospective case studies use the validated model to make predictions about the value of individual variables and the relationships between the variables in the model. For example, it is possible to model the effect of increasing patent length on the value of a patent. There are currently six case studies being conducted by IMPG that are related to the topic of this book: plant-derived vaccines' technology diffusion; Canada's access to medicines regime; the politics around blocking patents; knowledge flows; a comparative case study of traditional and indigenous knowledge in Brazil, Kenya and Canada; and a case study on Indonesia's phytomedicines industry.

   Case Study 1: Plant-Derived Vaccines

   Plant-derived vaccines involve the introduction of a gene into a plant species, such as a tomato or tobacco, to produce a vaccine for serious communicable diseases such as Hepatitis B. Plant-derived vaccines offer many benefits over their conventional counterparts, and may revolutionize the way that vaccines are delivered, particularly in developing countries. Challenges requiring scientific and technological innovation remain but, assuming they are overcome, regulatory obstacles (intellectual property, import/export, plant varieties legislation and environmental legislation) will determine the rate and extent of the technology's diffusion. This case study compares different scenarios under which plant-derived vaccines can be manufactured and distributed, and includes a semi-quantitative, dynamic simulation model developed to surpass educated guessing and speculation about new technology innovation and diffusion. The case study will be published in book form by John Wiley and Sons in 2010.

Case Study 2: Intellectual Property Governance and Non-State Actors: The Case of Bill C-9

This case study provides a greater understanding of how to bring non-state actors (civil society and industry) into decision-making around health care innovation policy. Specifically, the project identifies ways in which non-governmental organizations contributed constructively to policy development in relation to health research, development and commercialization. The case study is based on interviews with leading members of civil society, documentary and media analysis, and analysis of the political debates. It explores how civil society influenced the passage of Bill C-9 in Canada – called the Access to Medicines Regime – and contributed to health, innovation and patent policy. The case study deals specifically with an amendment to the *Patent Act* to help provide medications through compulsory licensing to developing countries in need. The case study follows the work of non-state actors from the Cancun Ministerial Meeting in 2003 that provided a mechanism through which actually to import these medications to their lobbying for Bill C-9, *An Act to Amend the Patent Act and the Food and Drugs Act*, which aims 'to facilitate access to pharmaceutical products in the developing world, in order to address public health problems, especially those resulting from HIV/AIDS, tuberculosis, malaria and other epidemics'. The Bill, the first of its kind among developed nations, became law on May 14, 2004. Only in the summer of 2007 did any country, however, actually make a request to use it. That country was Rwanda, which sought a first-line triple therapy for HIV/AIDS.

Case Study 3: The Politics of Blocking Patents

This case study examines how a small Utah biotechnology company's patent and business model erupted into an international debate about blocking patents. The company is Myriad Genetics and the technology is a genetic predisposition test for breast and ovarian cancer.

Based on academic literature, news reporting, and a workshop of the principal industry and governmental actors involved with the dispute, the case study illustrates how communication and institutional failures combined with a lack of trust combined to transform a poorly adapted business model into an international *cause célèbre*. The case study points to the need for a greater recognition that biotechnology patents exist within a different cultural and social context from most other inventions and that those who fail to account for this may pay a significant price.

Case Study 4: Knowledge Flows in Biotechnological Innovation
Knowledge and innovation flows can be tracked with new methods in bibliometrics and scientometrics, for instance, by looking at the fields of mouse genomics and stem cell research. To analyse the flow of knowledge from pure science via technology innovation into public discourse, a case study will explore research methods in three distinct fields: (a) citation analysis for scientific publications and patent literature; (b) word and concept analysis for summaries or full texts of scientific, patent, policy and media documents; and (c) social network analysis. Different sources of data for large-scale analyses, and verified or obtained appropriate licences for large-scale searches and downloads have been identified, including Medline/PubMed, ISI Web of Science, Elsevier's Scopus database, Delphion and national patent databases, Lexis/Nexis and Factiva for media and national government sites for policy documents, committee reports and political speech.

Case Study 5: Comparative Case Study of Traditional and Indigeneous Knowledge
As numerous international fora and organizations attempt to map out a strategy to recognize and protect traditional indigeneous knowledge, little attention has been paid to whether there is a coherent meaning to this knowledge that transcends the different cultures and societies involved. This case study remedies the gap by comparing the types, cultural meanings and legal protection offered to traditional and/or indigeneous knowledge in three places: Brazil, Kenya and Canada's North. The goal of the study is twofold. First, it seeks to determine whether there is a single overarching concept of traditional knowledge that applies in all of these places. Second, it attempts to determine whether, given the nature and meaning of such knowledge, it is best viewed through the lens of legal 'property' or as a component of sovereignty and autonomy of the peoples involved.

Case Study 6: Indonesia's Phytomedicines Industry
This case study explores how Indonesia can convert its advantages in plant diversity and in traditional medicine into the development of new medicines that address the country's essential health needs. Indonesia – and, by extension, similar developing countries – could take full advantage of modern biotechnology to meet its health concerns, but three obstacles have been identified: first, there needs to be better alignment of health research with health needs in the country. Second, trained managers are needed to move innovative products from the laboratory to the patient. Third, finding ways to stimulate the Indonesian market to accept and encourage locally invented and produced medications is a must.

3. *Workshops*
The dynamic simulation model and the case studies are supported and cross-referenced by means of international workshops at which members of the IPMG adopted the view that workshops provided an obvious and important dissemination opportunity, but equally they provided an opportunity to collect and validate data that would be used in the model and in the case studies. Accordingly, the IPMG has held many workshops, of which the event in Florence that led to this book was the fourth. Participants were invited to the workshop to examine the role of IPRs in innovation systems, and to provide their views about this issue from their disciplinary and regional backgrounds. There was considerable diversity of opinion at the workshop, and the collected wisdom and expertise were crucial to the methodology of IPMG's project insofar as they helped uncover deeply held assumptions about the role of intellectual property rights in innovation systems. The workshop was successful in not only generating this book, but also assisting IPMG members in their efforts to select, quantify and relate variables that are directly relevant to the issue of IPRs' role in innovation systems.

A similar workshop hosted by the IPMG was entitled 'Intellectual Property, Biotechnology Capacity and Development'. This workshop was held in Buenos Aires on September 25 and 26, 2006, and was hosted by the Argentinian Agencia Nacional de Promoción Científica y Tecnológica, and the Centre for Intellectual Property Policy. The workshop brought together policy-makers, non-governmental organizations, industry and academics to examine how developing countries can configure their intellectual property systems to attract and retain scientists and investment in local research and development. The focus was on policy options available to Latin America and the Carribbean to enhance their scientific infrastructure in the biotechnology field. These options include practices drawn from actual experience in developing and industrialized countries and new models put forward to overcome identified problems in innovation systems. Some of the novel mechanisms given special attention in this workshop are new venture capital structures for start-up technology firms, collaborative and open science mechanisms and novel reward systems for innovation. Perhaps the greatest issue identified in this workshop is how cutting edge social science research on innovation systems and intellectual property can be used to overcome the tremendous diversity in levels of innovation, diversity of resources, and social infrastructure in the region.

The Impact of IPMG

The IPMG has developed a conceptual model of how IP-related laws, practices and institutions actually work together to create or inhibit the production and dissemination of new knowledge in relation to biotechnological innovation. IPMG research demonstrates that: (a) IP laws do not operate in isolation from practices and institutions, and thus the implications of implementing particular legal rules cannot be assessed without a more comprehensive understanding of the relationship between rules, practices and institutions; and (b) while policy-makers generally rely on IP rules to achieve policy objectives concerning innovation, these objectives could be met through complementary or alternative practices and different institutional structures.

One of the challenges that the IPMG addresses is how to draw on this understanding of IP systems, better to develop understanding on how best to deploy IP laws, practices and institutions to meet social and economic goals. The IPMG distinuishes between 'old' IP and 'new' IP eras to highlight the changing attitudes toward the role of IP, the changing roles of institutions which rely on IP, and evolving IP practices. To make this transition, the final report of the IPMG, *Toward a New Era of Intellectual Property: From Confrontation to Negotiation*, describes a detailed framework of actions to be taken by governments, patent offices, universities and the scientific community (The International Expert Group 2008). These include: developing greater trust between actors; more and better communication; generating new models of IP; enhancing science, technology and engineering in the developing world; cross-cutting thinking about IP; and developing improved data sets and metrics. If followed, the framework of actions ought to increase short- to medium-term innovation levels; increase scientific infrastructure (particularly in developing countries and in economically or socially disadvantaged regions of developed countries); and increase access to technology.

A simulation model specific to the issue of IPRs' role in innovation has been developed. It incorporates the interactions amongst the selected variables, according to system dynamics modelling principles. At the time of writing, the model is being evaluated for its realism and accuracy using scenario analyses of policy changes that target specific variables to induce changes within the model. This evaluation includes checks for the model's internal validity (consistency with theory, links amongst variables of specific feedback loops, expert knowledge, empirical evidence from research), the model's external validity (consistent with historical data collected, gaps between calculated values from the model and historical data), and the sensitivity analysis of key parameters on overall model behaviour and performance on targeted policy parameters.

The motivations for the book are quite clear when one considers that policy and academic literatures often portray IPRs as *the* catalyst for biotechnology innovation. This role is contested by some commentators, but, as mentioned above, the problem is that decisive empirical data are missing. More important to the IPMG is that even if the relevant data exist, they need to be collected and understood within a rigorous framework that takes into account the complexities of the IP system in biotechnological innovation.

REFERENCES

Freeman C. (1987). *Technology and Economic Performance: Lessons from Japan*. London: Pinter.

Gold E.R., D. Castle, L.M. Cloutier, A.S. Daar and P.J. Smith. (2002). 'Needed: Models of Biotechnology Intellectual Property'. *Trends in Biotechnology* 20:327–9.

Gold E.R., W. Adams, D. Castle, G.C. de Langavant, L.M. Cloutier, A.S. Daar, A. Glass, P.J. Smith and L. Bernier. (2004). 'The Unexamined Assumptions of Intellectual Property: Adopting an Evaluative Approach to Patenting Biotechnological Innovation'. *Public Affairs Quarterly* 18:299–344.

The International Expert Group On Biotechnology, Innovation And Intellectual Property. (2008). *Toward a New Era of Intellectual Property: From Confrontation to Negotiation*. Available at www.cipp.mcgill.ca.

PART I

Intellectual property rights in innovation systems

Introduction

David Castle

It has become common parlance to describe innovation as taking place within a system. That is to say, innovation is thought to be a product of a system of integrated causes and effects, and the right way to analyse innovation is by thinking about a system. Gone from informed discourse about innovation is the caricature of the linear chain of events starting with the scientist-recluse and ending with with the monopolizing capitalist. The more sophisticated view of innovation systems does not, however, mean that we have answered all of the questions about how innovation works. Far from it in fact, as the chapters in this part reveal. Take, for example, the point that if innovation is a system-level phenomenon, it would be useful to know exactly how intellectual property rights contribute to the system. Yet as the contributions in this part point out, this question is not normally taken up directly, or at least not until now. Each of the chapters in this part finds its own way to this question, a fact which illustrates that IPRs play multiple roles in innovation systems, each of which can serve as an analytical starting point. In the first chapter, IPRs are the 'quanta' for the valuation of inventions, in the second chapter IPRs are policy tools for national and regional innovation system management, and in the third chapter IPRs are the source of tension about access to research tools and the impact on the development of genetic diagnostics and therapeutics.

Adam Holbrook's chapter begins with an account of what he calls the 'national system of innovation (NSI)' perspective on innovation systems and the role therein of intellectual property rights (IPRs). Holbrook's central premise is that one's view of innovation changes considerably if one abandons the Schumpeter conception of IPRs that restricts them to economic contexts. Holbrook's view, by contrast, is that IPRs can be understoond in a broader social context of innovation in which economic factors feature among other drivers of innovation. By looking at innovation through social and economic lenses, one can see, first of all, that firms are networked into private and public sector institutions which interact in a multitude of ways to diffuse, or block diffusion, of knowledge. From the standpoint of NSI theory, the central issue is to understand the formal and informal linkages are created between institutions, and how in turn those linkages modulate directly the flows of knowledge between institutions.

Second, Holbrook points out that if one takes a systems approach to thinking about IPRs in highly networked systems of innovation, it casts a different light on how we think about intellectual property. A strictly economic analysis of innovation would study the flow of capital in a national system. Holbrook, by contrast, suggests that if we think that IPRs embody economic and other social factors of innovation, then tracking their flow, as one would money, in an innovation system can be enlightening. Holbrook draws an analogy with data packet flows in the internet to suggest that IPRs are traceable packets of social and economic data in an NSI. IPRs may be characterized as the measureable 'quanta' of an innovation system because they are the basic aggregations of social and economic data exchanged in an NSI. Paradoxically, Holbrook argues, very well-defined IPRs are good for trading and stand out as 'units' of innovation because they are easily counted. But often the most valuable knowledge created in IPRs is the tacit knowledge – of what works and what does not work – that defies measurement in traded IPRs. The role of the NSI approach is to try to understand how knowledge flows from contexts in which there is somewhat diffusely defined invention to rigidly described IPRs.

Bjørn Asheim and his colleagues begin by noting that there are two large, but mostly independent, literatures on IPRs and innovation systems. Like Holbrook, they think that a national innovation systems perspective will bring to light the role of IPRs in innovation systems. Asheim et al. argue that an innovation systems approach is an important starting point for the discussion because one of the main reasons for supporting a national system of innovation is that it can bring the agencies of knowledge creation in close contact with private and public sector enterprise that will expand economic activity based on the knowledge. IPRs play an important role in generating the artificial scarcity necessary to drive the value of knowledge up, and to differentiate the value of different innovations that feed into a market system. IPRs, as Asheim and his colleagues point out, solve the knowledge valuation problem for markets. But IPRs do not themselves resolve issues about how different innovation systems ought to use the IPRs in pursuit of wealth creation, because the use of knowledge within different systems of capitalism defies a one-size-fits-all approach to integrating rules and practices about IPRs into an innovation system. Consequently, system-specific legislation, such as the Bayh-Dole Act, is unlikely to be usefully transferred to other innovation systems.

Within both the European Union and the United States there has been explosive growth in patenting activity in the last two decades for almost all industries. The biotechnology industry's growth spurt roughly corresponds to this period. The effects of biotechnology IPRs on the underlying science base and the production of applied science have been significant. Asheim et al. point out that the private receptor capacity in an innovation system is dependent on many things, but chief among them is being able to control the channelling of knowledge from public research institutions and keeping control of the valuation of that knowledge. An interesting dynamic between public and private institutions is created and maintained by IPRs insofar as they are part of the architecture for privately extracting value from otherwise public or 'open' science. Universities, having been at the centre of this social coordination of knowledge flows, have adapted their institutional mission along the way. Nowhere is this more obvious than in the changing role of US universities before and after the Bayh-Dole Act where university assignees on patents increased. In the European Union, where the assignees have tended to be industry even if the university is the inventor, there are initiatives afoot to change national laws more closely to resemble the Bayh-Dole Act. Asheim et al. consider the effects of Bayh-Dole on three countries (Germany, Switzerland and Denmark) within the broader European context. They conclude that the increase in patenting activity is largely a result of economic and institutional changes, rather than a sign of significant innovative activity in the biotechnology sector, and that the changes to rules about patent ownership that reflect the Bayh-Dole Act do not always have corresponding positive effects, as was the case in Denmark. In this respect, Asheim et al. end on the same theme as Holbrook regarding the highly context-dependent role of IPRs to innovations systems.

In the final chapter in this part, Koichi Sumikura considers the effects of IPRs, especially patents on disease genes, with respect to the development of molecular diagnostics and potential drug targets. Sumikura argues that there is a difficult balancing act that needs to be achieved to support the public's interest in having new gene-based diagnostics and therapeutics. On the one hand, public disclosure of research results increases access, while providing leading biotechnology and pharmaceutical firms with traditional intellectual property rights protection that will enable them to undertake potentially risky projects. The problematic situations where the balance is not struck are well known – access to diagnostics, as in the case of Myriad's BRCA tests, and licensing restrictions on research tools, particularly in cases where the intent is not to develop further commercializable technologies. Sumikura observes that Japanese university researchers have not been concerned until recently about the potential for IPRs to restrict their research activity. In a series of interviews of research scientists conducted by Sumikura a number of issues and events came to light that are eroding researcher confidence in there being a strict line demarcating research from commercial use of licensed technologies. For these researchers, the threat of IPR incursion into basic science is seen as a real problem that threatens their institutions, academic freedom and their individual research programs. Researchers, sensing the tide change in their own institutions and abroad, may become increasingly reluctant to take on new research projects intensively based on licensed technologies.

Sumikura considers the policy options that may address the concerns that have arisen in the literature about the negative impact of IPRs on research, and the concerns voiced by his study group. These include anti-monopoly provisions such as reach-through clauses and compulsory licences, but he notes that measures such as these can have the effect of destabilizing a patent system because they target a key provision – short-term monopoly rights. Other policy options, may be more attractive. For example, gene patents could be recognized in an IP system, but research exemptions would be made for genetic diagnostics or for research tools: a view upheld by many of the scientists Sumikura interviewed. Another option is narrowly to limit the effects of patent rights on genes, thereby keeping the scope of the patent rights narrowed to the claims made at the time of application. Again, these approaches may not enhance confidence in the patent system, and so Sumikura considers a coordinated solution involving integrated management of intellectual property, the development of research tool consortia, reciprocal access to research tools with individually held patents, and university institutional ownership policies that provide free access to research tools within the institution. Only through a careful balancing act between researcher interests, state laws, and the demanding realities of innovation systems can the public interest in timely diagnostics and therapeutics be met. For Sumikura, simple remedies to the problem do not seem to be possible. The implication is that a satisfactory solution will likely be a complex negotiation that reflects the intricate web of interests that sustain IPRs in biotechnology.

What is immediately striking about the three chapters in this part is that, despite being written from very different perspectives from industrialized countries that are not easily compared, there are important common themes running throughout. Notice, for instance, that in each chapter the place of IPRs in innovation systems is regarded as an important but under-studied issue. It is as if the descriptive project of saying what IPRs do for innovation systems had eluded the attention of the research community in general, and the authors of these chapters feel that they have to break new ground as they start to probe the question. Equally striking, however, is that each of the chapters takes on the issue by locating IPRs in national systems of innovation in order to be able to describe, within the boundaries of national research funding agencies, university systems, laws and culture. This strategy for tackling the problem generates equal suspicion among all of the authors of these chapters that generalizations stating the role of IPRs in innovation systems are not very likely to rise above the level of platitudes. For example, everyone knows that IPRs have a coordinating role between research universities and the private sector. But to go beyond the surface of this statement and answer, for example, a very complicated but relevant question about the impact of laws styled on the Bayh-Dole Act in other jurisdictions is quite difficult. Similarly, calling for reform in some generic conception of an IP system is an easy thing to do, but actually to see how reform would work in a specific country is quite different. Reform to the allowable scope of claims in a patent system might be feasible in Germany, but the same reform might not be feasible in Japan.

This part of the book attempts to answer the difficult question about the role of IPRs in innovation systems. No doubt the results are constrained by the countries considered and perspectives offered. Does that suggest that the responses are simply idiosyncratic? One way to broach this question is to look at the other option. Perhaps there is one true-for-all-of-time theoretical account of the role of IPRs in innovation systems that transcend the particular details of any one national innovation system. Perhaps there are even general principles one can observe in single cases. The lesson from this part, however, is that the details of how IPRs are implemented in an innovation system, the institutions and practices they foster and legitmate, and the character and depth of the innovation system give rise to context-sensitive detail that puts into doubt the prospects of a general theory of IPR–innovation system interaction. That is not to say that the general descriptive project is hopeless, but it is to emphasize that much more descriptive work needs to be done before much credence is given to normative projects calling for changes and reform in innovation systems and IP laws.

1. Are intellectual property rights quanta of innovation?

J. Adam Holbrook

INTRODUCTION

Policy makers in all levels of government are searching for means of understanding the role of innovation in the development of modern societies and for frameworks upon which they can construct their policies. In general, they are seeking economic frameworks; rarely do they use innovation policies as a lever to develop social policies. The reverse is not uncommon: education policies are sometimes used to foster a climate of innovation.

The problem is simply that most analysts do not differentiate between innovation in general, a *process* that produces new ideas and intellectual property rights (IPRs), and *specific* innovations which can best be described through IPRs. Frequently policy makers start with the work of Josef Schumpeter (1961) who identified five forms of innovation: new products, new processes, new markets, new resources, new organizations. This categorization of innovations examined the economic effects of specific innovations, not the process itself. Thus there are clearly policies and programs in the economic field that can be used to stimulate Schumpeterian innovation. Even so, in policy terms, the concept of innovation is usually confined to that of technological innovation, as defined by the OECD in the Oslo Manual:

> Technological Product and Process (TPP) Innovations comprise implemented technologically new products and processes and significant technological improvements in products and processes. A TPP innovation has been implemented if it has been introduced on the market (product innovation) or used within a production process (process innovation) [OECD 1997, sec. 15].

This, of course, is a narrower definition than that proposed by Schumpeter, but in the domain of public policy it is usually the basis of the framework used for policy formulation. Many academics claim that there should be a broadening of this definition to encompass organizational and service improvements.[1]

Schumpeter thought of innovation as a phenomenon comprised of discrete units: innovations. These innovations have the potential to alter the economic milieu in which they occur. But arguably, innovation is also a process. It has a dual nature; it is not only an economic phenomenon but also a social one. Everett Rogers (2003), in his book, *Diffusion of Innovations*[2] looked at how innovations are communicated, adopted and adapted. In particular he drew the distinction between an inventor, the individual who generates a new idea, and the innovator, who disseminates the idea to those who implement it. Innovation is as much a matter of communication as it is of invention. Thus, if innovation is regarded as a process, it is a process that is influenced by the environment in which it takes place.

Innovation (in general, a phenomenon) thus takes place in an environment in which discrete innovations appear. These particles of innovation can flow from one element of the economic and social fabric of a locality, a region, a nation or part of the world to another. Thus the flow of knowledge embedded in these particles can be described in terms of a 'system of innovation', either through people, or through financial flows that permit the creation of knowledge in the recipient institution. A good working definition is that proposed by Metcalfe:

> A system of innovation is that set of distinct institutions which jointly and individually contributes to the development and diffusion of new technologies and which provides the framework within which government form and implement policies to influence the innovation process. As such is a system of interconnected institutions to create, store and transfer the knowledge, skills and artefacts which define new technologies [Metcalfe 1995].

Discrete innovations are unique pieces of knowledge. Scientific and technical knowledge (of which IPRs are a subset) is a unique commodity in that, while it can be created, it cannot (usually) be destroyed, although knowledge is often forgotten. IPRs that are traded are (usually) well defined, such as patents or licences. Similarly they can be transferred, as IPRs, but the inventor of the knowledge (usually) remembers the knowledge he/she transfers to the recipient. Thus, while the inventor can sell his/her IPR, the inventor retains two important pieces of tacit knowledge:

- what worked, even if this cannot be exploited under the terms of the IPR transfer agreement, and,
- more importantly, what did not work. This knowledge is rarely codified and almost never sold, but nevertheless is extremely important for future development of the IP in question. What is usually not traded is the tacit knowledge of the original inventor – and it can

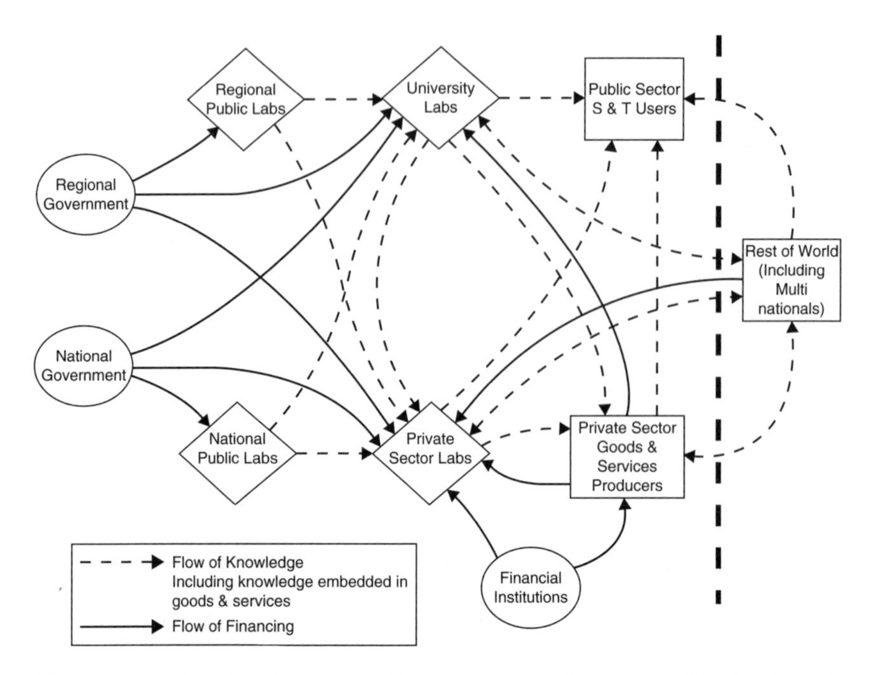

Figure 1.1 A national system of innovation (Holbrook and Hughes 1998)

be argued that often the knowledge of what did not work is more important than the knowledge of what did.

THE SYSTEMS OF INNOVATION APPROACH

A national system of innovation (NSI) describes the system by which knowledge flows among its constituent organizations and institutions, and relations among the organizations in the NSI. These flows can be quantified through financial flows or movements of people. These can be described in many ways: one visualization, based on the Canadian NSI, is shown in Figure 1.1.

The key characteristics of a NSI can be summarized as:

- Firms are part of a network of public and private sector institutions the activities and interactions of which initiate, import, modify and diffuse new technologies.
- An NSI consists of linkages (both formal and informal) between institutions.
- An NSI includes flows of intellectual resources between institutions.

- Analysis of NSIs emphasizes learning as a key economic resource and that geography and location matters (Holbrook & Wolfe 2000).

The NSI model of innovation tells us different things. On the one hand, the whole (the system) is much more than the aggregation of its parts. For instance, a national system of innovation is much more than the sum of its regional innovation systems. It also demonstrates that innovation from outside national boundaries is an important factor (indeed, in many countries external sources are the major source of innovations and IPRs: see the World Bank 2008). Thus one of the key elements of this type of policy analysis is the acknowledgement of the dependence of nations on sources of technology external to their NSI. A major policy objective of any national government must be to provide a suitable environment and mechanisms by which innovations can be brought into the nation and used for its social and economic benefit.

This model is important in that it addresses issues of systemic interaction and performance. In this sense, the OECD notes that:

> a country's innovation performance will depend not only on how it performs on each individual element of the NSI, but how these separate elements interact [OECD 2003].

The emphasis on the analysis of systemic failures was intended to shift policy analysis from promotion of state intervention by simple supply-side policies such as subsidies, to measures that ensure that the innovation system performs adequately as a whole. A key role for policy makers is 'bottle-neck analysis', to identify and try to rectify structural imperfections (Arnold 2004).

Most studies of innovation systems reflect on innovations that are created within that particular national system. But for the large part of humankind this is not so – most nations, particularly emerging economies, import IP through their purchases of IPRs.

While innovation, as defined by the OECD is 'new to the world', 'new to the nation' or 'new to the firm', there is often a fourth category: 'new to the market which you serve' (Holbrook & Hughes 2001). The innovator who brings new technologies to less developed regions of a country, or to less developed countries, is often, in both social and economic terms, far more important than the original inventor. Thus attempts to understand innovation systems, particularly in emerging economies, must focus on, and understand, how IPRs – codified innovations – are introduced into new markets (see Salazar & Holbrook 2004), and how these transactions are promoted.

Many inventions/innovations have occurred in one region or nation only to be ignored within that geographic socio-economic entity, taken up elsewhere, and developed to commercial success through a number of small, incremental (usually process) innovations. The example of the video-tape recorder is well known – invented in the United States, it was a useful piece of equipment for professional establishments until a number of incremental innovations in Japan made it possible to market the device as a consumer product. This example also provides evidence of another truth related to the analysis of systems of innovation: the input of innovations from abroad is probably the most important source of innovation in any economy, even in those economies which are highly innovative to start with.

It is also, in this context, worth considering the changing nature of production. When Schumpeter was writing, production was relative tightly integrated within single factory complexes. Today, an increasing variety of final consumer products result from the integration of a vast variety of component and modular products and services that have been produced off-site, very often by unrelated firms and often in a different country. Thus we may need to add 'architectural innovation' (Henderson & Clark 1990) to Schumpeter's original list. Wixted (2005) has shown that assembly of even technologically intensive 'products' (auto, aerospace and ICT in particular) rely heavily on imported parts. Wixted and Cooper (2007) have shown that, for ICT at least, the trend is towards greater reliance on imports. 'Products' is thus a vague term: what is a product for one company is an input for another.

INTELLECTUAL PROPERTY AND INTELLECTUAL PROPERTY RIGHTS

Intellectual property (IP) is a general term that can apply to almost any form of creative activity – from Leonardo da Vinci's *Mona Lisa* to a state of the art integrated circuit design. We tend to apply the concept of IP to the creation of codified knowledge, as opposed to the art of creating an object or tacit knowledge (or a skill). An artisan has tacit knowledge of his or her trade, and that knowledge is their IP. In terms of the NSI, IP is the knowledge held by any single actor (or group of actors) in the NSI and it can be tacit or codified. Intellectual property rights (IPRs) on the other hand are usually (not always) thought of as codified knowledge that, as a consequence of its codification, can be traded or transferred among actors.

In general one can describe four types of IPRs in the NSI:

1. IPRs coming from IP that is in the public domain, and thus accessible by anyone. These are IPRs, but there is no single owner or user (such as open source software). They can be tacit as well as codified.
2. IPRs coming from IP that is not in the public domain – for example, a patent. The IP thus defined by the corresponding IPRs can be traded for money or other value. The trade can be for exclusive use of the IP or for a more limited licence to use the IP.
3. IPRs coming from IP contained in a product or service that is traded. A piece of electronic equipment may contain several IPRs, and by buying the equipment or service the purchaser is, in effect, buying a limited licence, the use of which is limited to the use of the equipment (or service).
4. Tacit knowledge, ranging from learned skills (which could be self-taught or learned, for a fee, from someone else) to 'cultural' knowledge that is the property of a social group.

An IPR can retain its value even if it is not new. We also generally assume in these studies that IPRs are created and traded for positive purposes. But there are also many examples of patents that have been taken out to block other researchers' lines of enquiry, even if the originator has no intention of following up that particular line of enquiry. This type of IPR activity creates barriers to innovation, and has to be recognized as a type of bottle-neck blocking the transfer of IP within the NSI.

Foray has stated:

> Traditionally IPRs were considered one of the incentive structures society employed to elicit innovative effort. They co-existed with other incentive structures, each of which has costs and benefits as well as a degree of complementarity. We seem to be moving toward a new view, in which IPRs are the only means to commodify the intangible capital represented by knowledge. . . . [Foray 2002]

ARE INTELLECTUAL PROPERTY RIGHTS QUANTA OF INNOVATION?

As noted above, innovation must be seen through both economic and social lenses. Everett Rogers distinguished between inventors and innovators. He looked at innovation as a social activity, essentially the communication of ideas, the 'selling' of an idea, an invention, by an innovator to those who adopt the innovation. Thus innovations can have social value as well as economic value. Indeed, Rogers' visualization could be described as a wave radiating through society (or an NSI). This approach differs from the strictly economic view of innovation espoused by Schumpeter and his followers.

Suppose we think of knowledge as the 'fluid' coursing through an NSI. If we think of that system as being similar to the Internet, the digital packets of the Internet are replaced by particles of knowledge, defined in terms of specific IPRs. These IPRs can be thought of as blocks of knowledge, or as 'quanta' of knowledge flowing through the system of innovation. These quanta vary in economic and social value, which is how we might characterize individual units of IPRs. Quantum theory was developed by physicists to explain how light could behave simultaneously both as a diffuse wave and discrete stream of particles. Quantum physics links the energy of a quantum to the degree of precision to which it can be located. Simply put, one can measure the magnitude of a photon (the quantum of energy) or its position accurately, but one cannot do both at the same time.

The same might be said of particles of knowledge. Knowledge flows through the NSI like a fluid, but it can also be defined as discrete packets (IPRs). The more precise the description of the IPR, the more bounded it is and the more easily it can be valued or traded. If the knowledge content of an IPR can be more precisely defined, it has a higher degree of marketability (or higher value). If the knowledge is more diffuse (as in tacit knowledge), if it covers a greater area, and has a greater degree of imprecision over its boundaries, it is thus harder to value. But, as with quantum theory, it can be argued that one cannot have both a high degree of precision in defining the IPR and a wide span of the knowledge.

If we think of the particles of knowledge flowing through the NSI as quanta, we can then appreciate the roles of the various institutional actors in the system, and how they act on each IPR – quantum – of knowledge. We can search for bottle-necks and we can appreciate (using the analogy of wave-particle duality) how a quantum can both 'flow' and yet be traded as a discrete entity. This is not simply a theoretical construct – if Foray (2002) is correct that we are moving towards an NSI where IPRs are the only means to commodify knowledge, then we have to understand the systemic constraints on the quanta of knowledge flowing through the NSI and the limitations of an approach that sees IPRs only as particles.

ACADEMIC PAPERS AND PATENTS

How is a particular discovery modified by the environment in which it occurred? How can knowledge start out as something diffuse and yet wind up as a patent defined in legal language? In understanding the NSI for policy purposes we need to understand the process by which the NSI acts on a quantum of knowledge to transform it into an IPR, regardless of whether it has commercial value. As with any information in a physical transmission channel, the movement of quanta of knowledge is subject to the addition of noise and distortion. It can be amplified or attenuated by one or other of the many institutions it passes through. Usually the process is one of filtering. Thus the language of processing wave-like data can be applied to the policy process. Yet, while it starts as a wave, it ends as a packet, an IPR.

Patents are formal descriptions of IPRs, with the state granting exclusive privileges in return for disclosure, to stimulate innovation. Some patents (United States, and others with citations to previous patents) and some industries (chemicals, biotechnology, pharmaceuticals) can be analysed quantitatively. Patents with citations can be analysed statistically, for example, in the work of Francis Narin and CHI Research on patent bibliometrics.[3] There also are 'key' patents at the head of 'patenting trees' identifying major new IPRs and the spread of those IPRs (these too can be measured bibliometrically). But this focus on patents might be beside the point for industries relying on speed to market or trade secrets; patent analysis may not be indicative of what the industry is doing.

IPRs originating in the academic sector are made manifest in many ways. Both the system of academic publishing and the patenting system are formal methods of both publicizing and laying claim to IPRs. Academic papers contain IPRs, in that the authors are claiming priority of discovery. They are units of IPR, quanta, even though they may contain more than one individual piece of IP. These IPRs may be in the public domain,[4] but they still have the property of a 'right', even if it is only a 'moral' right (these IPRs have value, as, for example, in determining hiring and promotion). But the quanta generated are often very diffuse pieces of knowledge since the inventor does not offer specific uses for the knowledge or seeks to contain its use. These quanta are often easy to quantify and measure, for example the ISI Citation Index[5] is a useful method of measuring output by individual, by subject, by university, and other criteria. The science of bibliometrics is based on the analysis of publications and citations, and forms a rich field of data for looking at inter-relationships among generators of IPRs in the NSI.

The link between academic research and commercial IPRs is becoming more important. The Bayh-Dole Act in the United States is often seen as the defining policy statement, mandating that IP generated in the academic world should be moved expeditiously into the commercial world.[6] This policy of commercialization is contingent on accurate descriptions of IP generated in university laboratories. Yet the movement of these quanta of knowledge is often convoluted, as they must pass through a number of processes as they move from the original discoverer (often a graduate student working under the supervision of a faculty member) to a commercial enterprise. It is often difficult to describe exactly where an IPR came from or what exactly it applies to. The differences between academic IPRs and commercial IPRs, and indeed the transition from academic knowledge developed in a laboratory of a publicly funded institution to specific IPRs with commercial value, are possibly best understood if the IPRs are thought of as quanta and the process thought of as the quantum of knowledge moving from a high degree of uncertainty in both magnitude and breadth to one of higher precision in either value or coverage.

QUANTIFYING IPRs: MEASUREMENTS AND SURVEYS

If one takes the quantum theory analogy further, one can reflect on the fact that there is a direct, inverse, relationship between the physical size of a quantum and its energy:

> $\lambda E = Q$, where Q is Planck's constant, λ is the wavelength (size) of the quantum and E its energy.

Thus as its energy increases, its size decreases. As with the physical world, it can be argued that the more precisely an IPR can be defined the greater its marketability. Schumpeter argued that innovation is disruptive. Innovations move the state of an art from one level to another – a step function. But Rogers argues that the diffusion of an innovation follows an 'S' curve. It is no accident that the integral of a single wave is an 'S' curve, and that a step function is the limiting case of an 'S' curve. This argument by analogy creates a link between Rogers and Shumpeterian views of innovation. High impact innovations (and IPRs) are the commercial innovations that have value; slower and more incremental innovations pass through the NSI as waves, rather than as pulses, and have slower immediate effects, even if their long term effects are significant.

When it comes to understanding the role of IPRs in the commercial world, we need to be able to distinguish between relatively common incremental innovations and the relatively rare, disruptive Schumpeterian innovations. In practical terms, identification of narrowly defined, 'high energy' innovations may be useful in that it helps identify which IPRs are worth registering and protecting, and which consequently may have market value. A more diffuse innovation may be more difficult to commercialize simply because it is harder to define. This would allow for an understanding of innovation based on the continuum between Rogers' model and Schumpeterian innovation, and other possible determinants such as innovation in large organizations versus small organizations. (It may be possible to broaden the range of influences on innovation processes that are not necessarily accounted for in current innovation surveys (Salazar & Holbrook 2004). For example, there have been suggestions that there are gender influences on the nature of innovation: female innovators tend to work incrementally and cooperatively as opposed to more 'disruptive' individualistic male innovators. Innovation in large organizations (and the public sector) is usually the result of social processes, is incremental, and thus less 'catastrophic' and less disruptive than the Schumpeterian innovation which usually comes from an individual or a very small group.

Planck's equation implies that the product of size and energy is a constant. This is likely not the case for IPRs, but it does point to the need to understand which innovations are diffuse and which are more tightly defined. Thus innovation surveys should clearly ask respondents not only: 'Did you innovate?' but also 'How well-defined was your innovation?'.

CONCLUSIONS: THE POLICY IMPLICATIONS OF QUANTA OF IP – WAVES VERSUS PARTICLES?

The systems approach to innovation provides a conceptual framework for doing 'bottle-neck analysis', that is, the identification and rectification of structural imperfections of the transfer of IP. This allows the development of 'systemic' policies in addition to 'reinforcement' and 'bridging' policies, the last two being the bulk of innovation policies in both developed and developing countries. The understanding of regional systems of innovation becomes important because these have 'cultural' influences on how IP gets transferred within a particular region. Cluster promotion, a fashionable policy these days, can be considered a systemic innovation policy. But without an understanding of the 'whole' system of innovation the policy is unlikely to be successful.

There is an intriguing analogy one could explore with information communication theory – whether innovation system channel conditions affect the knowledge quanta moving through the channel, or whether, as some maintain, IPRs are neutral to the channel conditions. How does one quantify the transfer of tacit knowledge even though it cannot be codified (and thus quantified)? Are IPRs quanta because of the simple fact that they are codified, and thus can be traced through the innovation system, or are they simply artefacts of the legal rules within that system of innovation?

But there is an even more simplistic differentiation between particulate and continuous innovation. Policy makers often develop their innovation policies around two extremes: if IP is particulate, then policies should be put in place to encourage the development and registration of IPRs. On the other hand, if IP is a continuous (wave) phenomenon, then policy makers should try to develop the environment in which the creation of IP occurs – a more generalized approach to development of IPRs.

Policy makers have to reconcile the differences between open-source transmission of IPRs (such as open-source software) and the more traditional commercially restrictive transfer of IPRs in their system of innovation. They are faced with the same problem as physicists: how do you know if a quantum is a particle or a wave? Should regional/national innovation policies promote the generation of waves or particles?

There are many consequences of this policy dilemma. The traditional valuing of intellectual property as an IPR, with a finite boundary, provides the possibility for economic returns on its invention, transmission and exploitation. Much of our innovation policy is based on this (relatively) simple economic incentive, for individuals and institutions, to innovate. Equally evaluators of research performance have relied on statistics of the production of academic IP, the production of academic papers. Bibliometrics provide a great deal of useful information about the sociology of science, but it is arguable that using the production of academic papers to measure academic productivity is not a fair assessment of the quality of the work being performed by the researchers. The science of bibliometrics has been used to measure the performance of research teams in the academic world, sometimes with perverse results.

More recently some innovation policy makers have moved towards the promotion of the creation of a highly qualified workforce – a variation on the principle 'if you build it, they will come'. This is based on the assumption that if innovators can be trained, they will inevitably produce innovations. But will they produce incremental innovations or will they produce the 'killer applications', the disruptive innovative particles of Schumpetarian literature? Do many small incremental innovations when integrated over time have the same effect as a single highly energized innovation? Economic policy makers would prefer the latter, but it is questionable whether this ultimately leads to economic and social advancement.

But the dichotomy is greater than simply one of size – whether the innovations produced by a generally better educated workforce are more intense, and thus of greater economic value. One could also ask whether a better educated workforce produces a greater number of innovations, even if they are small ones. If a nation or region invests in the development of human capital it provides an environment in which innovations can flow through its system of innovation. These pieces of knowledge may be ill-defined and they may also be facilitative: the development of tacit skills enables the flow of knowledge from one individual to another. In such an environment is there a higher probability that less well defined pieces of knowledge, quanta, flow like ripples in a pond (waves) through the system?

Arguably this may be a situation that depends on the mandates of individual policy makers. In Canada, policy makers at the federal level operate on economic policy, and thus, of necessity, focus on the creation of well-defined IPRs. Federal policies are thus designed to foster innovation that results in specific, quantifiable, IPRs. But under the Canadian federal system, provinces have responsibility for social affairs, including education policies. Thus the creation of an environment in which innovation can occur is the domain of the provincial (that is, sub-national or regional) policy makers. The consequence is that federal policy makers must think of IPRs as being particulate, while their provincial/regional counterparts must think of the qualitative environment in which IP occurs.

Studies of systems of innovation help identify bottle-necks in the system of innovation. These bottle-necks restrict the flow of the quanta of knowledge. What are these bottle-necks? Are they social, legal, bureaucratic or inherent to the technologies that the quanta carry? The environment for the management of IPRs, such as the legal environment, may be one of the bottle-necks in the system of innovation (sometimes referred to as barriers to innovation). The flow of these quanta can be affected by a number of variables: the quanta vary in economic and social value over time, over place and among the various actors in the national system of innovation. Thus studies of IPRs need to look at at least these three dimensions.

Policy makers need to consider two completely different tracks: how to promote the generation of IPRs, narrowly defined quanta of innovation, but also how to promote the more diffuse concept of innovation, the concept of 'innovation'. Knowledge will flow through the system of innovation; the system itself must be prepared to accept the 'wave' of innovation.

NOTES

1. The latest version of the OECD's 'Oslo Manual' addresses some of these issues.
2. The most recent edition is the 5th edition published in 2003.
3. Now found at www.ipiq.com.
4. Leaving aside any consideration that the journal or book publishers may hold the copyright.
5. See www.isinet.com.
6. A summary can be found in *The Economist*, 24 December, 2005, at 109. However Mowery and Sampat (2005) argued that this model really only applied to the US, and not all OECD countries, and that even in the US the Bayh-Dole Act was 'neither necessary nor sufficient for much of the post-1980 growth in university patenting'.

REFERENCES

Arnold, E. (2004). 'Evaluating research and innovation policy: a systems world needs systems evaluation'. *Science and Public Policy*, 13(1), 3–17.

The Economist, 24 December, 2005, at 109.

Foray, D. (2002). 'Intellectual Property and Innovation in the Knowledge-Based Economy', *ISUMA*, Vol 3(1), 71–8 (see www.isuma.net)

Henderson, R.M. and Kim B. Clark (1990). 'Architectural Innovation: The Reconfiguration of Existing Product Technologies and the Failure of Established Firms'. *Administrative Science Quarterly*, Vol. 35(1), Special Issue: Technology, Organizations, and Innovation (Mar., 1990), 9–30.

Holbrook, J.A.D., and L.P. Hughes (1998). 'Measurement of Regional Systems of Innovation: Innovation in Enterprises in British Columbia', in J. de la Mothe and G. Paquet (eds), *Local and Regional Systems of Innovation*. Amsterdam: Kluwer Academic Publishers. pp. 173–90.

Holbrook, J.A.D. and L.P. Hughes (2001). 'Comments on the Use of the OECD Oslo Manual in Non-manufacturing Based Economies', *Science and Public Policy*, 28(2), 139–44.

Holbrook, J.A.D. and D. Wolfe (2000). 'Introduction: Innovation Studies in a Regional Perspective' in J.A. Holbrook and D. Wolfe (eds), *Innovation, Institutions and Territory – Regional Innovation Systems in Canada*. Montreal and Kingston: School of Policy Studies, Queen's University Press.

Metcalfe, S. (1995). 'The Economic Foundation of Technology Policy: Equilibrium and Evolutionary Perspectives' in P. Stoneman (ed.), *Handbook of the Economics of Innovations and Technological Change*. Oxford and Cambridge, Mass.: Blackwell. At 409–512.

Mowery, D.C. and B.N. Sampat (2005). 'The Bayh-Dole Act of 1980 and University–Industry Technology Transfer: A Model for Other OECD Governments?', *Journal of Technology Transfer*, 30(1/2), 115–27.

OECD. (1997). *Proposed Guidelines for Collecting and Interpreting Technological Innovation Data – Oslo Manual*. Paris: OECD.

OECD. (2003). *A Strategic View of Innovation Policy: A Proposed Methodology for Assessing Innovation Policy and Performance*. Paris: OECD, Working Party on Innovation and Technology Policy.

Rogers, E. (2003). *Diffusion of Innovations*. New York: 5th edition, The Free Press.

Salazar, M. and J.A.D. Holbrook (2004). 'A Debate on Innovation Surveys', *Science and Public Policy*, 31(4), 254–66 (a paper presented at a conference in honour of Keith Pavitt: 'What do we know about innovation?', SPRU, University of Sussex, November 2003).

Schumpeter, J.A. (1961). *The Theory of Economic Development*. Translated by Redvers Opie. New York: Oxford University Press.

Wixted, B. (2005). *Systems of Innovation Beyond Borders: Linked Clusters and the Role, Scale and Spatial Structure of Extra-Territorial Interdependencies*. Unpublished thesis, available at: http://library.uws.edu.au/adt-NUWS/public/adt-NUWS20060407.100831/index.html.

Wixted, B. and R. Cooper (2007). 'OECD Cluster Networks 1970–2000: An Input–Output Study of Changes in Interdependencies between 9 OECD Countries in ICT Industries', in R.J. Cooper, K.P. Donaghy and G.J.D. Hewings (eds). 2007. *Globalization and Regional Economic Modeling*. Heidelberg: Springer.

2. Intellectual property rights and innovation systems: issues for governance in a global context

Bjørn Asheim, Finn Valentin and Christian Zeller

INTRODUCTION: WHAT IS AN INNOVATION SYSTEM?

This chapter deals with IPRs in innovation systems. Even if both innovation systems and IPRs as such have been studied extensively, the specific problematic of this chapter has hardly been analysed before. This is even more surprising as the form and extent of IPR regulations potentially have big impacts on the functioning of an innovation system. According to Granstrand (2005), 'IPRs, particularly patents, play several important roles in innovation systems – to encourage innovation and investment in innovation, and to encourage dissemination (diffusion) of information about the principles and sources of innovation throughout the economy' (Granstrand 2005, p. 280). Thus, the extent, intensity and type of interactions between firms and universities, which represent the constituting relations of an innovation system, will obviously be affected by the way the IPRs are constructed. This is especially the case in science driven activities, such as biotech, which is the focus of this book, being the object of the majority of implemented IPR regulations. This chapter will investigate this problematic through a comparative approach, looking at how IPRs are implemented and their consequences for industry–university collaboration in the US and Europe, with a special focus on Germany, Switzerland and Denmark. While the analyses of the US and Germany and Switzerland give a general account of changes in the IPR regimes, the section on Denmark specifically looks at the consequences for industry–university coloration in biotech by the introduction of Bayh-Dole inspired legislation. This represents a contribution to the analysis of the importance of national variations in institutional frameworks which, in addition to sector differences (which in this chapter are handled by the differentiated knowledge based perspective), has been surprisingly little covered in the literature so far. The chapter starts with a short introduction to the innovation system concept as a national as well as a regional system.

The concept of the innovation system (IS) – originally developed by Bengt-Åke Lundvall – is a relatively new one, and was first used by Chris Freeman in his analysis of Japan's blooming economy (Freeman, 1987). The concept of a regional innovation system (RIS) appeared in the early 1990s (Cooke 2001), approximately at the same time as the idea of the national innovation system was becoming more widespread, thanks to the books by Lundvall (1992) and Nelson (1993). Characteristic of a systems approach to innovation is the acknowledgement that innovations are carried out through a network of various actors underpinned by an institutional framework. This dynamic and complex interaction constitutes what is commonly labelled systems of innovation (Edquist 1997), that is, systems understood as interaction networks (Kaufmann and Tödtling 2001). A set of variations on this approach have been developed over time, either taking as their point of departure either territories (national and regional) or specific sectors or technologies (Fagerberg et al. 2004).

The National Innovation Systems approach highlights the importance of interactive learning and the role of nation-based institutions in explaining the difference in innovation performance and, hence, economic growth across various countries. The rationale of having territorially based innovation systems (national and regional) is either the existence of historical technological trajectories based on 'sticky' knowledge and localised learning that can become more innovative and competitive by promoting systemic relationships between the production structure and knowledge infrastructure in the form of national or regional innovation systems (a policy of 'localised change' (Boschma 2004)), or the presence of knowledge creation organisations the knowledge of which could be exploited for economically useful purposes through supporting new emerging economic activity (a policy of 'structural change' (Boschma 2004)). The 'innovation system' concept can be understood in both a narrow and a broad sense. A narrow definition of the innovation system primarily incorporates the R&D functions of universities, public and private research institutes and corporations, reflecting a top-down model of innovation. A broader conception of the innovation systems includes 'all parts and aspects of the economic structure and the institutional set-up affecting learning as well as searching and exploring' (Lundvall 1992), and, thus, has a weaker system character. The formation of innovation systems must be understood in this context of creating a policy framework aiming at a systemic promotion of knowledge creation and learning, in which universities play a strategic role, as well as an efficient transfer to industry in order to secure the innovativeness and competitive advantage of nations, regions and firms (Freeman 1995; Cooke et al. 2000). According to Mowery and Sampat (2005) 'governments have sought to increase the rate of transfer of academic research advances to industry and to facilitate the application of these research advances by domestic firms since the 1970s as part of broader efforts to improve national economic performance. In this 'knowledge-based economy', according to this view, national system of higher education can be a strategic asset, if links with industry are strengthened and the transfer of technology enhanced and accelerated' (Mowery and Sampat 2005, 214).

IPRs IN REGIONAL INNOVATION SYSTEMS

When discussing the role of IPRs in innovation systems it is specifically the second rationale for innovation systems which is of relevance, that is, securing the exploitation of new knowledge creation at universities for economically useful purposes, applying a narrow definition of an innovation system. According to Mowery and Sampat (2005) this can be achieved by two types of policies: '(1) policies encouraging the formation of regional economic 'clusters' and spin-offs based on university research, and (2) policies attempting to stimulate university patenting and licensing activities' (Mowery and Sampat 2005, 225).

Mowery and Sampat (2005) emphasise that especially the first type of policy typically takes place at the regional level, in the context of RIS, seeking 'to spur local economic development based on university research, for example, by creating "science parks" located nearby universities campuses, support for "business incubators" and public "seed capital" funds, and the organization of other forms of "bridging institutions" that are believed to link universities to industrial innovation' (Mowery and Sampat 2005, 210).

An RIS is constituted by (1) the regional production structure or knowledge exploitation subsystem which consists mainly of firms, often displaying clustering tendencies, and (2) the regional supportive infrastructure or knowledge generation subsystem which consists of public and private research laboratories, universities and colleges, technology transfer agencies, vocational training organisations, etc. Thus, if the following two subsystems of actors are systematically engaged in interactive learning it can be argued that a regional innovation system is in place (Cooke 1998). From this it follows that clusters and RIS can (and often do) co-exist in the same territory. But whereas the regional innovation system by definition hosts several clusters, a cluster is not part and parcel of an RIS (Asheim and Gertler 2005).

According to Mowery and Sampat (2005) also 'the increased interests in

"Bayh-Dole type" policies is rooted in the motives similar to those underpinning policy initiatives that seek to create "high-technology" regional clusters' (Mowery and Sampat 2005, 228). As with the Bayh-Dole Act, the majority of these initiatives to support university-based research for the promotion of innovation and economic performance emphasise the importance of codification of knowledge as a precondition to imposing property rights on individual inventions, based on the premise that the impact of university research primarily happens through the production of new knowledge by new knowledge for commercialisation (for example patented discoveries) (Mowery and Sampat 2005). According to Mowery and Sampat (2005), 'in many respects, the Bayh-Dole act is the ultimate expression of faith in the "linear model" of innovation – if basic research results can be purchased by would-be developers, commercial innovation will be accelerated' (Mowery and Sampat 2005, 229).

The central role university research plays in contemporary economies with respect to the creation and diffusion of new knowledge has resulted in additional analytical frameworks – complementing the innovation system approach – pointing to the importance of strong links between universities, industry and government in knowledge-based economies. Two such concepts, the new interdisciplinary 'Mode 2' (Gibbons et al. 1994) and the 'Triple Helix' (Etzkowitz and Leydesdorff 2000) both argue that such interactions have increased (Mowery and Sampat 2005).

SPECIFICITIES AND CONTRADICTIONS OF KNOWLEDGE GENERATION

With the increasing socialisation of knowledge production the knowledge produced by the enterprises themselves increases. For this reason they depend more on 'intellectual common goods' in the form of generally available qualifications, information and knowledge (Jessop 2000). The contradiction between increasing socialisation of production and private appropriation inherent in capitalism appears even more obvious with knowledge production. The production of knowledge and new technologies is a process based on division of labour in complex systems and networks. Often innumerable people take part in this process. The classic contradiction between economic and business rationality, thus between privatisation of the benefit and externalisation of the expenses, becomes a particular guise. The enterprises aspire to free access to knowledge and information, and at the same time they want to reserve as much private property thereof as possible.

Knowledge production and its valorisation exhibit some characteristics which are crucial for the transformation of the regime of property rights

(Jessop 2000; May 2000; Husson 2001, 128; Thumm 2004, 531). In mainstream economics these traits are often described as market failure: First, intellectual activity enables considerable cumulative effects which bear much larger consequences than the productivity gains realised in material production. The benefits of science and information increase according to the number of people who use them. Knowledge develops and increases due to broad and free diffusion. These cumulative effects arise because codified information and knowledge can circulate extremely easily. They result from the collective and open character of intellectual activity ('public good' aspect and non-excludability of technological knowledge). Second, the production of scientific knowledge and of numerous new technologies requires very extensive, concentrated investments, similar to investments in fixed capital. Additionally, technological uncertainty investment leads to a general investment risk. The valorisation, however, can often be organised with only marginal additional costs. Knowledge-relevant information can be multiplied and used without large costs.

For these reasons, by using intellectual property titles (patents, copyrights) and technical safety devices (for example, copy protection of computer programs) firms seek to limit the uncontrolled diffusion of their products and artificially to create scarcity. Intellectual property titles are designed to render this artificial scarcity legitimate in the area of knowledge (May 1998, 69–70; Sell and May 2001, 472) and to exclude others from its use or force them to pay royalties. Intellectual property is a power instrument and contributes to a further accumulation of power. But, in contrast to the power which arises for owners of scarce material goods, scarcity of intellectual property must be produced artificially by legal regulations.

SECTORAL DIFFERENCES IN IPRs

As mentioned in the introduction, IPRs have potentially big impacts on the functioning of innovation systems. However, 'the importance of these roles varies across sectors (industries) and countries, and over time' (Granstrand 2005, 280). Granstrand underlines that especially with respect to patents the 'differences across industries or sectors are strikingly large' (Granstrand 2005, 282). Regions and nations display a large diversity when it comes to industrial structure, innovative capacity, competitiveness and economic growth. One way of analysing regional diversity within industrial sectors is to apply a differentiated knowledge base approach (Asheim and Gertler 2005; Asheim and Coenen 2005; Asheim et al. 2007). Despite the general trend towards increased diversity and interdependence in the knowledge generation process, Pavitt (1984) and others have argued that the innovation process of firms is also strongly shaped by their *specific* knowledge base, which tends to vary systematically by industrial sectors. Pavitt's typology has been further elaborated into a Synthetic-Analytical-Symbolic knowledge based typology, instead of the more narrowly defined traditional categories such as 'scientific', 'engineering' and 'artistic' knowledge bases, in order to capture the character of knowledge as output (Asheim and Gertler 2005; Asheim et al. forthcoming). More critically, this broader conceptual typology is intended to encompass the diversity of professional and occupational groups and competences involved in the production of various types of knowledge. As an ideal-type, synthetic knowledge can be defined as knowledge to design something that works as a solution to a practical problem. Analytical knowledge can be defined as knowledge to understand and explain features of the universe. Symbolic knowledge is knowledge to create cultural meaning through transmission in an affecting sensuous medium. As only the analytical (for example dominating pharmaceutical biotechnology) and synthetic knowledge (more important in agro-industrial and industrial biotechnology) bases are relevant with respect to biotech industry, the symbolic knowledge base will not be presented in more detail below.

Analytical Knowledge Base

This refers to industrial settings where scientific knowledge is highly important, and where knowledge creation is often based on cognitive and rational processes, or on formal models. Examples are biotechnology and nanotechnology. Both basic and applied research, as well as systematic development of products and processes, are relevant activities. Companies typically have their own R&D departments, but they also rely on the research results of universities and other research organisations in their innovation process. University–industry links and their respective networks are thus important and more frequent than in the other types of knowledge base.

Knowledge inputs and outputs are more often codified in this type of knowledge base than in the other types. This does not imply that tacit knowledge is irrelevant, since there are always both kinds of knowledge involved and needed in the process of knowledge creation and innovation (Nonaka et al. 2000; Johnson et al. 2002). The fact that codification is more frequent is due to several reasons: knowledge inputs are often based on reviews of existing studies, knowledge generation is based on the application of scientific principles and methods, knowledge processes are more formally organised (for example in R&D departments) and outcomes tend to be documented in reports, electronic files or patent descriptions. These activities require specific qualifications and capabilities of the people involved. In particular, analytical skills, abstraction, theory building and testing are more often needed than in the other knowledge types. The work-force, as a consequence, more often needs some research experience or university training. Knowledge creation in the form of scientific discoveries and technological inventions is more important than in the other knowledge types. Partly these inventions, under specific social and economic conditions, are susceptible to patents and licensing activities. Knowledge application is in the form of new products or processes, and there are more radical innovations than in the other knowledge types. Important routes of knowledge application are out-licensing activities of universities and publicly funded research organisations, and the creation of new firms and spin-off companies.

According to Granstrand (2005), 'patents are most likely to support the growth of knowledge-intensive industries in fields characterised by low ratios of imitation to innovation costs. Such low ratios are likely in areas with large-scale R&D projects, especially if the R&D results in highly codified knowledge, as in chemicals' (Granstrand 2005, 283). This will typically be the case in sectors or industries based on an analytical knowledge base, for example, the biotech industry, particularly that related to human health sciences and pharmaceuticals.

Synthetic Knowledge Base

This refers to industrial settings, where the innovation takes place mainly through the application of existing knowledge or through the new combination of knowledge. Often this occurs in response to the need to solve specific problems coming up in the interaction with clients and suppliers. Industry examples include plant engineering, specialised advanced industrial machinery and production systems, and shipbuilding. Products are often 'one-off' or produced in short series. R&D is in general less important than in the first type. If so, it takes the form of applied research, but more often it is in the form of product or process development. University–industry links are relevant, but they are clearly more in the field of applied research and development than in basic research. Knowledge is created less in a deductive process or through abstraction, but more often in an inductive process of testing, experimentation, computer-based simulation or through practical work. Knowledge embodied in the relevant technical solution or engineering work is at least partially codified. However, tacit knowledge seems to be more important than in the first type, in particular due to the fact that knowledge often results from experience gained at the workplace, and through learning by doing, using and interacting. Compared to the first knowledge type, there is more concrete know-how, craft and practical skill required in the knowledge production and circulation process. These are often provided by professional and polytechnic schools, or by on-the-job training.

The innovation process is often oriented towards the efficiency and reliability of new solutions, or the practical utility and user-friendliness of products from the perspective of the customers. Overall, this leads to a rather incremental way of innovation, dominated by the modification of existing products and processes. Since these types of innovation are less disruptive to existing routines and organisations, most of them take place in existing firms, whereas spin-offs are relatively less frequent.

However, this distinction refers to ideal-types, and most industries are in practice comprised of all three or at least two types of knowledge creating activities. For instance, biotechnology-based technology platforms or discovery tools also rely to a large extend on continuously improved engineering knowledge. The degree to which certain activities dominate is however different and contingent on the characteristics of the industry or phases of innovation processes within an industry (Moodysson et al., 2008).

'VARIETIES OF CAPITALISM' AND NATIONAL DIFFERENCES IN IPRs

Granstrand maintains that 'these intersectoral differences in the importance of IPRs have led several scholars to criticize the "one-size-fits-all" design of patent system' (Granstrand 2005, 283). This criticism becomes even more valid when national differences in institutional contexts, constituted by history, path dependence and institutional embeddedness, are taken into account. According to Mowery and Sampat (2005), the global diffusion of these policy proposals and initiatives – promoted among others by the OECD – display the classic problems of 'selective borrowing' from another nation's policies from implementation in an institutional context that differs significantly from that of the nation being emulated. . . . Indeed, emulation of Bayh-Dole could be counterproductive in other industrial economies, precisely because of the importance of other channels for technology transfer and exploitation by industry' (Mowery and Sampat 2005, 232–3).

Lam (2000) underlines that learning and innovation cannot be separated from broader social contexts when one analyzes the links between knowledge types, organisational forms and social institutions in order to meet the needs of specific industries in particular with respect to learning and the creation of knowledge in support of innovations. Soskice (1999) argues that different national institutional frameworks support different forms of economic activity, that is, that coordinated market economies (for example, the Nordic and (continental) West-European welfare states) have their competitive advantage in 'diversified quality production' (Streeck 1992), based on problem solving, engineering-based knowledge developed through interactive learning and accumulated collectively in the workforce (for example the machine tool industry based on a synthetic knowledge base), while liberal market economies (for example the US and UK) are most competitive in production relying on scientific based, analytical knowledge, that is, industries characterised by a high rate of change through radical innovations (for example IT, defence technology and advanced producer services). Following Soskice, the main determinants of coordinated market economies are the degree of non-market coordination and cooperation which exists inside the business sphere and between private and public actors, the degree to which labour remains 'incorporated' as well as the ability of the financial system to supply long term finance (Soskice 1999). While coordinated market economies on the macro level support cooperative, long-term and consensus-based relations between private as well as public actors, liberal market economies inhibit the development of these relations but instead offer the opportunity quickly to adjust the formal structure to new requirements using temporary organisations frequently.

These differences – due to the impact of the specific modes of organisation of important social institutions such as the market, the education system, the labour market, the financial system and the role of the state – both contribute to the formation of divergent 'business systems' (Whitley 1999), and constitute the institutional context within which different organisational forms with different mechanisms for learning, knowledge creation and knowledge appropriation have evolved (Casper and Whitley 2004).

Differences in capitalism between the liberal and the coordinated market economies can be observed when it comes to the ownership of IPR for inventions (either the academic institution or the researcher), the level of inter-university competition as well as the relative importance of private research universities. In Germany and Sweden, both typical examples of coordinated market economies, 'researchers have long had ownership rights for the intellectual property resulting from their work, and debate has centered on the feasibility and advisability of shifting these ownership rights from the individual to the institution' (Mowery and Sampat 2005, 232). This was changed in Denmark – another coordinated market economy, though with more liberalist traditions than Sweden and Germany – in 2000, when the Danish Law on University Patenting (LUP) became effective (see a later section for a discussion of its effects on university – industry cooperation within the pharmaceutical biotech industry).

With the exceptions of the university systems of the US and Britain, typical representatives of liberal market economies, inter-university 'competition' has been limited in most national systems of higher education. Inter-university competition played an important historical role in the evolution of US universities and their collaboration with industry; especially when university patenting began to change after the 1970s, as private universities expanded their share of US university patenting in the same period as the share of biomedical patents within overall university patenting grew (Mowery and Sampat 2005). A good illustration of the key role of a well-funded, high-quality research university for establishing a successful industry–university collaboration is MIT. Not surprisingly, research has found that a model design of Triple Helix based on MIT works less efficiently in different contexts with more average universities and regions (for example Australia and Sweden) (Cooke 2005).

While the US model of implementing IPR might function reasonably efficiently in the context of liberal market economies, these differences – based within a varieties of capitalism context – could lead to anomalies as well as dysfunctional situations and suboptimal allocation of resources in coordinated market economies. Mowery and Sampat (2005) mention several potential criticisms of, in particular the Bayh-Dole Act (Mowery and Sampat 2005, 230–32):[1]

> – Firstly, that the commercialization incentive resulting from the Bayh-Dole could shift the focus on university research away from 'basic' and towards 'applied' research.
> – Secondly, another potentially negative effect of a higher level of university patenting and licensing is a 'weakening of academic researchers' commitments to 'open science', leading to publication delays, secrecy, and withholding of data and materials'.
> – Thirdly, in view of the importance of the 'nonpatent/licensing' channels of interaction with universities in many industrial sectors, it is important that these channels are not restricted or impeded by the strong focus on patenting and licensing in many universities. Thus, the 'emulation of the Bayh-Dole Act is insufficient and perhaps even unnecessary to stimulate higher levels of university-industry interaction and technology transfer'.

FROM CLOSED TO OPEN INNOVATION – OR FROM OPEN SCIENCE TO PATENT-ENCLOSED INNOVATION?

The question of the role of intellectual property rights in innovation processes is controversial. On the one hand, intellectual property rights are justified as an answer to the market failure of technological knowledge which is an outcome of the specific characteristics of knowledge generation such as the inseparability of research expenditures and the burden of huge fixed costs for investors, the general investment risk that goes along with technological uncertainty, the 'public good' aspect, and the non-excludability of technological knowledge (Thumm 2004, 531). Thumm states that strengthening the patent system is likely to permit more trade in disembodied knowledge. Therefore, it is likely to facilitate the vertical disintegration of knowledge-based industries and to enable the entry of new firms that possess mainly intangible assets. Similarly Arora and Merges (2004) argue that a strong intellectual property regime supports vertical disintegration and the innovation activities of small firms. Chesbrough (2003a) describes the new configuration characterised by an increased vertical disintegration and a bigger share of extramural research activities, in a quite uncritical way as *open innovation.* On the other hand, David (2004) and Orsi and Coriat (2005) emphasise the questionable effects of the extension of intellectual property rights. This controversy recalls the crucial issue already raised by Dasgupta and David (1994), who argued that knowledge produced by academic institutions should retain the status of a 'public good'. Coriat, Orsi and Weinstein (2003) emphasise a finance-driven model of innovation linked to institutional changes and the move from 'open science' to 'patent intensive' science (Coriat and Orsi 2002; Coriat et al. 2003; Orsi and Coriat 2005). They emphasise that the rules of the game shaped by institutions such as financial markets, labour markets, the intellectual property regime, public research funding, the division of labour between academic and industrial research and the underlying accumulation regime must be considered.

New Industry Organization and Stronger Intellectual Property Rights

The discussion on these issues has to be put into the context of a far-reaching reorganisation of the industry organisation in the pharmaceutical and biotechnology industries. Due to the enormous differentiation of drug discovery technologies and in order to minimise their own risk, large pharmaceuticals outsource numerous research activities to biotechnology firms and academic research centres. They systematically observe technological development on a global scale and acquire promising substances and technologies. Thus, 'big pharma' relies heavily on appropriating externally produced knowledge. This analytical knowledge is generated in a few regional innovation arenas (Zeller 2004). Publicly financed institutions play a central role in this generation. Out-licensing of drug candidates and technologies has become an important source of income for biotech firms and even universities. Biotechnology companies often have a mediating role. They transform and develop basic analytical knowledge generated in publicly financed institutes. They then can further develop promising projects together with pharmaceutical companies, or out-license.

The notion of open innovation and the problematic of knowledge transfer through channels and pipelines need to be clarified. In the sense of Chesbrough, open innovation means that 'firms commercialise external (as well as internal) ideas by deploying outside (as well as in-house) pathways to the market'. The boundary between a firm and its surrounding actors has become more porous, enabling innovation to move easily between the interested players (Chesbrough 2003b, 37). However, the notion of 'open innovation' is misleading. Increased division of innovative activities, network membership and sharing of knowledge do not automatically mean augmented openness. On the contrary, exactly because the boundaries between a firm and its surrounding environment are more porous, intellectual property rights must be enforced as a consequence. In- and out-licensing are possible only when based on property rights. They have become bargaining chips for the exchange of technology between companies and for venture capital (Thumm 2002, 924).

An increased but highly selective openness can be observed in the sense of intensified knowledge transfer from universities to firms and between firms. However, science and knowledge generation have not become more open in the sense of a free dissemination of knowledge and information. On the contrary, the stronger IPR regime encloses knowledge more than before. There is even the danger that knowledge produced by academic institutions loses the status of a 'public good' (Nelson 2004; David 2004).

Open channels and closed pipelines of knowledge transfer closely interfere with one another (Owen-Smith and Powell 2004). Transactions are 'pipelines' when legally binding, confidential, contractual business is being transacted. But they rely on free contacts to 'open science'. Publicly funded research institutes are major magnets for profit-seeking biotech firms and for large pharmaceuticals exactly because they operate with relatively open science conventions. However, in doing so, 'big pharma' tries to absorb a good portion of that knowledge exclusively (Zeller 2004). This shows that closed and open science are inseparably related; closed corporate innovation enclosed by property monopolies profits from open science in universities. How the processes of knowledge generation, acquisition and subsequent valorisation interact depends on financial constraints, regulatory conditions and power relations within an industry, between firms, and even within firms.

Only codified knowledge can really be commercialised, either materialised in products or in the form of licences. The acquisition of knowledge through open channels is necessary in order to convert this knowledge into commerciable information. In contrast to the idealistic picture of open science and open innovation, in the regime of monopolised intellectual property rights, the economy is increasingly shaped by secretiveness and patent disputes. Thus, technological progress adopts a quite specific face (Perelman 2002, 43).

Enforcing Intellectual Property Rights and the Anti-commons Regime?

The effects of the strengthened and extended intellectual property rights are increasingly the subject of critical discussions. Various studies show that an extensive granting of patents can block the free usage and accumulation of knowledge and hence hinder innovation processes (Coriat and Orsi 2002, 1504; Perelman 2002; Rai and Eisenberg 2003; David 2004; Nelson 2004; Orsi and Coriat 2005).

First, this can happen if patents are granted too broadly and therefore close the gate on subsequent research in the same area. Second, there is the so-called anti-commons regime in biotechnology. If companies obtain private rights on DNA sequences, including fragments of a gene, before the corresponding gene, protein or the corresponding active substance has been identified, no one will be able to unify the rights or buy all licences. In such a case, various owners of fragmented goods are given the right to exclude all others, with the ultimate effect that the product will not be produced (Heller and Eisenberg 1998, 699). By creating the possibility of patenting gene fragments, regulatory authorities encouraged the race for private appropriation. However, the production of a recombinant protein drug and new genetic diagnostic test requires the combination of gene sequences. Third, innovation can be blocked if research tools, preliminary products for broad areas of research or key approaches are patented, and if the patent holder aggressively prosecutes unlicensed users or only grants one exclusive license (Nelson 2004, 464). Those who are willing to pay a large sum, generally speaking large companies, may be best capable of tackling those obstacles and creating monopoly-like situations with exclusive rights.

In biotechnology, primarily in the area of genomics, different firms have practised a systematic accumulation of patents. This has permitted them to enclose whole areas of drug targets, substances or technologies and, accordingly, to impose an 'immaterial toll' to block other interested parties if they are not willing to pay licensing fees. Entire cascades of products, such as the sequence of a protein encoded by a gene, the antibodies of a protein, gene vectors, host cells and genetically manipulated animals used in the preclinical trials, are often based on patents on gene sequences. Patent holders can thus block future, still unknown usages and obtain rents from royalties (Coriat and Orsi 2002, 1498; Orsi 2002, 74; Cassier 2003, 71).

Patents became a central valuation criterion not only for firms but also for academic research institutes. US universities increased their licence-based income from 186 million USD in 1991 to 1.3 billion USD in 2001. Many universities depend to a large extent on such income to finance their research expenditures (Ernst & Young 2005, 123). The tendency to patent extensively and on an increasingly broad scale can hinder innovation processes. Companies which finance academic research require reliable and confidential results. This again discourages open interaction between researchers. But it is exactly in this way that academic research in the biosciences can distinguish itself from the business-oriented research (Cooke 2004, 167).

CHANGES OF IPR REGIMES AND NATIONAL INNOVATION SYSTEMS IN THE USA AND EUROPE

All industrial sectors have experienced a drastic increase in the granting of patent during the last two decades. The United States Patent and Trademark Office (USPTO) granted 76,748 patents in 1985, 107,124 in 1991 and 221,437 in 2002. A similar development happened in Europe. There were 42,957 applications for patents in 1985, 60,148 in 1991 and 110,640 in 2002 at the European Patent Office (EPO) (Khan and Dernis 2006, 51ff). From 1990 to 2000, the number of patents granted in biotechnology rose by 15 per cent a year at the USPTO and 10.5 per cent at the EPO, against a 5 per cent a year increase in overall patents. The number of gene patents granted has risen dramatically since the second half of the 1990s. In 2001, over 5,000 DNA patents were granted by the USPTO, more than the total for 1991–95 combined. Similarly, the EPO estimates it has approved several thousand patents for genetic inventions (OECD 2002, 8). Applications to the EPO for biotechnology patents rose from 2,453 in 1991 to about 6,200 in 2000, and slightly declined to 5,876 in 2002 (Khan and Dernis 2006, 34). The European Commission described this trend towards increased patenting activity as the 'pro-patenting era' (European Commission 1999, 14). This trend is expected to continue in the future, as different surveys have confirmed (Thumm 2004, 529).

This inflation in patenting does not necessarily reflect an increase in inventive activity. It also expresses that patents are used for reasons other than the traditional appropriation function. This is generally known as 'strategic patenting' (Thumm 2003, 34; 2004, 529). This is not surprising in a context where knowledge and know-how in the form of patents have become a strategic commodity of firms. Kortum and Lerner (1999) explained this pro-patenting era with changes in management of innovation involving a shift towards more applied activities. Moreover, firms are more conscious of the importance of intellectual property rights. An increase in the bargaining power of companies and higher product market competition are the most important factors underlying this trend in patenting (Thumm 2003, 15).

These explanations need to be complemented with a broader social and economic perspective. The explosive expansion of intellectual property monopolies is less a result of technological breakthroughs than of far-reaching economic and institutional changes. A new regime of intellectual property rights consisting of far-reaching institutional changes emerged. It was accompanied by a changed role for universities and publicly funded research, and cannot be separated from the ascent of concentrated financial capital, changes in financial markets and the entry of pension funds into venture capital. These complementing and reinforcing *institutional complementarities* strongly influence national and regional innovation systems (Coriat 2002a; Orsi 2002; Coriat et al. 2003, 240).

The patent system is an important element of national innovation systems and of innovation policies in states. However, OECD countries are currently converging with respect to their policy designs in science and technology policies, which reflect similar constraints imposed by liberal policies. The US is the main model of reference for most changes and reforms (Braun 2004, 132). This convergence happens also in the field of intellectual property rights. In the context of globalisation of markets and the increasing importance of patents as means to secure investments the international patent system continuously increases its importance at the expense of national patenting procedures. The changes in the intellectual property regime in the US in the past three decades are crucial: first, because of the relevance of the US economy to global dynamics, and second because of the pioneer role of US policy. In the course of the implementation of the TRIPs (Trade Related Aspects of Intellectual Property Rights) agreement, the US was able to enforce its 'philosophy' of intellectual property rights almost on a global scale (UNDP 1999; May 2000, 68ff; Coriat 2002b, 184).

Below, first the major changes of the IPR regime in the US are presented. Then the measures towards homogenisation of intellectual property rights in Europe and the creation of a European IPR regime are described. Finally, it is shown how Germany, Switzerland and Denmark adapted their intellectual property rights regime to this changing framework.

BAYH-DOLE ACT AND CHANGING LEGAL PRACTICE IN THE USA

Technologies developed over the course of the 1970s permit the modification of genetic substances. In 1980, the US Supreme Court for the first time granted a patent for a genetically engineered micro-organism to General Electric (*Diamond v. Chakrabarty*). This decision marked a decisive turning point in US patent policy. Henceforth, inventors could enforce a monopoly claim on life-forms and gene sequences. Such patents unleash a considerably larger monopoly effect than one which monopolises only a production process (Coriat and Orsi 2002, 1498; Orsi 2002, 74; Cassier 2003, 69)

A comprehensive change in the property rights regime then occurred: first, the judges decided that discoveries and not just inventions could be patented. Second, the directives of the USPTO in 1995 and the US courts also authorised patents which could be a basis for different further developments, even if their use could not be proven at the time of the patent application. The renunciation by the courts of industrial 'utility' as a criterion enabled the granting of patents on inventions the uses of which were in the very early stages of research (Orsi 2002, 71f). Thus scientific insights became the objects of systematic privatisation. Since research results can also be patented now at an early stage of the innovation process, it has become possible to block research activities which rely on these insights. Generally, the scope and reach of patents have been greatly extended.

These institutional changes were conducted in light of a change in the universities' role. In the 1970s and 1980s, universities were increasingly assigned the task of contributing to the re-establishment of the US economy's international competitive position and technological leadership in certain areas. A cornerstone of this development was the *Patent and Trademark Amendments Act*, known as the Bayh-Dole Act, in 1980. It gave universities and public research institutes the chance to own property rights on results from research financed by federal money. The passage of the Bayh-Dole Act, which could thus be viewed as 'one part of a broader shift in US policy toward stronger intellectual property rights' (Mowery and Sampat 2005, 228), was strongly promoted by US research universities active in patenting. Previously, universities once understood their mission to be the practice of open sciences and the elaboration of publicly accessible knowledge. Thus since 1980, universities have been able to patent their findings and subsequently commercialise them, be it by new start-up firms or through licensing of patents to other firms (Argyres and Liebeskind 1998, 435f; Coriat and Orsi 2002; Orsi 2002; Nelson 2004, 462).

HOMOGENISATION OF THE REGULATORY FRAMEWORK IN EUROPE

Although later, Europe is experiencing the same development towards stronger intellectual property rights. Two changes in the institutional landscape are to be mentioned: first, in the context of the creation of a single European market, the harmonisation of intellectual property rights systems became more important. Second, the implementation of the TRIPS reinforced intellectual property right protection. Different national systems of intellectual property rights are in this political and economic context considered as non-tariff trade barriers.

Intellectual property laws are usually nationally based, whereas competition is transnational. The national, European (European Patent Convention, EPC) and international (Patent Convention Treaty, PCT) patent rights exist in parallel within Europe. The use of the European patent system has risen tremendously during the last decade and has largely eclipsed the number of national patent applications in Europe (Thumm 2002, 921).

The existing European patent system is based on the 1973 Munich Convention on the European Patent, in accordance with which the European Patent Organisation was established. The European Patent permits a patent application in an official language (English, French, German) of the European Patent Office (EPO) in a unified procedure to receive patent protection in all designated member states of the European Patent Convention of 1973. The European Patent Convention promoted administrative ease but did not introduce a single patent in the sense of enforcement with a material Europe-wide patent law (EPO 2006; Thumm 2002, 921).

A European Community patent, covering with one right the entire territory of the European Union, has not yet been implemented. The proposal for a Community patent, made in 1975 (Convention for the European Patent for the common market) and again in 1989 (Agreement relating to Community patents), has still not been adopted. The Member States were reluctant to establish a European Community Patent because of the high costs of translation into all the official EU languages and because Member States still consider the granting of patents as a matter for national sovereignty (Thumm 2002, 921). Currently, the Community Patent is still the object of controversy. However, the European Commission launched a new initiative to move towards the implementation of a Community Patent on January 16, 2006 (European Commission 2002, 2006). This renewed effort is in line with the life sciences and biotechnology strategy demanding explicitly a European Community Patent adopted in January 2002 which is an element of the competitive agenda of the EU decided at its Lisbon summit in 2000.

There is still a possibility of different interpretations by national laws and national courts in the European patent system. But during the 1980s, European industry increased its pressure to unify the European regulation for biotechnological inventions and to obtain patents on life forms. Directive 98/44/EC of the European Parliament and of the Council of 6 July 1998 on the legal protection of biotechnological inventions, adopted after a 10-year debate, was a further step toward harmonisation (EP and EC 1998). The EU countries should have integrated it into their national law by July 2000. But the Directive still continues to be debated in some countries. The Directive does not create any specific patent law for biotechnological inventions. It rather makes adaptations and amendments which the national legislators must implement. National patent laws remain the essential basis for the legal appropriation of biotechnological inventions (Thumm 2002, 924; 2003, 44).

Heavily discussed in European countries was the patentability of DNA sequences. The Directive excludes the human body and the discovery of one of its parts (for example a gene) or parts of it from patentability. However, a part of the human body (for example a human hormone, a human gene or nucleotide sequence) that is derived from genetic research and isolated from the human body by means of a technical procedure is patentable, even if the isolated part is completely identical to the natural part in the human body. In this respect, there is a convergence of patent policy in Europe and in the US. Both patent offices provided a large number of patents for gene sequences (Schöllhorn and Büchel 2002, 41; Thumm 2002, 926).

A major difference between the patent systems in the European countries and the US concerns the timing of the patent application. The first to invent system in the US requires for novelty that an invention must not have been in public use or on sale or patented or described in a printed publication prior to one year before the US filing. The European system is a first to file patent system. The researcher applying first for a patent has priority over those having made the same invention but who arrived at the patent office later. In the US the researcher can profit from a so-called grace period of 12 months during which she/he can publish her/his results without making a later patent application impossible. The grace period is criticised in particular by industry representatives because it creates a period of uncertainty mainly in complex and competitive fields such as biotechnology. However, universities, research institutes and firms collaborating with public institutes also favour such a grace period in Europe (Schöllhorn and Büchel 2002, 44; Thumm 2002, 923).

Bayh-Dole inspired legislation steps up the role of universities in taking out patents on the inventions of their employee scientists. Preparations for this legislation have not included systematic studies of the manner and the extent to which university scientists, prior to the legislation, were involved in commercial patenting. Nor was any consideration given to how this involvement could be affected by such legislation.

In the wake of this wave of legislation across Europe several recent studies demonstrate that university invented patents are far more prevalent than university owned patents, for example in Italy (Balconi, Breschi and Lissoni 2004), Finland (Meyer 2003) and Germany (Schmoch 2000). The same pattern is identified for single large universities, for example, the Louis Pasteur University (Llerena, Matt and Schaeffer 2003; Saragossi and van Pottelsberghe de la Potterie 2003). A recent study offers a useful overview of these findings (Crespi, Geuna and Nesta 2005) and presents results from a large analysis of 9,000 EPO patents across six European countries, identifying one inventor in each. The sample also includes a small segment of 294 inventor contributions from university scientists, which gave rise to only 85 university assignments.

University-invented, company-owned patenting turns up as a far more prevalent mode of academic contribution to technological invention than is the mode in which universities are assigned patent rights. For the overall contribution of academia to the technological performance of European companies it emerges as an important issue if that contribution is negatively affected by Bayh-Dole inspired legislation, and if, in that case, adequate new mechanisms appear as substitutes. These issues are addressed below based on an empirical study on the effects of the Danish version of Bayh-Dole.

STRONG FEDERALIST SUPPORT POLICY AND REGULATORY ADAPTATION TO THE INTERNATIONAL CONTEXT IN GERMANY

For a long period Bayer and Hoechst were among the leading chemical and pharmaceutical companies in the world. This traditionally strong position was lost in the 1990s. No longer is there a German pharmaceutical company among the global top ten. Only specialised pharmaceutical firms such as Schering (acquired recently by Bayer) and large firms still owned by families such as Boehringer Ingelheim and Merck KGaA were able to defend their strong position. The biotech industry in Germany also emerged in the early 1990s, more than 10 years later than in the US. The rise of the biotech industry in Germany has been characterised by the economic and political ambition to catch up compared to the US and Britain.

On the federal level, three political measures substantially influenced the evolution of biotechnology in Germany: first, the establishment of the genomics centres in Berlin, Heidelberg, Cologne and Munich between 1984 and 1989, second the adoption of gene technology law in 1990 and its amendment in 1993, and, third, the BioRegio contest organised by the BMBF (Federal Ministry for Education and Research) in 1995. This BioRegio contest of 1995 played a crucial role. Actors in regions were invited to submit proposals and to describe how commercialisation of biotechnology in their region could be promoted. The winners, the three most organised regional organisations, Initiativkreis Biotechnologie München, BioRegio Rheinland and BioRegio Rhein-Neckar-Dreieck, each received €25 million over five years to invest in biotechnology. BioRegio Jena in East Germany was awarded a further grant of €15 million. This competition activated a start-up dynamic in these winner regions and also in other regions. The BioRegio contest was an expression of a deliberate policy to promote the commercialization process of the technological potential in order to improve national competitiveness. Due to the federalist structure of Germany this policy was implemented on a regional level, institutionally using the rivalry between regions in Germany (Adelberger 1999, 7; Giesecke 2000, 218; Reiss 2001, 143; Zeller 2001: 66).

The BioRegio contest was followed by a series of further promotion programs of the BMBF. BioProfile supported the specialisation of established bioregions. BioChance launched in 1999 funded certain highly risky R&D projects conducted by small and medium-sized biotech companies. Until 2005 it supported young firms translating their biotechnology knowledge into new products totalling €50 million. Starting in 2003 BioChancePlus offered €100 million for project submissions to small and medium biotechnology firms. By means of the BioFuture contest the BMBF will finance young scientists in biotechnology and related fields totalling 75 million Euro until 2010 (Kaiser and Prange 2004, 402; Ernst & Young 2006b, 97).

In the context of a so-called innovation offensive, the BMBF started a High-tech Masterplan in 2003 which aimed at better access to venture capital, the creation of a competitive tax environment and new collaboration models between public research and small enterprise. This plan was less specifically directed to biotechnology- than to technology-based companies in general. Additionally, on the level of the Bundesländer, especially Bavaria, the biotechnology industry was promoted by different cluster initiatives implemented by the Bundesländer (Ernst & Young 2004, 111; 2006b, 100).

In parallel the German government began to adapt its regulatory framework, especially in the field of intellectual property rights, to the general trend in Europe which had began even earlier in the US. In June 2003 the German government decided to transpose European Community Directive 98/44 into national law, and in March 2004 the German parliament started consultations on the Directive. The biotechnology sector strongly supported the swift implementation of the European Biopatent Directive (Reiss and Hinze 2004, 56).

In Germany, university patents were traditionally regulated by the Inventions made by Employees Act (ArbNErfG). In order to promote freedom of research it included a so-called 'professor privilege', according to which the university inventor and not the university was the owner of the patent. This regulation was criticised by some practitioners and industry representatives. They argued that the research institutions had no incentives to support patent applications (Giesecke 2000, 211). In 2002 the ArbNErfG was amended, transferring responsibility for patent application from the inventor working at a university to the institution. The aim of this amendment consisted of encouraging patenting activities in universities. Accordingly, since February 2002 an inventor, if he wants to publish his results, has to present the invention to the employing research institution, which has to prove the need for applying for a patent. The institution can decide whether it wants to file a patent application or leave the invention to the inventor for application. Inventors receive 30 per cent of the profits (Reiss and Hinze 2004, 43, 55f). The abolition of the 'professor privilege' encouraged the universities to create their own structures for the commercialisation of inventions and patents. However, an inventor can abstain from publication and therefore impede or prevent a patent application. This possibility creates uncertainty for firms interested in collaborating with universities. This shows that, depending on the specific societal context, a regulatory amendment can provoke results which were not originally intended.

However, the German biotech industry remained relatively weak compared to those in Britain and Switzerland, despite strong government support and the considerable public financing of the biotech sector. This raises the question of the extent to which such a promotion policy, applied in specific historical, economic and societal contexts and trajectories, can be an adequate means to improve so-called national competitiveness.

FEDERALIST LIBERAL SCIENCE POLICY AND ADAPTATION OF FAVOURABLE REGULATORY CONDITIONS IN SWITZERLAND

Switzerland has traditionally applied a very liberal style in determining technology policy. Although Swiss governments, both at the federal and canton level, have always fostered the promotion of academic research, undertaking technological research was, for ideological reasons, mainly left to the business sector. Federal administration has traditionally been weak (Braun 2004, 107f). Swiss government strategy consists in developing excellent research in some selected priority areas. Applied technological research is mainly concentrated in a few multinational enterprises. Particularly, in the field of the pharmaceutical and biomedical industries large transnational companies are responsible for the main part of research expenditure. The Swiss federal government has not conceived a more active technology policy. It focuses on providing a good infrastructure and favourable legal conditions. The innovation system worked almost without extra-university government-financed research institutes that could have been used for technology policy purposes. Switzerland has a long tradition of niche specialisation and the development of high-value added products. This strategy has been highly successful including remarkable results in basic research (Arvanitis et al. 2003, 37ff, 65, 81; Braun 2004, 109f).

The universities (Basel, Zürich, Lausanne and Geneva) as well as the Federal Institutes of Technology (ETH Zürich and EPF Lausanne) have been crucial for the creation of new ventures and technology transfer. Novartis and Roche, two major pharmaceutical companies both located in Basel, played another central role in the formation process of the Swiss biotechnology industry. Therefore, the large majority of biotech firms in Switzerland are concentrated in the same regions.

In line with the policy mentioned above, which consisted in strengthening selected technological fields, the Swiss National Science Foundation (SNF) launched the Swiss Priority Program Biotechnology (SPP Biotechnology) in 1992 with public funds. A total of six research modules in biotechnology and complementary activities in continuing education, information, communication, technology assessment and technology transfer were designated to receive state support over a period of 10 years, ending in December 2001. The SPP Biotechnology had the objective of promoting strategic, applied biotechnology research in Switzerland. This program contributed to the creation of 18 new companies. The program budget included SFr100 million allocated by the Swiss Federation. Additionally, some SFr40 million cause from industry, as well as more than SFr60 million of venture capital throughout the duration of the program (Carrin et al. 2003, 42). One outcome of the SPP was the creation of the technology transfer office, Biotectra, in 1996. Unitectra, the technology transfer organisation of the universities of Berne and Zürich and the SNF's SPP Biotechnology was founded in 1999 and is the successor institution of Biotectra (Carrin et al. 2003, 28). The other universities created their respective technology transfer organisations. In 2001, the SNF launched the first 14 National Centres of Competence in Research (NCCR) which represented a new instrument of research and technology policy after completion of the SPP. In the field of life sciences, four NCCRs have been started. They have focused on research in genetics, neurosciences, structural biology and molecular oncology (Carrin et al. 2003, 25f, 42).

The CTI (Commission for Technology and Innovation) assumes an important role in the biotech promotion policy. The CTI has been a traditional key instrument of the Swiss federal government's technology policy. In 2000, it became the 'federal agency for applied research and development'. The CTI promotes applied research and development projects through public–private partnerships and supports start-ups. The CTI Start-up initiative begun in 1995 boosts entrepreneurship at the junction between universities and industry. It supports the commercialization process of products and especially companies in their start-up phase. It selects new firms for venture capital financing and grants them the 'CTI Start-up label' which qualifies them to access CTI support and venture capital. This label also helps to convince potential investors investing in a project. Through its Life Sciences section, the CTI backs applied R&D projects based on a public–private partnership model (50/50 funding as a basic rule) and facilitates a transfer from SNF to CTI funding. For example, it supports follow-up projects stemming from the National Centres of Competence in Research (NCCR) (Carrin, et al. 2003, 21, 62; Ernst & Young 2006a, 10f).

In contrast to Germany, Switzerland does not recognise a so-called 'professor-privilege'. According to the Swiss Law of Obligation (Obligationenrecht/OR), inventions made by employees in the course of their employment belong to the employer. Consequently, the universities and research organisations are the applicant and owner of the patents claiming such inventions.

The government combined the liberal attitude with a careful adaptation of legal conditions. Although Switzerland is not a member of the European Union, its government has very consciously harmonised its regulatory framework in strategically important fields with the framework of the EU. In keeping with this pattern, in 1999 the parliament engaged the Federal Council to adapt the patent law to the European Biotech Directive, passed by the European Parliament and Council in 1998. This revision aims to bring the patent law into conformity with the EU Directive on the legal protection of biotechnological inventions (Thumm 2003, 4). Mainly, the pharmaceutical and biotechology industries have lobbied for the revision. This revision also partially takes into account the interests of academic research in biotechnology. The draft revision of the patent law adopted by the Swiss Federal Council in November 2005 defines two major issues (Ernst & Young 2006a, 16f):

Patentability of genes: According to the EU directive 98/44, isolated genes as such are patentable. In Switzerland, however, naturally occurring genes would be excluded from patenting under the new law. This is a restriction of patentability that goes beyond the practice applied by the EPO. However, it would continue to be possible to patent 'derived sequences' such as the cDNA produced by PCR (Polymerase Chain Reaction), under the condition that at least one function of these sequences is known. This means that the properties and applications of sequences that are derived from gene sequences must already be described in the patent application. Adding them later is no longer possible. This is intended to prevent speculative patent applications.

Patentability of human beings and their body parts: In agreement to European Directive 98/44, the human body as such and human physical components in their natural environment shall also be excluded from patenting. The subject of a patent is a technical teaching as to how human beings can utilize nature in a new way for commercial purposes. This technical beneficial effect makes the discovery an invention under patent law. However, isolated and perhaps technologically modified components of the human body outside their natural environment (such as isolated and possibly genetically modified blood cells) are patentable. Explicitly excluded in the new patent law and also corresponding to Directive 98/44/EC are cloning of human organisms, chimeras with human germ cells, modification of human germ line cells or unmodified human embryonic stem cells.

Role of Patents in Swiss Biotechnology

There is not much empirical evidence available on the impact of intellectual property rights in biotechnology. In the context of the revision process of the patent law the Swiss government requested the Swiss Federal Institute of Intellectual Property to analyse the impact of patents on biotechnological innovations and the economic implications of patents (Thumm 2003). The number of patent applications rises with firm size. However some very active small companies are strongly dependent on patenting and commercialisation. Small companies consider patents to be highly important for acquiring venture capital. They are also more innovative in terms of patent applications per employee in research and development. Similarly, research institutes consider their patents to be important for obtaining financing research and development. For most companies having participated in this survey patents are very important for collaborations with other companies and for timing their scientific publications. Particularly, large companies pay attention to patents when they examine possibilities for collaborations or mergers (Thumm 2003, 19, 23; 2004, 534). Not surprisingly and in line with the economic structure of the country the patent applications of Swiss firms are internationalised. This is expressed in a high number of triadic patents (patent applications filed with the European Patent Office, the US Patent and Trademark Office and the Japanese Patent Office). Switzerland takes the eighth place in triadic patent families, and measured in patent families per million inhabitants Switzerland is the leading country, together with Sweden (OECD 2003). Swiss biotech firms normally apply also for European and US patents (Thumm 2003, 17ff).

EFFECTS OF BAYH-DOLE INSPIRED LEGISLATION ON THE BIOTECH INNOVATION SYSTEM IN DENMARK

This section takes a closer look at some of the implications which Bayh-Dole inspired legislation (discussed in previous sections) has had for the coherence of innovation systems (discussed in the first section of this chapter) in Denmark, focusing on the effects on industry–university (I–U) collaboration in biotech research (Valentin and Jensen 2006).

In the 1950s both Denmark and Sweden enacted legislation explicitly excepting academic scientists from the general principle whereby employers hold the right to inventions produced by their employees. Referred to as the 'teacher's exception' (similar to what in Germany was referred to above as the 'professor's privilege', this principle has been maintained in Sweden until now, but was changed in Denmark on January 1, 2000 in the Law on University Patenting (LUP).[2] The Act transfers to universities the ownership of inventions made as part of the work of employees. That also pertains to inventions resulting from collaborative work with third parties (for example firms), but in these cases the university may renounce the right to the inventions made by the project.

Sweden has maintained its teachers exception, but is currently considering reforms in this area. Hence, comparing post-LUP patterns in I–U collaboration in Denmark with Sweden offers possibilities of a quasi-controlled experiment. Such a comparison benefits from the broad similarities between the two countries, and from the fact that regulations of university IPR uniformly affect all academic research, unmodified by national variations between for example, private v public universities, or by variations at lower levels of government (länder or states).

This comparison is particularly informative if referring to a field in which I–U collaboration plays a significant role. That is the case in biotechnology where a number of studies have documented the strong reliance of firms on knowledge transfer from academic science (Santos 2003; Powell 1998; Liebeskind, Oliver, Zucker and Brewer, 1996; Fuchs 2003). Biotech is also a field in which the two countries are strikingly similar, particularly when we focus on the segment of Dedicated Biotechnology Firms (DBFs) specialised in drug discovery. This segment emerged in the two countries at the same time during the 1990s, and has grown to the same number of firms, with quite similar patenting profiles. Denmark has 51 DBFs, of which 48 have filed patents. In Sweden 41 of the total of 44 DBFs have filed patents.

The data by which the study compares Denmark and Sweden are extracted from all the 1,087 patents filed by these firms. Inventors listed (by name and area only) on each patent front page were identified and their organisational affiliation was established[3] as per the date of patent application. These 3,640 inventor identifications permit the calculation of the share of university scientists for each patent, and of aggregated shares for each country, for specific periods.

The effects of LUP on the share of university scientists in these patents are examined, defining the 'event date' with a lag of one year (that is, January 2001), since the law respects collaborative arrangements established prior to its enactment.

Tests are made for LUP-related shifts (1) in the share of domestic scientists (that is, Danish university scientists contributing to patented inventions assigned to Danish DBF), and (2) in the share of non-domestic university scientists. For both share differences between pre- and post-LUP levels, comparing Denmark and Sweden, are tested by Difference-in-Difference (DD) regressions. Furthermore, to enhance the interpretation of DD findings, tests are made for LUP-related shifts in trends, of shares of university scientists, aggregated separately for the two countries by quarters.

These tests bring out the following:

- Since the mid 1990s the number of DBF patents in Sweden and Denmark as a whole has increased steeply, as has the number of academic scientists contributing to inventions in absolute number. An increase is also observed in academic inventors as a share of all inventors.
- Throughout the 1994–2004 period, academic involvement in Swedish patents was notably above what was seen in Danish patents. However, the Danish pattern, until the introduction of LUP, quite systematically converged towards the higher Swedish level.
- Compared to the Swedish control group, the DD regression identifies a drop of 12.6 per cent in the share of domestic academic inventors behind the Danish patents, specifically attributable to the event. In the trend analyses this appears as a reversal of the previous convergence towards the higher Swedish level into a sharp downward trend.

- Non-domestic inventors on the whole are involved with notably lower shares as compared to domestic inventors, and also in this respect the Danish level is below that of Sweden. However DD regressions identify a significant increase in Danish non-domestic shares of 13.7 per cent, as compared to the Swedish control group, specifically for the post-LUP period. The trend analysis in this case offers only tenuous results, but they suggest a steep increase in the Danish post-event involvement of non-domestic inventors. Consistent yearly increases were observed over the four post-event years, bringing the level from 4.9 per cent to 11 per cent of all Danish inventor contributions.

Taken together these findings strongly indicate a Danish pattern, significantly distinct from the Swedish counterpart, of a post-LUP decrease in the involvement of domestic academic scientists. At the same time, non-domestic university scientists to a notable extent substitute for their domestic counterparts in Danish inventor teams.

To see this decline in university involvement in context the study examines whether Danish academics in the life sciences instead redirect their inventiveness into university-owned patents, as indeed was one of LUP's key objectives. The number of domestic academic inventor contributions to university-owned patents, in relevant IPC categories, filed following the LUP is identified. This number may be subtracted from the number of inventor contributions equivalent to the 12 per cent post-LUP decline detected in the DD regression. On this basis the study estimates the extent to which university owned patents provide a substitutive outlet. The two variations of this estimate presented in the study indicate substitutions at rates of 19 per cent and 26 per cent. University-owned patents, in other words, are very far from providing a substituting outlet for the inventive potential of university scientists previously mobilised for company-owned drug discovery patents. By far the largest part of this academic inventive potential simply seems to have been rendered inactive as an effect of LUP.

To examine the causes behind observed LUP effects the study is underpinned by interviews with academics and DBFs selected from the patent data on which the study is based, identifying explanations consistent with quantitative results from other studies (D'Este and Patel 2005). The interviews bring out that collaborations typically are concerned with exploratory research, referred to by some firms as 'pre-project research'. DBFs enter such collaborations well aware that it is highly uncertain if and how results will be commercially relevant, and that they become so only based on an unknown body of subsequent in-house R&D. Firms tend to find this set-up inconsistent with the LUP principle that universities *ex ante* own the rights to resultant inventions, and reduce the involvement of Danish academic scientists accordingly. For their part, academic scientists typically collaborate in exploratory research to deploy some specialised research experience, which collaboration may be enhanced by offering better research funding, access to an interdisciplinary environment or to the specialised research facilities of the industrial partner. The academic scientist collaborates on such explorative research, in other words, based on skills and motivations which would not allow her to pursue equivalent patenting on her own.

This explanation clarifies not only why LUP has brought about a decline in collaboration within exploratory research. It also explains why LUP would not necessarily have similar effects in university research operating in 'down-stream' fields closer to technology. Down-stream fields are less demanding in terms of complex, post-discovery development, so here academics are better positioned to invent on their own, without relying on clues and information from industrial partners. Furthermore, in cases where such downstream issues are addressed in collaborative projects, they also lend themselves more easily to *ex ante* allocation of IPRs.

Implications for the Danish Innovation System in Biotechnology

Several studies document that research collaborations in biotechnology are established with a strong preference for partners from the same country, even the same region (Allansdottir, Bonaccorsi, Gambardella, Mariani, Orsenigo, Pammolli and Riccaboni 2002; Coenen, Moodysson and Asheim 2004). The advantage of proximate research relationships is not derived from superior qualities of partners who just happen to be local. Rather it comes from the fact that proximate relations tend to be embedded in networks in which actors have repeated interactions and learn about each other over time via multiple channels (Powell 1998; Pyka and Küppers 2003). In this way networks become architectures capable of retaining and transmitting vastly richer information about each actor, as compared to arm's-length relationships to partners who are distant (in the sense of not being part of the network) (Reagans and McEvily 2004). That is why networks offer superior capacity for searches, allowing actors with complex agendas to access the types of complementarity which give rise to effective research partnering (Valentin and Jensen 2002). However, depending on which activities are carried out and their respective knowledge bases, such networks can also have a global reach, as is the case with epistemic communities (Moodysson et al. forthcoming).

For the issues in this chapter the implication of this argument is that an important part of the value emerging from industry–academia collaborations lies in the quality of the network through which either side may undertake effective search so as to identify 'the right complementarity at the right time'. Danish DBFs have no advantage above that of DBFs from other countries when it comes to the search in the global 'market' for academic collaboration. But they do have an advantage in the search in the Danish academic setting, since there are strong indications that they are particularly well connected with Danish universities. The authors behind the LUP analysis summarised above in another study identify all founders and board members affiliated with all Danish and Swedish DBFs through their first year of existence (Valentin and Jensen 2005). This study shows that the vast majority of founders and board members are recruited from Danish organisations.

Founder teams involving Danish university scientists established more than half of Danish DBFs. Similarly academics were present on more than half of the boards that took firms through their first year of business. These compositions of founder teams and boards make them highly effective in subsequent search into the academic potential for research collaboration.

These figures bring out the particular connectivity which Danish DBFs have into Danish academia. In turn this connectivity is a key asset for scientists from both the academic and the industrial side when they look for the complementarity of skills and agendas which is so important for making university–industry research collaboration effective and useful for both commercial and scientific objectives. The chapter demonstrates, consistently with this argument, that this composition of founders and boards matters for the ability of Danish DBFs to establish the diversity of inventor collaborations which in turn affects their commercial performance (Valentin and Jensen 2005). These observations substantiate that research networks are an important part of the Danish innovation system in biotechnology, and that it matters significantly for the inventiveness and competitiveness of firms. It is therefore cause for concern that LUP, as an unforeseen consequence, induced Danish biotech firms to disengage themselves from the national research network, substituting instead a search in the global market for academic research partners. It signifies that LUP, as an unintended side effect, seems to have induced an erosion of parts of the national innovation system of considerable value for Danish science-based competitiveness.

CONCLUSIONS

IPR as a regulatory institution in innovation systems can play a decisive role with a potentially significant impact on the governance of the innovation systems. In this chapter we have shown that across industrial sectors there has been a dramatic increase in patents granted during the last two decades both in the US and in Europe. In Europe, Switzerland is the leading country measured in patent families per million inhabitants, followed by Sweden. This reflects the strength of the pharmaceutical biotech industry in these two countries in contrast to Germany, the biotech industry of which has remained weak compared to those of Britain and Switzerland.

However, this big increase in the use of IPRs is more a result of economic and institutional changes than of technological breakthroughs. A new regime of IPRs emerged accompanied by a change in the role of universities and publicly funded research, and thus, had major impacts on the functioning of regional and national innovation systems. The US has been the model of reference for most of these changes and reforms led by the introduction of the Bayh-Dole Act in 1980. Also in Europe Bayh-Dole legislation has been influential, for example in stepping up the role of universities taking out patents on the inventions carried out by their employed scientists. However, company-owned, university-initiated patenting represents a more important academic contribution to technological invention than university-owned patent rights. Moreover, the Danish example showed that the introduction of a Bayh-Dole inspired law has had negative impacts on university–industry collaboration in contrast to the situation in Sweden, which still deserves the principle of 'teachers exception'.

As shown in the chapter the consequences of introducing IPR differ according to sectors, knowledge bases, phases in the innovation process, and political-institutional frameworks. A patent based IPR makes a 'better fit' in certain sectors and knowledge bases than in others, and works better in specific political environments than in others depending on the types of institutional complementarities present. There may, for example, be a real problem with the current IPR system favouring radical innovations when biotechnology develops from early-phase, hi-tech to more generally diffused multi-tech. All this provides strong arguments against a global generic patent system.

NOTES

1. The Bayh-Dole Act policy can, however, have disadvantageous effects even in the US. However, the US system is a global system and only works reasonably well because it can attract values from other parts of the globe.
2. The Act on Inventions at Public Research Institutions of June 2, 1999 may be accessed at http://www.videnskabsministeriet.dk/cgi-bin/doc-show.cgi?doc_id=14206&leftmenu=

LOVSTOF. An English translation is available at http://www.videnskabsministeriet.dk/cgi-bin/doc-show.cgi?doc_id=20047&doc_type=22&leftmenu=1.
3. The methodology for inventor tracking is described in Valentin and Jensen (2006).

REFERENCES

Adelberger, K.E. 1999, 'A Developmental German State? Explaining Growth in German Biotechnology and Venture Capital', *BRIE Working Paper 134*, 24.

Allansdottir, A., A. Bonaccorsi, A. Gambardella, M. Mariani, L. Orsenigo, F. Pammolli and M. Riccaboni. 2002, 'Innovation and competitiveness in European biotechnology', Enterprise Papers 7, Brussels: Enterprise Directorate-General, European Commission.

Argyres, N.S. and J.P. Liebeskind. 1998, 'Privatizing the Intellectual Commons: Universities and the Commercialization of Biotechnology', *Journal of Economic Behavior & Organization*, 35(4), 427–54.

Arora, A. and R.P. Merges. 2004, 'Specialized Supply Firms, Property Rights and Firm Boundaries', *Industrial and Corporate Change*, 13(3), 451–75.

Arvanitis, S., H. Hollenstein and D. Marmet. 2003, 'Die Schweiz auf dem Weg zu einer wissensbasierten Ökonomie: eine Bestandesaufnahme'. *Studie im Auftrag des Staatssekretariats für Wirtschaft.* Bern: Staatssekretariat für Wirtschaft (seco). 186 Supplement.

Asheim, B. and M. Gertler. 2005, 'The Geography of Innovation: Regional Innovation Systems', in J. Fagerberg, D. Mowery and R. Nelson (eds), *The Oxford Handbook of Innovation.* Oxford: Oxford University Press, 291–317.

Asheim, B. et al. (2007), 'Constructing Knowledge-based Regional Advantage: Implications for Regional Innovation Policy', *International Journal of Entrepreneurship and Innovation Management*, 7, 140–55.

Balconi, M., S. Breschi and F. Lissoni. 2004, 'Networks of Inventors and the Role of Academia: an Exploration of Italian Patent Data', *Research Policy*, 33(1), 127–45.

Bathelt, H., A. Malmberg and P. Maskell. 2004, 'Clusters and Knowledge: Local Buzz, Global Pipelines and the Process of Knowledge Creation', *Progress in Human Geography*, 28(1), 31–56.

Boschma, R. 2004 'Rethinking Regional Innovation Policy: The Making and Breaking of Regional History', in G. Fuchs and P. Shapira (eds), *Rethinking Regional Innovation and Change: Path Dependency or Regional Breakthroughs?* Dordrecht: Kluwer International Publishers, pp. 249–720.

Braun, D. 2004, 'From Divergence to Convergence: Shifts in the Science and Technology Policy of Japan and Switzerland', *Swiss Political Science Review*, 10(3), 103–35.

Carrin, B.J., Y. Harayama, J.A.K. Mack and Z. Zarin-Nejadan. 2003, 'Science–technology–Industry Network – The Competitiveness of Swiss Biotechnology: A Case Study of Innovation', *RIETI Discussion Paper Series 04-E-007*: Geneva, Tohoku, Neuchâtel: Research Institute of Economy, Trade and Industry.

Casper, S. and R. Whitley. 2004 'Managing Competences in Entrepreurial Technology Firms: a Comparative Institutional Analysis of Germany, Sweden and the UK'. *Research Policy*, 33, 89–106.

Chesbrough, H.W. 2003a, 'The Era of Open Innovation', *Sloan Management Review*, 44(3), 35–41.
Chesbrough, H.W. 2003b, 'The Logic of Open Innovation: Managing Intellectual Property', *California Management Review*, 45(3), 33–58.
Coenen, L., J. Moodysson and B.T. Asheim. 2004, 'Nodes, Networks and Proximities: On the Knowledge Dynamics of the Medicon Valley Biotech Cluster', *European Planning Studies*, 12(7), 1003–19.
Cooke, P. 1998, 'Introduction: Origins of the Concept,' in H. Braczyk, P. Cooke and M. Heidenreich (eds), *Regional Innovation Systems*. London: UCL Press 2–25.
Cooke, P. 2001, Regional Innovation Systems, Clusters, and the Knowledge Economy. *Industrial and Corporate Change*, 10(4), 945–74.
Cooke, P. 2004, 'The Molecular Biology Revolution and the Rise of Bioscience Megacentres in North America and Europe', *Environment and Planning C: Government and Policy*, 22(2), 161–77.
Cooke, P. 2005, 'Regionally Asymmetric Knowledge Capabilities and Open Innovation. Exploring "Globalisation 2" – A New Model of Industry Organisation', *Research Policy*, 34(8), 1128–49.
Cooke, P. et al. 2000, *The Governance of Innovation in Europe. Regional Perspectives on Global Competitiveness*, London: Pinter.
Coriat, B. 2002a, 'Le nouvau régime américain de la propriété intellectuelle. Contours et caractéristiques clés', *Revue d'économie industrielle* 99(2), 17–32.
Coriat, B. 2002b, 'Du "Super 301" aux TRIPS: la "vocation impériale" du nouveau droit américain de la propriété intellectuelle', *Revue d'économie industrielle* 99(2), 179–90.
Coriat, B. and F. Orsi 2002, 'Establishing a New Intellectual Property Rights Regime in the United States: Origins, Content and Problems', *Research Policy*, 31(8–9), 1491–1507.
Coriat, B., F. Orsi and O. Weinstein. 2003, 'Does Biotech Reflect a New Science-based Innovation Regime?', *Industry and Innovation*, 10(3), 231–53.
Crespi, G., A. Geuna and L. Nesta. 2005, 'Labour Mobility from Academia to Business', DRUID Summer Conference, Copenhagen Business School, June 27–29, http://www2.druid.dk/conferences/viewpaper.php?id=2773&Cf=18
Dasgupta, P. and P.A. David. 1994, 'Toward a New Economics of Science', *Research Policy*, 23(5), 487–521.
David, P.A. 2004, 'Can "Open Science" be Protected from the Evolving Regime of IPR Protections', *Journal of Institutional and Theoretical Economics*, 160(1), 9–34.
D'Este, P. and P. Patel, 2005, 'University – Industry Linkages in the UK: What are the Factors Determining the Variety of Interactions with Industry?', DRUID Summer Conference Copenhagen Business School, June 27–29, http://www.druid.dk/uploads/tx_picturedb/ds2005-1447.pdf.
Edquist, C. 1997, 'Introduction', in: C. Edquist (ed.) *Systems of Innovation: Technologies, Institutions and Organisations*. London: Pinter.
European Parliament and European Commission. 1998, Directive 98/44/EC of the European Parliament and of the Council of 6 July 1998 on the legal protection of biotechnological inventions, http://eur-lex.europa.eu/smartapi/cgi/sga_doc?smartapi!celexapi!prod!CELEXnumdoc&lg=en&numdoc=31998LØØ44&model=guichatt
European Patent Office. 2006, European Patent Convention. http://www.european-patent-office.org/legal/epc/index.html, accessed 8 May 2006.

Ernst & Young (ed.). 2004, 'Per Aspera Ad Astra "Der steinige Weg zu den Sternen"'. *Deutscher Biotechnologie-Report*. Mannheim: Ernst & Young.
Ernst & Young (ed.). 2005, 'Kräfte der Evolution'. *Deutscher Biotechnologie-Report*. Mannheim: Ernst & Young.
Ernst & Young. 2006a, *Swiss Biotech Report 2006*. Basel, Zürich: Ernst & Young.
Ernst & Young (ed.). 2006b, 'Zurück in die Zukunft'. *Deutscher Biotechnology-Report*. Mannheim: Ernst & Young.
Etzkowitz, H. and L. Leydesdorff. 2000, 'The Dynamics of Innovation: from National Systems and "mode 2" to a Triple Helix of University-Industry-Government Relations'. *Research Policy*, 29, 109–1023.
European Commission. 1999, 'Strategic Dimensions of Intellectual Property Rights in the Context of Science and Technology Policy', *ETAN Working Paper*. Prepared by an independent ETAN Expert Working Group for the European Commission Directorate General XII - Science, Research and Development Directorate AP - Policy Co-ordination and Strategy, Ftp://Ftp.cordis.europa.eu/pub/etan/docs/ipr-seminar-report.pdf.
European Commission. 2002, 'Life Sciences and Biotechnology – A Strategy for Europe'. Communication from the Commission to the European Parliament, the Council, the Economic and Social Committee and the Committee of the Regions, COM (2002) 27.
European Commission. 2006, Consultation and public hearing on future patent policy in Europe: The European Commission, Internal Market. *Webpage* 16 Jan. 2006, 3 July 2006, 16 Aug. 2006. http://ec.europa.eu/internal_market/indprop/patent/consultation_en.htm.
Fagerberg, J. et al. 2005, *The Oxford Handbook of Innovation.* Oxford: Oxford University Press.
Freeman, C. 1987, *Technology Policy and Economic Performance: Lessons from Japan*. London: Pinter.
Freeman, C. 1995. 'The "National System of Innovation" in Historical Perspective'. *Cambridge Journal of Economics*, 19, 5–24.
Fuchs, G. 2003, 'Biotechnology in Comparative Perspective', London: Routledge.
Gibbons, M. et al. 1994, *The New Production of Knowledge – The Dynamics of Science and Research in ConTemporary Societies*, London: Sage (reprinted 1999).
Giesecke, S. 2000, 'The Contrasting Roles of Government in the Development of Biotechnology Industry in the US and Germany', *Research Policy*, 29(2), 205–23.
Granstrand, O. 2005, 'Innovation and Intellectual Property Rights', in J. Fagerberg et al. (eds), *The Oxford Handbook of Innovation*. Oxford: Oxford University Press, 266–90.
Heller, M.A. and R.S. Eisenberg. 1998, 'Can Patents Deter Innovation? The Anticommons in Biomedical Research', *Science*, 280, 698–701.
Husson, M. 2001, *Le grand bluff capitaliste.* Paris: La Dispute.
Jessop, B. 2000, 'The State and the Contradictions of the Knowledge-driven Economy', in J.R. Bryson; P.W. Daniels, N. Henry and J. Pollard (eds), *Knowledge, Space, Economy*. London and New York: Routledge, 63–78.
Johnson, B. et al. 2002, 'Why All this Fuss about Codified and Tacit Knowledge?', *Industrial and Corporate Change* 11(2), 245–62.
Kaiser, R. and H. Prange. 2004, 'The Reconfiguration of National Innovation

Systems – the Example of German Biotechnology', *Research Policy*, 33(3), 395–408.
Kaufmann, A. and F. Tödtling. 2001, 'Science-Industry Interaction in the Process of Innovation: The Importance of Boundary-Crossing Between Systems'. *Research Policy*, 30, 791–804.
Khan, M. and H. Dernis. 2006, 'Global Overview of Innovative Activities from the Patent Indicators Perspective', *STI Working Paper* No. 2006/03. OECD.
Kortum, S. and J. Lerner. 1999, 'What is Behind the Recent Surge in Patenting?', *Research Policy*, 28(1), 1–22.
Liebeskind, J.P., A.L. Oliver, L.G. Zucker, and M.B. Brewer. 1996, 'Social Networks, Learning and Flexibility: Sourcing Scientific Knowledge in New Biotechnology Firms', *Organization Science*, 7(4), 428–43.
Llerena, P., M. Matt and V. Schaeffer. 2003, 'The Evolution of French Research Policies and the Impacts on the Universities and Public Research Organizations', in Aldo Geuna, Ammon Salter and W.E. Steinmueller (eds), *Science and Innovation. Rethinking the Rationales for Funding and Governance*, Cheltenham: Edward Elgar, 147–68.
Lundvall, B.-Å. 1992, *National Innovation Systems: Towards a Theory of Innovation and Interactive Learning*. London: Pinter.
May, C. 1998, 'Thinking, Buying, Selling: Intellectual Property Rights in Political Economy', *New Political Economy*, 3(1), 59–78.
May, C. 2000, *A Global Political Economy of Intellectual Property Rights. The New Enclosures?*, London, New York: Routledge.
Meyer, M. 2003, 'Academic Patents as an Indicator of Useful Research? A New Approach to Measure Academic Inventiveness', *Research Evaluation*, 12(1), 17–27.
Moodysson, J., L. Coenen and B. Asheim. (2008), 'Explaining Spatial Patterns of Innovation: Analytical and Synthetic Modes of Knowledge Creation in the Medicon Valley Life Science Cluster', *Environment & Planning A*, 40, 1040–56.
Mowery, D. and B. Sampat. 2005, 'Universities in National Innovation Systems', in J. Fagerberg et al. (eds), *The Oxford Handbook of Innovation*, Oxford: Oxford University Press, 209–39.
Nelson, R. 1993, *National Innovation Systems: A Comparative Analysis*. Oxford: Oxford University Press.
Nelson, R. 2004, 'The Market Economy, and the Scientific Commons', *Research Policy*, 33(3), 455–71.
Nonaka, R.T. et al. 2000, 'SECI, Ba and Leadership: a Unified Model of Dynamic Knowledge Creation', *Long Range Planning* 33, 5–34.
OECD. 2002, *Genetic Inventions, Intellectual Property Rights and Licensing Practices. Evidence and Policies*. Paris: OECD.
OECD. 2003, *Compendium of Patent Statistics*. Paris: OECD.
Orsi, F. 2002, 'La constitution d'un nouveau droit de propriété intellectuelle sur le vivant aux Etats-Unis: Origine et signification économique d'un dépassement de frontière', *Revue d'économie industrielle*, 99(2), 65–86.
Orsi, F. and B. Coriat. 2005, 'Are "Strong Patents" Beneficial to Innovative Activities? Lessons from the Genetic Testing for Breast Cancer Controversies', *Industrial and Corporate Change*, 14(6), 1205– 21.
Owen-Smith, J. and W.W. Powell. 2004, 'Knowledge Networks as Channels and Conduits: the Effects Spillovers in the Boston Biotechnology Community', *Organization Science*, 15(1), 5–21.

Pavitt, K. 1984, 'Sectoral Patterns of Technical Change: Towards a Taxonomy and a Theory', *Research Policy*, 13, 343–73.
Perelman, M. 2002, *Steal this Idea. Intellectual Property Rights and the Corporate Confiscation of Creativity*. New York: Palgrave Macmillan.
Powell, W.W., 1998, 'Learning from Collaboration: Knowledge and Networks in the Biotechnology and Pharmaceutical Industries', *California Management Review*, 40(3), 228–40.
Pyka, A. and G. Küppers (eds). 2003, *Innovation Networks. Theory and Practice*, Cheltenham: Edward Elgar.
Rai, A.K. and R.S. Eisenberg. 2003, 'Bayh-Dole Reform and the Progress of Biomedicine', *American Scientist*, 91(1), 52–9.
Reagans, R. and B. McEvily. 2004, 'Network Structure and Knowledge Transfer: The Effects of Cohesion and Range.', *Administrative Science Quarterly*, 48(2), 240–68.
Reiss, T. 2001, 'Success Factors for Biotechnology: Lessons from Japan, Germany and Great Britain', *International Journal of Biotechnology*, 3(1/2), 134–56.
Reiss, T. and S. Hinze. 2004, *The Biopharmaceutical Innovation System in Germany. OECD Case Study on Structure, Performance, Innovation Barriers and Drivers*. Stuttgart: Fraunhofer IRB Verlag.
Santos, F.M. 2003, 'The Coevolution of Firms and their Knowledge Environment: Insights from the Pharmaceutical Industry', *Technological Forecasting and Social Change*, 70(7), 687–715.
Saragossi, S. and B. van Pottelsberghe de la Potterie. 2003, 'What Patent Data Reveal about Universities: The Case of Belgium', *Journal of Technology Transfer*, 28(1), 47–51.
Schmoch, U. 2000, 'Wissens- und Technologietransfer aus offentlichen Einrichtungen im Spiegel von Patent- und Publikationsindikatoren', in Ulrich Schmoch, Georg Licht and Reinhardt.Michael (eds), *Wissens- und Technologietransfer in Deutschland*, Stuttgardt: Frauenhofer IRB Verlag, 17–37.
Schöllhorn, A. and D. Büchel. 2002, *Patentes Wissen. Der Schutz des Geistigen Eigentums für biotechnologische Erfindungen*. Basel: Interpharma.
Sell, S. and C. May. 2001, 'Moments in Law: Contestation and Settlement in the History of Intellectual Property', *Review of International Political Economy*, 8(3), 467–500.
Soskice, D. 1999, 'Divergent Production Regimes: Coordinated and Uncoordinated Market Economies in the 1980s and 1990s', in H. Kitschelt et al. (eds), *Continuity and Change in Contemporary* Capitalism, Cambridge: Cambridge University Press, 101–34.
Streeck, W. 1992, *Social Institutions and Economic Performance – Studies of Industrial Relations in Advanced Capitalist Economies*. New York: Sage Publications.
Thumm, N. 2002, 'Europe's Construction of a Patent System for Biotechnological Inventions: An Assessment of Industry Views', *Technological Forecasting and Social Change*, 69(9), 917–28.
Thumm, N. 2003, 'Research and Patenting in Biotechnology. A Survey in Switzerland No. 1 (12.03). Swiss Federal Institute of Intellectual Property: Bern. 97, http://www.ige.ch/e/jurinfo/documents/j10005e.pdf.
Thumm, N. 2004, 'Strategic Patenting in Biotechnology', *Technology Analysis and Strategic Management*, 16(4), 529–38.
United Nations Development Programme (UNDP) 1999, *Human Development Report*. New York, Oxford: Oxford University Press.

Valentin, F. and R.L. Jensen. 2002, 'Reaping the Fruits of Science', *Economic Systems Research*, 14(4), 363–88.

Valentin, F. and R.L. Jensen. 2005, 'Entrepreneurship, Networks and Strategising in Biotech Start-ups', 21st EGOS Colloquium – 'Unlocking Organizations', June 30 – July 2, 2005. Freie Universität Berlin in cooperation with Wissenschaftszentrum Berlin.

Valentin, F. and R.L. Jensen. 2006, 'Effects on Academia-Industry Collaboration of Extending University Property Rights', Research Centre on Biotech Business, *Biotech Business Working Paper*, http://ideas.respec.org/a/kap/jtecht/v32y2007:3p251-276.html.

Whitley, R. 1999, *Divergent Capitalism: The Social Structuring and Change of Business Systems*. Oxford: Oxford University Press.

Zeller, C. 2001, 'Die Biotech-Regionen München und Rheinland. Räumliche Organisation von Innovationssystemen und Pfadabhängigkeit der regionalen Entwicklung', in R. Grotz and L. Schätzl (eds), *Regional Innovationsnetzwerke im internationalen Vergleich*. Münster: Lit-Verlag, 59–82.

Zeller, C. 2004, 'North Atlantic Innovative Relations of Swiss pharmaceuticals and the Importance of Regional Biotech Arenas', *Economic Geography*, 80(1), 83–111.

3. Intellectual property rights policy for gene-related inventions – toward optimum balance between public and private ownership

Koichi Sumikura

INTRODUCTION

This chapter discusses the use of patented genes in genetic diagnosis and the use of patented research tools in research and development (R&D) activities as cases illustrating typical problems with regard to intellectual property rights policy for gene-related inventions. Taking these problems into account, the chapter attempts to identify issues concerning possible measures to promote the public ownership of research results while granting ownership to frontrunners in technology. These measures include the application of anti-monopoly laws, compulsory licensing, new legislation allowing exemption from the effect of rights and new legislation allowing the limitation on the effect of rights. All these measures may be effective in maintaining the balance between the public and private ownership of technology. However, problems exist with these measures as they fail to provide the stability of rights and are not consistent with the relevant international conventions. Therefore, it is hoped that measures will be worked out to enhance access to patented inventions without changing the existence of the patent or the effect of the patent rights.

One such measure considered in this chapter is the establishment of a patent distribution mechanism whereby patented inventions in a certain field of technology are gathered and managed by a single organization to facilitate the conclusion of license agreements for individual technologies. For instance, the establishment of a 'research tool consortium' to collect research tool patents and offer them for use in academic research activity would facilitate and enhance the use of patented inventions in academic research activities.

The results of a survey of life science researchers, conducted by the author of this chapter, show that there is a substantial need among these researchers for such a mechanism. The same survey also found that many researchers hope for a mechanism based on the principle of reciprocity with respect to patented inventions owned by individuals. Under this mechanism, a researcher offering his or her own patented invention for free would be entitled to use patented inventions of other researchers for free. Using the results of this survey as a reference, this chapter considers specific aspects of such a research tool consortium.

GENETIC RESEARCH AND MEDICINE

There is no question that numerous genetic research projects are now being conducted at a broad range of institutions, such as universities, public research institutions, hospitals with research functions, and companies, and their research achievements are contributing greatly to the development of medicine. The identification of a disease gene leads to the development of a new method of genetic diagnosis. Meanwhile, by screening compounds that interact with proteins having certain genetic coding, candidate molecules for the development of new drugs can be identified.

Under the patent systems in Japan and many other countries, genes are subject to protection by patent as an 'invention of a composition of matter'. When a new gene is identified and the function of the gene sequenced, the polynucleotide chain having this particular base sequence will be patented, provided the designated patentability conditions of novelty, utility and non-obviousness, and description requirements for patent specifications are fulfilled.

Gene patents thus granted have influence on genetic diagnosis as well as on the development of drugs using the patented genes. But the impact is more direct on genetic diagnosis than on the development of drugs, as genetic diagnosis using a patented gene cannot be made without permission from the patent holder. In comparison, in case of developing a new medicine using a patented gene, even if there is any influence of the patent with regard to the use of the medicine, it is indirect. In rare cases, genes themselves are used as a medicine. However, the general approach within the drug discovery industry today is to screen the compounds that interact with proteins having certain gene coding and to develop new drugs based on the screened compounds. Thus, the exercise of gene patent rights has no influence on the use of new drugs as a final product. But it should be noted that the screening of compounds involves the use of various research tools, such as cell lines, disease-model mice, analysers, research reagents and research kits, and thus the degree of ease with which such patented research tools are used can influence the development of drugs and even the improvement of medicine as a whole.

One important factor that must be taken into account in considering a patent system for genetic research is that R&D on drugs and diagnostic methods is being undertaken in today's economic society, where decisions on R&D investments are made based on market principles. Corporations are able to conduct R&D by investing massive amounts of money because they are entitled to exclusive rights to their inventions under the patent system. Were it not for the patent system, they would make little effort to develop new drugs or diagnostic methods and there would be little improvement in the quality of medicine. However, there are also instances where patents inhibit the use of genetic diagnosis and the development of new drugs.

Therefore, in order to ensure smooth access to newly-developed drugs and diagnostic methods while promoting the improvement of the quality of medicine through the development of new drugs and diagnosis methods, it is necessary to encourage downstream R&D activities by making research results public under certain conditions or by imposing some limitations on the scope of private ownership, but without changing the basic framework of an intellectual property rights protection system that allows frontrunners in R&D to obtain private ownership of research results. In other words, an appropriate 'balance between public and private ownership' must be realized. Based on this recognition, this study attempts to sort out what measures are available to realize such an appropriate 'balance between public and private ownership', to point out problems with each of the measures, and to present feasible policy measures.

WHERE THE PROBLEMS LIE: TWO TYPICAL CASES

Genetic Diagnosis

One example in which the existence of a gene patent resulted in the reduction of access to medicine is the case of Myriad Genetics, Inc. and its patents for the BRCA1 and BRCA2 breast and ovarian cancer genes.[1] Using these genes, Myriad Genetics developed a method for diagnosing a predisposition for breast cancer and obtained patents in Japan, the United States, Europe, Canada and many other countries, for the genes and for the method for using the genes in diagnosis. Three patents were granted in Japan,[2] eight in the US,[3] and four in Canada.[4] Where such patents exist, a license must be obtained from Myriad Genetics in order to provide genetic diagnosis for breast cancer. The company, however, is pursuing a policy of providing genetic diagnosis on its own instead of licensing its patents to other parties, setting up a genetic diagnosis business called BRACAnalysis. With respect to the BRCA genes, four patents were granted in Europe, too.[5] However, after a series of objections were filed against the BRCA patents, the European Patent Office (EPO) revoked one of the patents in May 2004 and substantially reduced the scope of rights of three other patents in January and June 2005.[6]

In Canada, Myriad Genetics holds a license agreement with MDS Laboratory Services to provide BRACAnalysis testing across the country. After acquiring Canadian patents for the BRCA1 gene in October 2000 and the BRCA2 gene in April 2001, Myriad Genetics began to request licensing fees from medical institutions providing genetic diagnosis for breast cancer. Prior to this, provincial governments had been offering genetic testing for breast cancer using BRCA genes at a cost of about 1,200 Canadian dollars per sample. But Myriad Genetics demanded that they pay 3,850 Canadian dollars in licensing fees. Responses have varied among provinces; some provincial governments continue to provide BRCA genetic testing by agreeing to pay the licensing fees to Myriad Genetics; other governments have temporarily halted the testing; while still others continue to provide diagnosis without paying licensing fees, in effect, ignoring the patents. In any event, it is clear that the existence of the patents is curtailing access to this particular diagnostic technology. Thus, the Myriad Genetics case is frequently referred to when discussing the negative effect of the gene patent system. In Japan, opposition to the BRCA patents is not as great as it is in North America and Europe. But potential risks always exist for the occurrence of similar cases in which the presence of a gene patent impedes access to medical services.

Research Tools

In addition to the aforementioned problem concerning the direct obstruction by patent rights of access to medicine, another problem in which a patent granted to a certain upstream technology hampers the R&D of downstream technologies and prevents the improvement of medicine, must be taken into account in considering intellectual property rights policy for gene-related inventions.[7]

Currently, in Japan and other developed countries, massive amounts of research funds including government funds are being poured into basic research conducted by universities and public research institutions in the basic life sciences. This academic research is funded without explicit expectations that application development is a direct goal. From basic academic research, a number of important innovations that will ultimately lead to the development of new medicines are being created, but indirectly. As discussed below, however, the existence of patents for research tools can result in impeding the promotion of academic research, making it necessary to form a proper system.

In gene-related R&D, various research tools such as cell lines, transgenic mice and vectors are used, and the presence of these tools has now become indispensable for conducting research. A 'research tool', as referred to here, is defined as a tool (including both composition of matter and methods) that is used in a laboratory, not for the purpose of improving itself, but to achieve a certain research goal. In cases where a certain research tool is patented, a license must be obtained from the patent holder, in the same way as with other patented inventions.

Article 69(1) of the Japanese Patent Law stipulates that the 'effect of the patent rights shall not extend to the working of a patented invention for experiment or research'. The 'experiment or research' as referred to therein, according to a study by Keiko Someno,[8] falls into one of three categories: (1) patentability research, (2) function research, and (3) experiments for the purposes of improvement and/or development. It is held that the use of research tools falls into none of these three categories. That is, in deciding whether the use of a certain patented research tool in research activities constitutes the infringement of the patent, academic institutions such as universities are no exception, and it does not matter whether the research activities are conducted for a commercial or non-commercial purpose, or whether they are in the stage of basic research or application development. This position was adopted by the Industrial Structure Council, an advisory body to the Ministry of Economy, Trade and Industry, during its discussion in 2004,[9] and coincides with the position held in the US and Europe.

For instance, in a case where a certain enzyme used for gene amplification is patented, the use of the gene of the enzyme to synthesize, isolate and extract the enzyme in large volumes, after which the mass-produced enzyme is used in gene amplification experiments in a laboratory, would constitute the infringement of the patent.[10] On the other hand, it is interpreted that research for the enhancement of the enzyme's gene amplifiability does not constitute the infringement of the patent because it falls under the category of experiments for the purpose of improvement and/or development.

Until now, researchers have been able to use research tools virtually free from worry about the possibility of patent infringement, so long as they are engaged in academic research. It has been common practice within the research community for researchers freely to use research tools for academic purpose, provided patents for such tools are owned by an academic institution or individual researchers belonging to such an institution.[11] Also even in cases where a certain research tool is patented by a private-sector company, the risk that researchers may face litigation as a result of using the tool in academic research has actually been extremely low; as private-sector companies see little merit in exercising their rights against the use of their patented research tools in academic research.

Interviews with life science researchers, conducted by the author from 2004 to 2005, revealed that anxiety among researchers in using research tools in academic research has risen to an unprecedented level. As reasons behind this, the following factors can be cited:

- With universities becoming more forthcoming in acquiring patents for their research results and selling them to industry, companies are coming to see universities as partners on an equal footing to pursue profits together in the market. On the other side of the same coin, universities are being recognized as competitors in businesses in which the possibility is growing for universities and their affiliated researchers to be sued by a patent-holding company for the infringement of a patent.
- Corporate patent strategies, particularly in case of companies holding patents for research tools (many of which are US or European venture businesses), are becoming stringent. For instance, a definite distinction is made between the use for academic purposes and the use for commercial purposes, or reach-through provisions designed to restrict the handling of resulting research results, such as those under which a patent-holding company would be entitled to receive any license resulting from research using its patented research tool and those obliging a company using the patented tool to report the research results to the patent holder prior to announcement to the public.
- In cases where researchers recognize the existence of a patent, researchers are often unable to get a clear-cut answer to their inquiry about the specific way of handling the patent; a company holding the patent or its agent would give a vague answer that leaves the possibility of the right being exercised in the future, saying that the use of the patented tool would not be an issue for now but there is no telling about the future.
- As a consequence of the growing awareness regarding patents among university researchers, competition among university researchers over academic priority may evolve into patent conflicts.
- As mentioned above, the Industrial Structure Council has confirmed that a license needs to be obtained for the use of patented research tools even when they are used in research for academic purposes, and

this position has been widely propagated among those engaged in academic research. And in fact, as a consequence of such development, many researchers, who were using research tools patented by others without feeling any anxiety, are now suddenly becoming quite concerned, having recognized that the long-established common practice concerning the use of patented tools, in fact, has no legal foundation.

- In the US, it has been established through a series of court precedents that a certain degree of exceptions for experimental use should be allowed. However, the ruling in *Madey v. Duke University*,[12] in which the scope of exceptions was one of the contentious points, narrowly interpreted the 'exceptions for experimental use' and said that research activities conducted by the university do not fit the definition of the exceptions for experimental use. Though opinions are divided about the degree of applicability of this particular ruling, the fact that university research activities have been judged as not meeting the definition of exceptions for experimental use has found its own way to spread out and come to be shared by researchers in the field of life sciences.
- In universities and public research institutions, a mechanism has been established for transferring their inventions to private-sector companies. However, the reality is that neither the intellectual property offices of individual universities nor technology licensing organizations (TLOs) have the capacity to manage the use of patents held by others in research activities.
- In cases where research conducted by teaching staff is regarded as part of their duties and it is policy that any invention resulting from their research activities will be subject to institutional ownership, any patent infringement litigation that may arise from academic research will be filed directly against the institution concerned. Therefore, universities need to know what research tools are being used by their researchers and check for the existence of any patent for each of these tools. As mentioned above, however, it is virtually impossible for the institutions to carry out this task. Thus, it is expected that, in many universities, the task of checking for the existence of any patent and then acquiring the relevant license, should the need arise, will be placed on the individual teacher. Should this become a reality, it would impose a substantial burden on the individual teacher.
- Meanwhile, there is always the risk of litigation because it is difficult for individual teachers to check patent status and acquire a license for each of the research tools used in research.

Because of factors such as those described above, it is feared that researchers may become reluctant to make research plans requiring the use of new research tools. Whether such research tools are actually patented or not cannot be determined without checking. However, irrespective of whether or not certain research tools are patented, the abovementioned situations are imposing constraints on the promotion of research activities. Measures of some sort must be implemented to change this.

MEASURES TO REALIZE AN APPROPRIATE BALANCE BETWEEN PUBLIC AND PRIVATE OWNERSHIP

What can be said about both the two problems discussed in the previous section is that it is necessary to create a mechanism in which achievements resulting from R&D efforts will be made public to the greatest degree possible, and at the same time provide an opportunity to recoup R&D investments by allowing inventors to claim exclusive rights in their achievements under the patent system. In other words, it is necessary to find and realize the optimum 'balance between public and private ownership'. Taking these two issues into consideration as the major points of concern, this section presents and examines policy measures to solve the problems.

'Non-granting of patents', which is the decision not to grant any patent to genes or research tools, would be a possible policy option if thinking only of public ownership. Indeed, as discussed above, Europe has opted for non-granting by rejecting certain BRCA patents. Of course, a non-granting decision would not pose any big problem if the decision were made on a specific case and were reasonable in the light of patent requirements. However, if the non-granting option, a trump card that should be kept in reserve, becomes a policy tool that is used in all time, it presents a huge problem. This is not a reasonable policy, because under such a situation those who have undertaken R&D would be unable to recoup their investments. Thus, non-granting of patents is not included in the policy categories discussed hereunder. In the explanation below, policy measures are sorted out from the standpoint of seeking partial public ownership, or public ownership under specific conditions, while acknowledging a certain degree of rights to frontrunners.

Application of Antimonopoly Legislation

One way of seeking the optimum balance between public and private ownership is through the application of competition policies. The exercise of intellectual property rights is treated as an exception under anti-monopoly and antitrust laws.[13] However, such exceptional treatment is not granted automatically because certain acts by patent holders, such as demanding unreasonably expensive licensing fees by taking advantage of the lack of alternatives for their patented technology, may be considered an abuse of rights. The anti-monopoly provisions that are potential options include reach-through provisions, essential facility theory and compulsory licensing.

In light of anti-monopoly and antitrust laws, reach-through provisions are designed to extend the scope of rights to cover areas where patent rights have yet to be established. Similar cases include a clause under which the licensee (patent user) is obliged to allow the licensor (patent holder) to claim ownership in any improvements made by the licensee on the patented inventions (the so-called 'assign-back' clause) or another clause under which the licensee is obliged to allow the licensor to obtain an exclusive license to use any improvements made by the licensee (the so-called 'exclusive grant-back' clause). Both of these schemes 'fall under the definition of unfair business practice (as defined in item 13 "transaction based on restrictive conditions" in the general specifications of unfair business practices set forth by the Japanese Fair Trade Commission) and are likely to be judged illegal'.[14] On the other hand, such rules do not necessarily apply to reach-through provisions included in a license agreement for patented research tools, a scheme for handling 'derivative inventions' (coined by the author) derived from the use of research tools, because such provisions are, as is apparent from the definition provided above, not stipulated as an improvement on the invention. Yet, reach-through provisions are no different from assign-back and grant-back clauses, in that they impose restrictive conditions on transactions, and it is considered reasonable to make judgements by citing rules for improvement inventions. Thus, imposing an assign-back or exclusive grant-back obligation on licensees with respect to derivative inventions goes beyond the scope of exceptions granted for the exercise of intellectual property rights and may be deemed to fall under the definition of unfair business practice.

As such, it is considered possible to set clear rules, by incorporating the existing criteria and standards for the handling of rights in a license agreement with respect to derivative inventions made through the use of research tools. Of course, an agreement will not present any problems even if it includes reach-through provisions, provided that reasonable licensing fees are set and that the licensee is able to save on the initial cost of obtaining the license by arranging the agreement to allow the licensor to earn fees contingent on success. This type of agreement is beneficial to both parties.[15]

What is called the 'essential facility theory' can be cited as one way to construct logical arguments for finding a certain patent to violate anti-monopoly legislation, for instance, in cases where a gene patent unfairly prevents the implementation of genetic diagnosis or a research tool patent unfairly prevents the implementation of R&D. This is a theory that interprets a certain patent as violating anti-monopoly legislation if license refusal results in obstruction of access to an essential facility. The theory, though not yet clearly defined for the Japanese system, is being applied in the US to prohibit the monopoly of physical infrastructure such as electric cables and has been applied in some court cases in Europe.[16] However, it seems that, even internationally, no clear consensus exists about the 'essentiality' and rational criteria have yet to be formulated. Under these circumstances, should this theory be arbitrarily applied to cases where the exercise of intellectual property rights is interpreted as preventing the public ownership of technology useful for human beings, the stability of rights would be undermined and the intellectual property rights system would exist only in name.

Another way of achieving the optimum balance between public and private ownership is to establish a compulsory licensing right for patents used in genetic diagnosis and research tool patents. This is a system set forth under the patent laws of many countries. It is generally provided that the government may determine that a patent holder should be forced to allow the use of its patented technology by third parties if and when the presence of the patent is hampering the proliferation of technology deemed necessary for public health and welfare. In Japan, Article 93 of the Patent Law stipulates that should a party be seeking the use of a patented invention deemed necessary specifically for public interests, the party may request adjudication by the Minister of Economy, Trade and Industry if it fails to obtain the relevant license through its negotiation with the patent holder. In principle, a patent holder can freely decide to whom it will allow the use of its patented inventions. However, once adjudication proceedings have been carried out in accordance with the designated procedures, the patent holder, regardless of its own intentions, must issue a license (compulsory licensing) if so decided by adjudication. For such a license issued in accordance with the adjudication decision, the patent holder is entitled to receive a license fee, but cannot itself decide the amount of the fee by negotiating with the patent user.

Article 31 of the Agreement on Trade-Related Aspects of Intellectual Property Rights (TRIPS) also stipulates that compulsory licensing may be permitted only if the proposed user has made efforts to obtain authorization from the right holder on reasonable commercial terms and conditions, and if such efforts have not been successful within a reasonable period of time. Moreover, the Article also states that a member country may waive such requirement and immediately allow compulsory licensing in the case of a national emergency, in other circumstances of extreme urgency or in cases of public non-commercial use.

Like the aforementioned essential facility theory, this compulsory licensing system appears to be an attractive scheme to cope with situations in which the presence of intellectual property rights is obstructing the public ownership of a useful technology. Indeed, in September 2001, France's Minister of Research and Minister of Health and Social Protection issued a joint statement calling for applying the compulsory licensing system to genetic diagnosis.[17] With respect to patented inventions used as a research tool, Strandburg (2004) calls for enabling patent holders to issue a license on their own terms and conditions for a period of five years, and applying compulsory licensing if they fail to offer a license on reasonable terms and conditions after the initial five-year period.[18]

Compulsory licensing rights have rarely been invoked in developed countries because such a move would destabilize the rights stipulated by the law. Japan is no exception; there has not been a single case subjected to adjudication, including adjudication in case of non-issuance of licenses (Article 83) or adjudication in cases where the use of a patent held by a third party is involved (Article 92). Arguments have been made calling for the application of compulsory licensing rights as a means to facilitate the proliferation of medical services, including drugs and diagnostic methods. However, the frequent application of compulsory licensing rights could discourage efforts to develop new drugs and diagnostic methods. Therefore, any problems that may arise from the existence of a patent, except in case of emergency, should be dealt with by other means, such as the purchase or assumption of payment of licensing fees by the government or an international organization.

New Legislation Allowing Exemption from the Effect of Rights

Another approach is establishing exceptions to the effect of rights by means of legislation or, in certain cases, by changing the interpretation of the law. Possible measures include setting forth provisions that acknowledge gene patents but designate genetic diagnosis as an exception to the effect of such patent rights, and provisions that acknowledge research tool patents but designate academic research at universities and public research institutions as an exception to the effect of such patent rights.

With respect to the latter scheme, active discussions are being carried out in Japan and internationally on whether 'experimental exceptions' to academic research or non-commercial research at academic institutions should be permitted. As discussed above, it is generally perceived in Japan that the use of patented research tools does not fit the definition of an

Table 3.1 What should be made exempt from the effect of patent rights?

a.	**All** research activities should be **within the scope** of the effect of patent rights.	13%
b.	All research activities at **academic research institutions** such as universities and research institutes of an incorporated administrative agency (IAA) status should be **outside the scope** of the effect of patent rights whereas all research activities at **private sector companies** should be **within the scope** of the effect of patent rights.	16%
c.	In addition to criteria specified in **b.**, research activities at **academic research institutes** such as universities and research institutes of an IAA status should be **outside the scope** of the effect of patent rights so long as such research activities are carried out for **non-commercial purposes**, but should be **within the scope** of the effect of patent rights if they are for **commercial purposes** (for instance, if research activities are being conducted jointly with a private sector company or if a patent application has been filed for any achievement in the research).	56%
d.	Criteria specified in **b.** should apply only to cases where a patent has been obtained in **research financed by public funds** such as research grants provided by the central government.	6%
e.	Criteria specified in **c.** should apply only to cases where a patent has been obtained in **research financed by public funds** such as research grants from the central government.	7%
f.	Others	3%

exception provided for under Article 69(1) of the Patent Law, and thus any party wishing to use the tools must obtain a license from the relevant patent holder. However, a survey of life science researchers at universities and public research institutions (hereinafter referred to as 'Life Science Researcher Survey'),[19] conducted by the author and others in 2004, found that 56 percent of the respondents believed that non-commercial research at academic research institutions should be exempt from the effect of patent rights, whereas 13 percent believed that there should be no exemption from the effect of patent rights, and 16 percent that all the research activities conducted at universities and other academic institutions should be exempt (see Table 3.1).

In creating a system that allows for certain exemptions from the effect of rights, possible categories for differentiating between what falls under the effect of patent rights and what does not include 'non-commercial research and commercial research', 'academic research and non-academic research', and 'basic research and applied research' (with the former category being exempt from the effect of patent rights and the latter being subject to the effect of patent rights). However, it is difficult clearly to distinguish between 'non-commercial research and commercial research' because many of the technologies developed in basic research activities at universities are now being transferred to industry, and something that started out as research for non-commercial purposes may become a commercial research project from a certain stage onward. The distinction between 'non-academic research and academic research' would be made based on the nature of the organization undertaking the research. But then, a problem would arise as to how to handle research activities that are conducted at academic research institutions but in collaboration with a private-sector company. Although it is difficult to define 'basic research and applied research', it is possible to differentiate between them by establishing criteria for each type of research, such as whether the research is at the stage that is before or after clinical testing or of screening. Because differentiation is based on the stage of research development, there is a drawback in that it is difficult for those engaged in upstream research to recoup the cost of R&D investment.

In addition to the abovementioned difficulties in setting guidelines, trying to designate exemptions by means of legislative measures is just as arbitrary as the aforementioned essential facility theory and compulsory licensing, and the social demerit caused by impairing institutional predictability poses a serious obstacle.

New Legislation to Allow Limitations on the Effect of Rights

A final policy option for research tools is the limitation on the effect of rights by introducing new legislation. This option is slightly more moderate than the one described in the foregoing section, under which patent holders would not be able to exercise their rights under certain conditions. Normally, frontrunner scientists who have discovered a new gene are granted a substance patent for the gene, which entitles them to exercise rights over any use of the gene. In recent years, however, gene sequencing and function prediction have become easier than in the past, presenting the problem that the scope of rights obtained by frontrunners is too broad for the actual degree of scientific contribution by them.[20] Granting of a patent in a way to cover a disproportionately broad area compared to the actual result of research would hamper the subsequent development of circumventing inventions. That is, with other researchers becoming discouraged from attempting new challenges, progress towards innovation would be slowed.

To address this problem, a notable legislative step was taken in Germany. The revised Patent Act, which was approved by the Bundestag, the lower chamber of the federal parliament of Germany, on December 3, 2004, and then by the Bundesrat, the upper chamber, on December 17, 2004, was a domestic legislative step to implement EU Directive 98/44/EC. What is characteristic of the revised law is that if the patent is a substance patent obtained by discovering the functions of a naturally-occurring human gene, it limits the effect of patent rights to the use of the functions disclosed at the time the application for the patent was filed.[21] The formation of this system is quite reasonable and intended to define the scope of privatization in accordance with the degree of scientific contribution, making public ownership possible for the remaining areas of the scope. However, this measure is meant to provide special treatment only for technologies of a specific area and may contradict Article 27(1) of TRIPS, which stipulates that patents must be granted without discrimination as to the field of the invention.

Apart from legislating new law, it is possible to limit the scope of the effect of rights by applying stricter screening criteria or changing standards for court decisions. In the stage of granting a patent, this can be done by applying stricter criteria for the written description and enablement requirements, as well as for the support and clarity requirements for written descriptions of claims. Also, in the stage of interpreting claims, that is, in court decisions, the scope of the effect of rights can be limited by changing the criteria for applying the doctrine of equivalents to narrow the scope within which a new invention is judged as an equivalent to a certain patented invention.

FEASIBLE SOLUTIONS

Although being intended to promote the facilitation of genetic diagnosis using patented genes, as well as of R&D activities using patented research tools, the policy measures discussed in the preceding section have, as pointed out earlier, their respective deficiencies. Applying the essential facility theory, establishing compulsory licensing rights and legislating new laws to set exemptions from the effect of patent rights would impair the stability of rights and even undermine the credibility of the patent system. Thus, none of them can be deemed realistic in practical application. Meanwhile, imposing a certain limit on the effect of patent rights with respect to certain subject matters, as practiced in Germany, has the possibility of causing a problem in maintaining consistency with the requirements under TRIPS, and thus it is unlikely this measure will spread to many countries. In addition, should such arbitrary application of the patent system to a certain field take root, it would adversely affect other fields. Therefore, this section attempts to propose a measure for systematically solving the problem in a way that is different from those discussed thus far. This measure aims to increase the degree of public ownership by facilitating access to patented inventions while at the same time returning benefits to patent holders, without making any changes to the existence and effect of patent rights.

Integrated Management of Patents

Even where the existence of patent rights is presupposed, it is still possible to balance the goals of both public and private ownership of patented inventions by having patent holders issue a non-exclusive license under reasonable terms and conditions, instead of an exclusive license to a single licensee. Furthermore, the degree of public ownership can be increased by having patent holders issue a non-exclusive license to any applicant in a non-discriminatory manner, because such non-discriminatory issuance would effectively provide prospective users with inexpensive access to the patented invention. In other words, provide a license under reasonable and non-discriminatory (RAND) conditions.[22] Patent holders, for their part, can also expect greater revenue by providing a license in this manner because they would be collecting fees, although small in amount per license, from a number of licensees.[23]

By offering a multiple number of patented inventions in a specific field for use under RAND conditions, that is, creating a bundle of licenses available under RAND conditions, the accessibility to inventions in that specific field can be effectively enhanced. Consequently, work and procedures for concluding a licensing agreement can be reduced compared to those required when such patented inventions are individually offered, thereby creating a situation where patented inventions become publicly available to a greater extent. The integrated management systems that are being used to create bundles of patents include the following:

- *Patent pools.*[24] One example of a patent pool is the MPEG-LA, which handles a series of essential patents including MPEG-2, a public standard for image compression technology. The MPEG-LA collects patents relating to specific public technology standards and provides a portfolio license for them under RAND conditions.
- *Patent platforms.*[25] Under the 3G Patent Platform, a licensing regime handling patents for third-generation mobile communication systems, an essential patent holder and a licensee are, in principle,

to enter into a 'Framework Agreement' as a standard license agreement. However, a license agreement that is different from the standard license agreement (for instance, a cross license agreement agreed to by the parties concerned) may be concluded if so wished by both the patent holder and the licensee.

- *Patent clearinghouse.* This is the equivalent of a convenience store for patents, where patented inventions in a certain field of technology are gathered and a license agreement for each of them can be concluded in a simple manner. The Public Intellectual Property Resource for Agriculture (PIPRA), an organization set up in the US for the purpose of facilitating the use of basic tools in the field of agriculture-related biotechnologies, is considered to be one form of a patent clearing house.[26]

Creating a mechanism for facilitating access to patented inventions by taking into account the characteristics of existing organizations such as those listed above would make it possible to promote the public ownership of R&D achievements with respect to patents used for genetic diagnosis and gene-related research tool patents.

From now on, genetic research will make further advances and eventually almost all human genes, which form the foundations of research, will come to be known. When this happens, a patent will be obtained each time a disease-causing mutation of a specific gene is identified, which would easily bring about a situation where a multiple number of entities held patents for a multiple number of mutations of a single gene. In order to facilitate the provision of genetic diagnosis under such a situation, it is desirable to have a mechanism in which all such patents for mutations are gathered at and provided by a single organization. Furthermore, this mechanism can be an extremely convenient tool for genetic diagnosis providers, if the mechanism is designed in such a way that users, by accessing the organization, can easily search databases for gene patents held by the organization and conclude the necessary license agreement.

The same can be said about research tool patents; if basic research tool patents for life sciences were gathered under a single organization, it would help facilitate R&D activities both at universities and private-sector companies. Moreover, in certain fields of inventions where both academic institutions, such as universities and private-sector companies, can become licensees, as is the case with research tools, it would be necessary to prepare two different sets of terms and conditions for licensing; one applicable to academic institutions and the other to private-sector companies.

The discussion thus far has been made in consideration of the two cases, that is, genetic diagnosis and research tools. In the context that

Table 3.2 Is a research tool consortium necessary?

a	Yes, it is.	72%
b	No, it is not.	4%
c	Not sure.	24%

academic institutions, such as universities, can be licensors and licensees, it is expected that research tools will require a more complex mechanism for patent distribution than was necessary for genetic diagnosis. Thus, a mechanism for facilitating the distribution of patented research tools will be discussed below.

Aspects of a Research Tool Consortium

A mechanism for the integrated management of patents can be referred to as a 'research tool consortium', and by focusing on this perspective, the optimum embodiment of a 'balance between public and private ownership' can be examined. The envisioned function of a research tool consortium is to gather all patents necessary for R&D and services in a certain specific field so that a licensing agreement can be concluded easily. A portfolio license, such as the one provided by the MPEG-LA, is rarely offered. In many cases, a convenience store-like system (called a patent clearinghouse), under which users would pick up only those products they need from the assortment available, would better fit the reality. The operator of a research tool consortium needs to negotiate, in principle, with all the patent holders and set terms and conditions for the licensing of their patents beforehand. On the other hand, licensees using the consortium would not have to negotiate with each patent holder. Thus, in cases where there are many licensees and the operator of a research tool consortium has an established capacity for concluding license agreements, transaction costs for society as a whole can be reduced.

The necessity of a consortium-style organization has been acknowledged by 72 percent of the respondents to the Life Science Researcher Survey discussed earlier (see Table 3.2). The Survey (see Table 3.3) showed that 55 percent of those who responded felt it would be desirable for the scope of patented inventions handled by the consortium to encompass all the necessary patents, regardless of whether they are owned by private-sector companies or universities and public research institutions. Many of those holding research tool patents are bio-ventures in the US and Europe, most of which specialize in providing research tools. In order to incorporate patents held by such private-sector businesses, it would be necessary

Table 3.3 What is the most desirable scope of patented inventions handled by the research tool consortium?

a. Patented inventions owned by all sorts of organizations, including universities and private-sector companies (including those in overseas), should be handled.	55%
b. Only those owned by academic institutions such as universities should be handled.	31%
c. Only those derived from research financed by public funds such as research grants from the central government should be handled.	9%
d. Only certain types of fundamental patented inventions should be handled.	2%
e. Others	2%

to negotiate and conclude a license agreement with each of them to enable the consortium to provide their patents.

On the other hand, the question posed in Table 3.3 also found that many people (31 percent) expressed the view that a consortium should be created based only on tools held by universities and other academic institutions. This is thought to be a realistic view. Because it is conceivably difficult to gather patented inventions for major research tools by negotiating with all the patent holders, these respondents feel it is desirable to launch a consortium covering patents held by some universities and public research institutions in the first phase and then in the next phase, build on that foundation by calling on even more organizations to participate.

Inventions under Individual Ownership

In the Life Science Researcher Survey, respondents were asked under what situation they would be willing to offer their patents to the consortium if they were patent holders. It was found that 46 percent of the respondents were in favor of a system based on the principle of reciprocity; in other words, they would offer their patented inventions for free use if they were entitled to the free use of patented inventions owned by others. They exceeded the 27 percent of the respondents who replied they would provide their patents if they received sufficient financial compensation (see Table 3.4). There were very few people who did not want to offer patented inventions under any conditions. Based on these findings, it is possible, at least as far as university researchers are concerned, to gather a certain number of patented inventions under a consortium if rules are set in such a way that researchers offering their own patented inventions and having

Table 3.4 If there is a 'research tool consortium', would you offer your patented inventions to the consortium?

a. I would offer my patented inventions for free if I could use other patented inventions for free.	46%
b. I would offer my patented inventions for free provided that the amount of fees is sufficient.	27%
c. I would not offer my patented inventions under any condition.	2%
d. I don't know.	20%
e. Others	5%

given promises to licence them out under RAND conditions are entitled to use other patented inventions held by the consortium.

If a situation emerges, however, where those offering only useless tools are allowed to use all the other tools, it will cause a sense of unfairness within the consortium, which is supposed to be an entity established based on the principle of reciprocity. This could lead to an increase in the number of those who seek to take but give little, called free riders, and if this situation continues, the whole scheme would cease to function. Therefore, in order for this sort of scheme to succeed, there must be a mechanism to attract many researchers and provide them with incentives to contribute good tools. One possible way to achieve this is to implement a system in which each use of a research tool is reported to the secretariat of the consortium, and the secretariat announces a ranking of the most frequently used research tools. This would encourage researchers to participate and offer good inventions to the consortium. On the other hand, venture businesses specializing in providing research tools would see no merit in joining a consortium based on the principle of reciprocity. Thus, as mentioned above, it is necessary for the operator of the consortium to negotiate with venture businesses and obtain licenses to use their inventions.

Inventions under Institutional Ownership such as Those Owned by Universities

Following the incorporation of Japanese national universities in April 2004, many universities and public research institutions have come to hold ownership of inventions made by researchers affiliated with them. However, many inventions that had existed before university incorporation belong to the individual researchers.[27] Therefore, the situation today is that inventions under individual ownership and those under institutional ownership co-exist.

With respect to inventions under institutional ownership, such as those owned by universities, an invention made by a researcher affiliated with a certain university is theoretically available for free use by other researchers of the same university in research activities conducted within the university.[28] This system can be expanded to cover a group of universities so that researchers affiliated with these universities can use each other's research tools either for free or at minimal cost, while at the same time allowing external researchers, such as those who are not affiliated with any of the group universities, to use research tools by paying reasonable fees and by completing simple procedures. This is the basic concept of a consortium under institutional ownership.

Possible Policy Measures

The foregoing sections have discussed the idea of research tool consortiums. As to how to implement this idea, however, there are few policy actions available because decisions on conditions for licensing and participation in a research tool consortium are made, in principle, by individual patent holders. Specifically, possible measures are limited to:

- financially assist the establishment of such consortiums and support the development of human resources,
- call on universities and individual researchers to participate, and
- make it a rule to have all the results derived from government-funded research projects placed under a consortium.

It would be reasonable for the government, as a condition for providing research grants, to make it a rule that those engaged in government-subsidized research activities must agree to allow any research results to be licenced out under RAND conditions or provide their patented research tools to a consortium to be offered out under a portfolio license.

On the other hand, however, such policy actions are not binding for research activities not financed by the government. Thus, it is hoped that by setting up and operating certain consortiums, the practice will gradually expand to the whole research community.

SUMMARY

This chapter has taken up genetic diagnosis using patented genes and R&D using patented research tools as cases illustrating typical problems with regard to intellectual property rights policy for gene-related

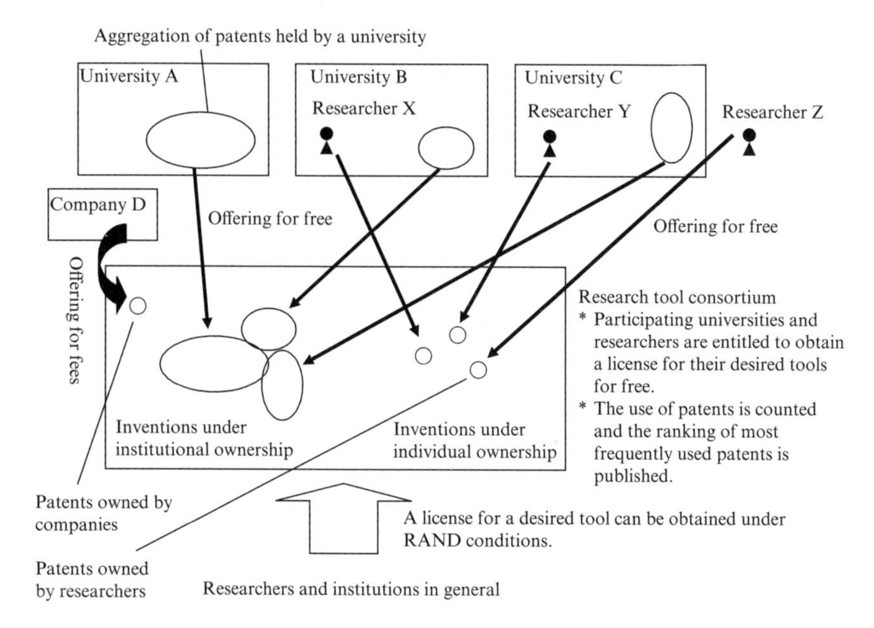

Figure 3.1 Conceptual picture of a research tool consortium

inventions. Taking these problems into account, the chapter has examined the feasibility of several possible options, including the application of anti-monopoly laws, compulsory licensing and new legislation to allow exemptions from the effect of rights or to limit the effect of rights, as a means to promote the public ownership of research achievements while giving exclusive rights to frontrunners. However, problems exist with these measures as they fail to provide the stability of rights and are not consistent with the relevant international conventions. Thus, it is necessary to work out measures that can enhance accessibility to patented inventions without making any changes to the existence of the patent or the effect of the patent rights.

As one way to achieve that end, the establishment of a consortium for integrated management of patents was examined, using cases such as patent pools as references. Figure 3.1 shows the characteristics of a research tool consortium, which was considered as one form of consortium. Specifically, the research tool consortium has the following characteristics:

- Participating universities have a collection of their institutionally-owned patents placed under the management of the consortium for free. Participating researchers have their individually-owned patents placed under the management of the consortium for free.

- The operator of the consortium builds its capacity for license negotiations, and then, negotiates and obtains the necessary licenses for certain patents held by private-sector companies and others, if the operator finds it necessary to make these patents available through the consortium.
- Participating universities and researchers are entitled to use any patents managed by the consortium for free or at a discount in return for offering their patents to the consortium for free. Different terms and conditions are set depending on whether use is by universities (in cases where all researchers affiliated with the university use a patent) or by individual researchers.
- Researchers, universities and other research institutions that are not offering any patents to the consortium may receive a license for patents held under the consortium by paying fees (under RAND conditions in many cases). Here again, terms and conditions are different depending on whether use is by individual researchers or for use by universities and other institutions.
- A license is provided for commercial use under different terms and conditions from those applicable to non-commercial use.
- For each research tool managed under the consortium, the number of times the tool is used is counted and a ranking of the most frequently used tool is announced.
- In cases where a research tool is to be created under a government-subsidized research project, those engaged in the project must agree to have the tool placed under the management of the consortium as a condition for receiving research grants.

It has been pointed out that Japanese researchers are feeling great concern about the possible existence and exercise of patent rights in using research tools for academic research. However, if most of the major research tools were managed under a consortium, such as the one described above, the researchers would be able to concentrate on their research without worrying about having to handle the rights for using research tools.

Patents used in genetic diagnosis (patents for disease-related mutations) are different from research tool patents, in that universities are not end users. However, it is believed that the use of such patents can be facilitated by creating a mechanism very similar to the one used for research tool patents, to manage all the related patents in an integrated manner. As suggestive as this idea might be, as a mechanism for facilitating the use of patents the idea is still at a preliminary stage. It must be refined further by studying concrete issues such as pricing and the planning of license agreements.

NOTES

1. For more details see Williams-Jones (2002) and Paradise (2004).
2. The three patents are Patent No. 3241736 (substance patent for BRCA1), Patent No. 3455228 (substance patent for BRCA2), and Patent No. 3399539 (diagnosis method using BRCA1).
3. See Paradise (2004).
4. Ibid.
5. Ibid.
6. On May 18, 2004, EP 699754 for BRCA1 was revoked. In January 2005, two BRCA1 patents (EP 705902 on January 21 and EP 705903 on January 25) were reduced in the scope of patent rights with regard to specific gene probes but not their diagnostic applications. On June 29, 2005, the scope of patent rights of EP 785216 for BRCA2 was reduced to cover only the use of a specific mutant of the BRCA2 gene. The information is based on the EPO website (http://www.epo.org/about.us/press/releases/archive/2005/29062005.html.
7. Pioneering research works about this problem include Nakayama (2003) and Walsh, Arora and Cohen (2004).
8. See Someno (1988).
9. See 'Issues Concerning Smooth Use of Patented Inventions', a November 2004 report compiled by the Working Group on Patent Strategy Plan, Patent System Subcommittee, Intellectual Property Committee, Industrial Structure Council, available at http://www.jpo.go.jp/shiryou/toushin/shingikai/pdf/strategy_wg_prob/00.pdf.
10. However, this does not constitute an infringement of the patent provided that the enzyme has been purchased from a distributor formally licensed by the patent holder and that it is used for the purposes designated under the agreement concluded at the time of purchase.
11. In the case of research tools for which high reproducibility is difficult to realize, such as cell lines and mice, the general practice within the research community has been for a patent-holding researcher to offer his or her patented research tool to other researchers for free use and these researchers, in return, make reference to this when publishing their research results.
12. *Madey v. Duke University*, 307 F. 3d 1351 (Fed. Cir. 2002). Dr John Madey, a former Duke University professor, sued Duke University for infringing his patents by continuing to use certain equipment that had been installed in the university and contained devices covered by his patents even after his resignation from the university. The university lost the case.
13. In Japan, Article 21 of the Anti-monopoly Law stipulates this.
14. See Yamaki (2002), at 202.
15. Similar ideas are provided in a June 2002 report compiled by a Fair Trade Commission study group examining issues on patents in new fields from the viewpoint of competition policy. See a news release at http://www.jftc.go.jp/pressrelease/02.june/02062603.pdf.
16. See Basheer (2004).
17. Press Release, Institut Curie, 'The key dates' (Sept. 26, 2002), available at http://www.curie.fr/upload/presse/keydates.pdf.
18. According to Strandburg (2004).
19. The survey was conducted at the Kobe Portopia Hotel in August 2004, during the conference of the 'Four Genome Domains' study group under a government-subsidized scientific research project in priority areas. A total of 172 individuals responded to the questionnaire. Of the 132 respondents whose affiliation could be identified, 130 were from a university or a public research institution, and only two were from a private-sector company. Of the 127 respondents whose title could be identified, 28 were university professors, 56 were researchers of other status (associate professors, assistants, post doctors, etc.), and the remaining 43 were students.

20. See Sumikura (2001) and (2002).
21. This is based on Section 1a (4) of the revised Patent Act of Germany, which stipulates: '[i]f the subject-matter of the invention is a sequence or partial sequence of a gene, the structure of which is identical to the structure of a natural sequence or partial sequence of a human gene, then its use, for which the industrial applicability has been described in concrete terms . . . , is to be included in the claim'.
22. RAND conditions is a well-established term that is used in 'patent declarations', a document submitted by a company holding a patent to a standards organization, if the company's patent rights conflict with the standards that are being set.
23. As detailed in Watanabe and Sumikura (2002), the Cohen and Boyer patent for gene recombination, an invention that is believed to have generated the greatest amount of licensing fees for a technology transferred from an American university, has been licensed out to a number of licensees on a non-exclusive basis.
24. See Sumikura (2003) and Ozaki and Kato (1998).
25. See Kato (2001).
26. See Atkinson (2003). The PIPRA website is: http://www.pipra.org/.
27. Prior to the incorporation of national universities, inventions made by national university researchers were placed under either national or individual ownership. According to a 2002 report compiled by a working group of the Ministry of Education, Culture, Science and Technology on intellectual property, however, those under individual ownership accounted for 86.4 percent of inventions that had been subjected to deliberation by national university invention committees, an organ commissioned to decide the ownership of inventions.
28. In relation to this, each university should create rules about how to return benefits to individual researchers for the use of their inventions within the university.

REFERENCES

Atkinson, R. et al. (2003), 'Public Sector Collaboration for Agricultural IP Management', *Science.* 301: 174–175.

Basheer, Shamnad (2004), 'Block Me Not: Genes as Essential Facilities?' *Chizaiken Kiyo 2004*, 82–87.

Kato, Hisashi (2001), 'Dai-San-Sedai Idotai-Tushin no tameno Patento Purattofomu Raisensu (Patent Platform License for Third-Generation Mobile Communications)', *Chizai Kanri,* 51: 559–569.

Nakayama, Ichiro (2003), 'Nichibei-Hikaku kara Mita Tokkyoken to "Jikken no Jiyu" no Kankei ni tsuite: "Shiken/Kenkyu no Reigai" no Hensen to Kadai (Relationships between Patents and the "Freedom of Experiment" from the Viewpoint of Comparison Between Japan and the United States: The Development and Future Challenges of "Experimental Use Exceptions")', *AIPPI* 48: 436–472.

Ozaki, Hideo; and Hisashi Kato (1998), 'MPEG Patento Potoforio Raisensu (MPEG Patent Portfolio License)', *Chizai Kanri,* 48: 329–337.

Paradise, J. (2004), 'European Opposition to Exclusive Control over Predictive Breast Cancer Testing and the Inherent Implications for U.S. Patent Law and Public Policy: A Case Study of the Myriad Genetics' BRCA Patent Controversy', *Food and Drug Law Journal.* 59: 133–154.

Someno, Keiko (1988), 'Shiken/Kenkyu ni okeru Tokkyo-Hatsumei no Jisshi (I) (Implementation of Patented Inventions in Testing and Research (I))', *AIPPI* 33: 138–143.

Strandburg, K. (2004), 'What Does the Public Get? Experimental Use and the Patent Bargain', *Wisconsin Law Review*, 73: 81–155.

Sumikura, Koichi (2001), 'Seimei-Kogaku to Tokkyo no Shin-Tenkai: Genomu-Tanpakushitsu Kaiseki to Tokkyo (New Developments in Biotechnology and Patents: Genome and Protein Sequence Analysis and Patents)' and Aita, Yoshiaki; Hirashima, Ryuta; and Sumikura, Koichi "Sentan-Kagaku-Gijutsu to Chiteki-Zaisanken (Advanced Science Technologies and Intellectual Property Rights)" Chapter 1, Japan Institute of Invention and Innovation.

Sumikura, Koichi (2002), 'Idenshi Tokkyo (Gene Patents)', *Associe*, 9: 60–70.

Sumikura, Koichi (2003), 'Sentan-Kagaku-Gijutsu ni okeru Tokkyo Puru no Katuyo (The Use of Patent Pools in the Advance Science Technology Field)', Vol. 1 and 2, *Bio Industry*, 20(2): 42–52, 20(3): 55–62.

Walsh, J.P., A. Arora and W.M. Cohen (2003), 'Effects of Research Tool Patents and Licensing on Biomedical Innovation', in W.M. Cohen and S.A. Merrill (eds), *Patents in the Knowledge-Based Economy*. The National Academies Press, Washington, DC: 285–340.

Watanabe, Toshiya and Koichi Sumikura (2002), *TLO to Raisensu Asosieito (TLO and Licensing Associates)* Tokyo: BKC.

William-Jones, B. (2002), 'History of a Gene Patent: Tracking the Development and Application of Commercial BRCA Testing', *Health Law Journal*. 10: 121–144.

Yamaki, Yasutaka (2002), 'Q&A Tokkyo-Raisensu to Dokusen-Kinshi-Ho (Q&A: Patent Licensing and the Antimonopoly Law)', Commercial Law Center, Inc.

PART II

Intellectual property management in biotechnology

Introduction

Karen L. Durell

Undoubtedly, intellectual property management ('IPM') is quickly becoming an important aspect of corporate strategies in a growing variety of markets. In fact, the term IPM has achieved almost buzzword status and is applied by groups as diverse as law firms to private-research organizations. Such indicators show that IPM is moving to the forefront of corporate interest, including in the biotechnology sphere. Of course, as a term becomes more broadly used, its meaning can be diluted. This part examines what IPM means in light of innovators' need effectively to manage intellectual property (IP). The following three chapters will look at aspects of IPM and attempt to offer a broad-view approach to the issues related to its adoption in IP intensive innovation systems. Jointly, these chapters provide us with indications about the state of the art in IPM, and draw important linkages between it and IP. A common theme that runs through the chapters is the fact that although IPM is a frequently used term, not everyone who utilizes it ascribes it an identical meaning. IPM is not accompanied by an authoritative definition, and is therefore open to interpretations that are context and user dependant. Although each author will discuss IPM individually, we accept as a general statement that IPM refers to a strategic application and ordering of intellectual property rights ('IPRs') within an organization. The management strategies used by organizations evolve to facilitate the utilization of IPRs to achieve a specific, defined goal set by an organization. As the nature of institutions applying IPM will differ greatly one to another, so will the goals set by each. This in turn means that the IPM strategy will also differ. The result is that IPM can have many faces, and for this reason it can be hard to give a consistent definition, or a consistent set of illustrations drawn from the practical uses of IPM in the organizations.

The first chapter is by Patrick Sullivan, a leader in IPM and one of the first people in the world to develop and implement IPM strategies. Sullivan provides us with a primer on IPM offering a consistent grounding in the issues that supports all subsequent discussion of the topic. As a starting point, Sullivan points out that IPRs are tools. Although the popular discourse often attributes certain outcomes to individual IPRs, the reality is that how one exercises one's right of IPR ownership is the outward manifestation of the strategic interest held by the IP owner. In this respect, IPRs are tools for strategic management, rather than being ends in themselves. As Sullivan makes clear in his chapter, an important starting point for IPM is to understand that IPRs are not the foreground, but will instead be subsumed in broader interests that structure the organization's overall management strategy.

Sullivan also explains the ways that one can find IPRs implemented. He acknowledges the many values that may be attributed to IPRs and explains that the value which IPRs represent to a particular corporation will be directly influenced by the goals of the organization. For example, IP can be managed in a way that is an explicitly offensive strategy, or the converse can be true and the approach will be defensive. Organizations will test the waters with new IP to find its value within their strategic purposes, which is one way that organizations can gain a sense of the accountable value of the IP assets. This is a key concept of IPM. Each firm that engages in IPM will design its strategies to promote its internal goals. Varying goals necessitate varying applications of IPR tools, so that no two corporations are likely to apply identical IPM strategies. For this reason, IPR tools will have different values to different corporations. For example, the IP may have revenue value, it may have a role in cost-reduction or strategic positioning value. There can also be regional differences if, for example, one takes the North American context in which IP is sometimes treated as if it were a substitute for cash, whereas in the European Union IP is often characterized in a social networking vein. Thus, IPRs that are not particularly useful to the internal workings of corporation A may glean significant value for the organization if they are licensed out to corporation B, that has a great need of the IPR. Thus, IPM is not merely about remaining internally focused. External relationships between corporations can be vital.

These themes continue through the chapters that follow. Sharon Oriel, who has been involved in corporate IPM for many years, including 30 years at Dow Chemical, offers an example of the difficulty that is posed when an IPM term must be defined. She notes that confusion occurs even amongst IPM professionals, and yet in order for effective discourse to occur it is crucial that IPM terms be commonly understood. For the particular context in which she was anchoring her remarks – the biotechnology industry – this turns out to be very important because biotechnology firms tend to have a common profile, and that means the culture of a firm or the entire sector will have a significant influence on the IMP strategy that will be applied. Oriel characterizes biotechnology firms as being small, start-up organizations with a single product with no interest in diversification but which are keenly aware of the need to establish competitive advantage. In biotechnology firms, science tends to be paramount, but the business aspects of the firm take a back seat.

What impact on IPM do these traits of biotechnology firms have? First, because of their small size and relative newness, biotechnology firms are often principally concerned with cost management issues. The result is that biotech IPM often emphasizes small-scale IPRs (for example, national patents rather than global patent portfolios achieved through PCT patent filings). Players within this sector are also often engaged in heated disagreements about issues of ownership and ethics relating to IPM strategies and thus, although IPMs applied in this sector are all likely to be small in scale, they will be diverse as regards the level of enforcement and retention of IPRs.

A further consideration for the average biotechnology company involves considerations of value. As Oriel states, a biotech firm is responsible for delivering value to its stakeholders. This is another instance where clear definitions are crucial. The shift in terminology away from 'shareholder' is deliberate. It is indicative of Oriel's view that many companies see themselves as having not only obligations to shareholders, but also obligations to a broader community of interests. In this respect, there is an expansion beyond the narrow group of parties who hold shares in a company to the wider group of parties who have an interest in the company's ability to achieve success. In a limit case, one can perhaps imagine groups such as CAMBIA, or some other kind of open source organization, explicitly inverting the order from stakeholders to shareholders.

In the final chapter in this part, Karen Durell and Richard Gold move away from the specifics of particular IPM strategies that may be adopted, which is the main focus of Sullivan and Oriel. Instead, Durell and Gold focus on the scope of the strategies themselves. The authors identify three different general IPM strategies – firm-centric, firm-networked and non-firm. Durell and Gold offer examples of the present application of firm-centric and firm-networked IPM, which are common. They argue that IPM is poised to expand to include a non-firm approach, an option that is not bound to function within either a firm internally, or inter-firm relationships.

A non-firm approach has the potential to shift the foundation of IPM goals to a wider sphere of organizations, those that participate in a common market or sphere of innovation, which in turn has the potential to spur innovative uses of intangibles. Actions undertaken within a system of innovation, rather than a single firm, cause the organizations within the system to set a common goal and to create relationships that will aid in the attainment of that goal. An implication of moving IPM beyond the boundaries of the traditional firm may be that the pool of decision-makers would expand to include policy-makers at the industry and government levels. This change could widen the range of IPM tools available, so as to encompass public policies, industry practices, international treaties and the regulatory environment. The goal of dynamic IPM may be stated as competition through cooperation. Moreover, non-firm IPM can alter the value of IPRs, as their use and potential for bundling are augmented in a wider sphere of management. Durell and Gold provide several examples, from the biotechnology sphere as well as other sectors, of firm-centric and firm-networked approaches. These are used to compare and contrast existing IPM approaches with the potential represented by non-firm IPM.

In summary, these chapters provide an overview of IPM in biotechnology and a suggestion of the future progression of strategies in this field. As a relatively new corporate tool IPM has much room for growth and the authors attempt to point out how it may flourish in the biotechnology sector. As a final comment on the part, it is worth noting that several common themes ran through all of the chapters. Foremost amongst these is the observation that the category of intangible assets includes more than just intellectual property. For this reason it is crucial that an IPM strategy identify which intangibles it will address – for example tradeable IP, or non-tradeable business practices and culture. All of the speakers also agreed that firm culture has a significant impact upon IPM and will shape the strategies as it influences the goal of an entity. Linked to this issue is the fact that IPM should remain open to new methods and not become static. Open source initiatives may be an example of a progression which can be integrated into IPM and may therefore merit further consideration in firms previously not inclined to take open source seriously as an option. Particularly as non-firm entities become more important players, it will be important for firms to consider the full range of options they have available for valuing and managing their intellectual property assets.

4. Fundamentals of intellectual property management

Patrick H. Sullivan

The field of intellectual property (IP) management has developed significantly since the mid-1980s. IP has moved beyond the narrow purview of the corporation's legal department into the broader arenas of business tactics and strategy. Any conversation about the role of intellectual property in a business or an industry, such as biotechnology, demands that participants be aware of the range of opportunities and practices that underlie IP and IP management.

The rich and well-established business literature on IP management falls into two major groups: IP as a legal asset and IP as a business asset.[1] These two views of intellectual property are quite different in nature, and are typically held by different sets of people within a corporation. Those who hold the 'legal asset' view of IP, usually attorneys or people with legal training, are interested in providing legal protection for the organization's innovations. Their view is primarily (although not entirely) defensive: like builders of fortifications, their goal is to create a safe and strong set of walls to keep out attackers who threaten to steal the organization's intellectual treasures. These people seek to create the strongest possible legal barriers to competitors and, with only rare offensive forays against individual infringers, position the organization behind strong defenses. On the other hand, those for whom IP is primarily a business asset tend to focus on maximizing the organization's revenues and profits. For them, the management of IP is primarily an offensive activity: they use IP as one of several springboards for increasing the firm's earnings. The information on IP and IP management in this chapter presents both of these perspectives.

WHY IS IP IMPORTANT?

Intellectual property is an intangible. It represents a set of rights and has no physical substance. As an intangible, it falls within the set of 'assets' that have been called the firm's 'intellectual capital'. Intellectual capital or intangibles (the terms may be considered synonomous) now represent a major portion of the market capitalization value of a firm. In recent years the number of companies the value of which lies largely with their intangibles has increased dramatically.

In a study of thousands of non-financial, publicly traded firms in the Compustat database over a 20-year period, Dr Margaret Blair of the Brookings Institution reported a significant shift in the makeup of the value of company assets.[2] In 1978, her study showed 80 percent of the average firm's value could be attributable to tangible assets on its balance sheet, with 20 percent of the company's value due to other factors. Ten years later, in 1988, the makeup had shifted to 45 percent tangible assets and 55 percent other factors. By 1998, only 30 percent of the value of the firms studied was attributable to the value of their tangible assets, while a stunning 70 percent was attributable to other factors. Blair concluded that the value not explained by the firm's tangible assets was associated with the firm's intangible 'assets'.

An Accenture study written in 2000 found similar patterns in the evolving size of the 'other factors' component of market capitalization, but came to different conclusions as to the explanation. The Accenture study concluded that the amount of a firm's market capitalization not accounted for by its tangible assets represented the firm's assets and activities *not under active management*.[3] Significantly, the Accenture paper did not identify intangibles as the major component of firm value; rather, it only defined the non-tangible component as those value-generating items and activities *not under active management*. This would include, but not be restricted to, the firm's intangibles. It could also include factors such as investor expectations.

The factors that contribute to a firm's market capitalization are complex and difficult to pin down. They are largely made up of investors' expectations of future revenue in addition to any value that they may attribute to the firm's assets, both tangible and intangible. But if more than 70 percent of investor expectations relate to future revenue, with the value of the firm's intangibles playing a significant role in those expectations, then it is fair to say that intangibles (including intellectual property) have assumed an increasing portion of the value of enterprises over the past two decades.

THE DEVELOPMENT OF IP LAW

Intellectual property represents the legal rights to intellectual activity in the industrial, scientific, literary and artistic fields. Intellectual property laws aim to safeguard creators and other producers of intellectual goods and services by granting them certain time-limited rights to control the use(s) made of their creations. Intellectual property is traditionally divided into two branches: 'industrial property' and 'copyright'.[4]

Until the latter half of the nineteenth century, there were no international agreements on legal rights for the owners of intellectual creations. Countries enacted their own laws to suit their individual interests and needs for protecting intellectual activity. As a result, it was difficult for an individual or a firm to obtain legal protection for intellectual works in another country because countries' laws were so diverse. A patent application, for example, had to be made at approximately the same time in multiple countries in order to prevent a subsequent publication in one country from invalidating the novelty of the invention in the others.

The spurt of technology development in the latter half of the nineteenth century made it imperative that a more internationally oriented and harmonious set of intellectual property laws be developed in order to facilitate the flow of technology as well as international trade. Under the auspices of the Empire of Austria-Hungary, the Congress of Vienna for Patent Reform was convened in 1873 and elaborated a set of principles on which an effective and useful patent system should be based.

The Vienna Congress was followed by an International Congress on Industrial Property in Paris in 1878, which recommended that an international diplomatic conference be called to determine the basis for uniform legislation in the field of industrial property. A diplomatic conference in Paris in 1883 produced the Paris Convention for the Protection of Industrial Property. Initially, two secretariats were created (one for industrial property and one for copyrights), but in 1893 the two secretariats were merged into one. The Paris Convention has been revised from time to time since 1883. Each of the revision conferences ended with the adoption of a revised Act of the Paris Convention. Although this string of 'Acts' still has historical interest, the great majority of the world's countries are now parties to the latest Act, that of Stockholm in 1967.

The Stockholm Act had several important accomplishments. First, it created the World Intellectual Property Organization (WIPO) as a specialized agency of the United Nations. The mission of WIPO is to promote through international cooperation the creation, dissemination, use and protection of works of the human mind for the economic, cultural and social progress of all mankind. It seeks to create a balance between protecting the moral and material interests of creators on the one hand, and providing worldwide access to such creativity on the other.

The Act also provides that:

> intellectual property shall include rights relating to:
> – literary, artistic, and scientific works

- performances of performing artists, phonograms, and broadcasts
- inventions in all fields of human endeavor
- scientific discoveries
- industrial designs
- trademarks, service marks, and commercial names and designations
- protection against unfair competition

. . . and all other rights resulting from intellectual activity in the industrial, scientific, literary or artistic fields.[5]

WIPO defines intellectual property protection as including the following:

Patents – A patent is the right granted by the State to an inventor to exclude others from commercially exploiting the invention for a limited period, in return for the disclosure of the invention, so that others may gain the benefit of the invention.

Copyrights – Copyright law protects the owner of rights in artistic works against those who 'copy' or who take and use the form in which the original work was expressed by the author. Copyright law protects the *form* (that is to say the creativity in the choice and arrangement of words, musical notes, colors, shapes, and so on) of the expression of ideas, not the ideas themselves.

Trademarks – A trademark is any sign that individualizes the goods of a given enterprise from the goods of competitors.

Industrial Designs and Integrated Circuits – 'Industrial design' refers to the creative activity of achieving formal or ornamental appearance for mass-produced items.

Geographic Indications – 'Champagne', 'Cognac', 'Roquefort', 'Tequila', and 'Darjeeling' are some well-known examples of names associated with products of a certain nature and quality. One common feature of all those names is their geographic connotation, the function of designating existing places, towns, regions, or countries. These are defined as indications of source and appellations of origin.

Protection Against Unfair Competition – Unfair competition is defined as any act of competition contrary to honest practices in industrial or commercial matters. Examples of unfair competition include (but are not limited to) 'creating confusion', 'discrediting', and 'misleading' activities.

INTANGIBLES AND INTELLECTUAL PROPERTY

It is generally agreed that the intangibles of an organization are representations of several different key forms of knowledge. During the 1990s, a

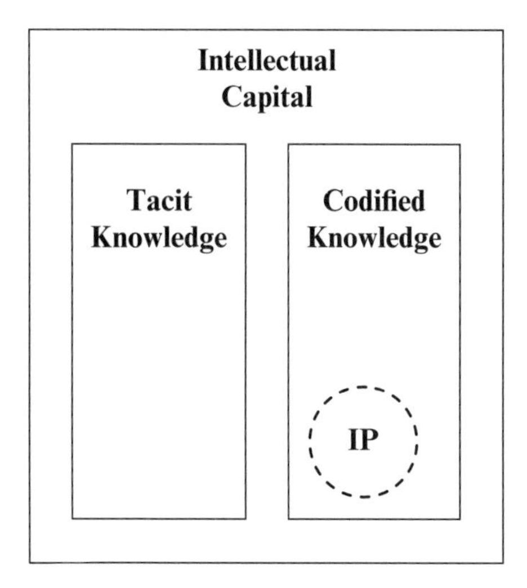

Figure 4.1 Elements of intellectual capital

community arose of those interested in the management and exploitation of these several forms of knowledge. It became known as the 'intellectual capital' community (the term 'intellectual capital' has largely been replaced in recent years by the term 'intangibles'). This community gave rise to the ICM Gathering, a group that defined the elements that constitute both tacit knowledge and codified knowledge.[6] As shown in Figure 4.1, intellectual property is a subset of a firm's codified knowledge.

In the simplest terms, intangibles are the firm's knowledge, know-how and relationships. Intangibles may be tacit or codified. When they are tacit, they reside within the mind(s) of company employees and other stakeholders. When they are codified, they have been committed to some form of media – typed into a computer, drawn on a blueprint, written on a piece of paper, or painted on a canvas. Owners may decide to protect some of their codified knowledge in the form of patents, copyrights, trademarks or whatever additional forms of legal protection are available in their country.

While the 'box' model of Figure 4.1 proved useful for visualizing the key elements of the value extraction process, particularly where IP was involved, it was not particularly helpful for managers who wanted to create value directly from tacit knowledge. For this purpose, the 'three-ring' model of intellectual capital proved more meaningful (see Figure 4.2).

This model of intellectual capital, while frequently used by companies interested in creating more in-house knowledge, is less useful for managers concerned with IP because it subsumes intellectual property into the

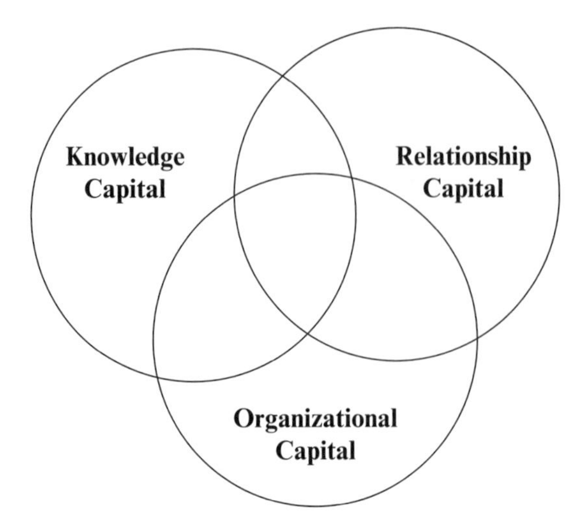

Figure 4.2 A value-creation view of intellectual capital

organizational capital of the company, where it is less visible and where its relationships with the other forms of knowledge intangibles are not as clear.

THE VALUE OF INTELLECTUAL PROPERTY TO THE ORGANIZATION

Most of us think of value in dollar terms. But firms can extract many kinds of value from their intellectual property, including value that is not measurable in dollar terms. We refer to dollar value as 'direct value' and to other kinds of value as 'indirect value'. Both kinds of value may be the result of either defensive or offensive activity.

- *Direct value*: value from activities that unambiguously link the IP with an identifiable flow of cash. Direct value can be measured in currency (for example, dollars, yen, euros).
- *Indirect value*: identifiable value, often intuitively obvious and sometimes compelling, that is not associated with a specific transaction (such as a license or a sale).
- *Defensive activity*: prepares the firm for invasive action by individuals or groups outside the firm. It entails the development of assets or resources that will help repel or neutralize intrusive activities that threaten the firm.

- *Offensive activity*: targets individuals or groups outside the firm. Its purpose is to advance the organization's ability to achieve its strategic goals or to implement its strategy. Offensive activity frequently concerns revenue or profit generation.

OFFENSIVE AND DEFENSIVE ROLES FOR IP

At a meeting of the ICM Gathering in 2001, in a discussion of value measurement and valuation, the participating companies agreed that accounting-based measurement was not helpful. Accounting-based measurement of value is most helpful when reporting externally on the firm's financial assets because the accounting framework is well-known. Unfortunately, accounting-based measures of intangible value are notoriously inaccurate and misleading. For the internal management of intangibles, Gathering companies developed alternative methods of valuation, methods that required an understanding of the *kinds of value* that intangibles provided to the organization. The participating organizations identified a spectrum of different kinds of business value of IP to their firms. These fell under two major headings: defensive value and offensive value.

Whereas the kinds of business value available under the *defensive* heading were small in number, the kinds of *offensive* value intangibles could provide was much more extensive (see Table 4.1). From the list of the kinds of value provided in Table 4.1, companies may identify which kinds of value are most consistent with their organization's business activity and business strategy. Knowing the kind of value they seek from their intangibles, companies may measure the degree to which their intangibles provide what is sought. In addition, they can then devise methods and processes for its extraction and delivery. Because most companies do not know how many different kinds of value their IP can provide, they tend to underspecify (or not specify at all) what they want from their intangibles.

Once a company knows the kinds of value it wishes to extract, the task of measuring that value becomes much more meaningful.

IP and Business Strategy

Whether a company's IP strategy is or should be offensive or defensive in nature depends in large part on the company's business strategy and the role intellectual property is expected to play in that strategy. Every company should have a vision of the company it wishes to become in the future. That vision establishes long-term goals that incorporate a set of operationally meaningful statements, which in turn affect employees'

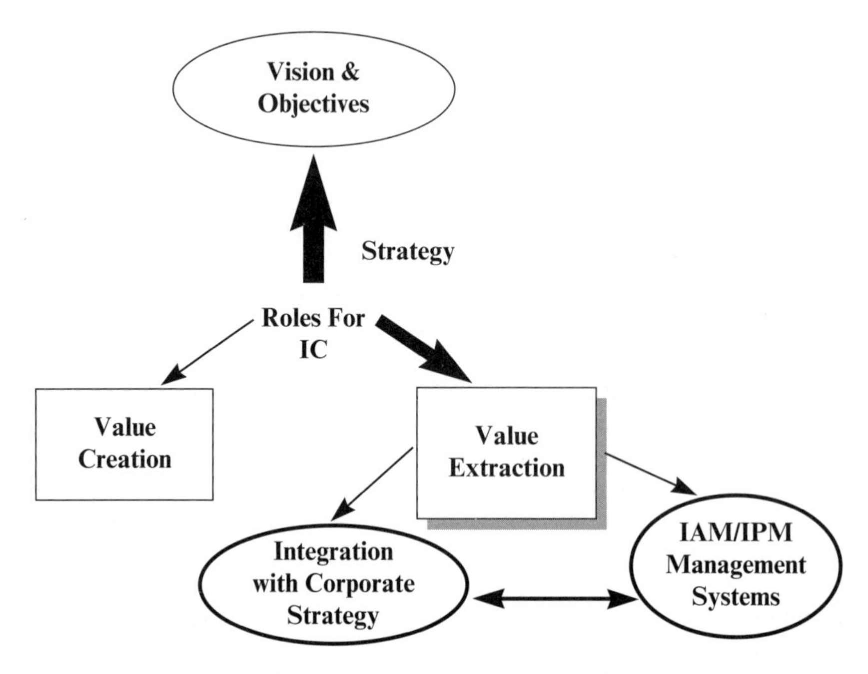

Figure 4.3 Relationships among vision, strategy and intellectual capital

day-to-day activity. For knowledge companies, the firm's strategy, the set of decisions about how to make progress toward the vision, includes the role the firm's intellectual capital is expected to contribute (see Figure 4.3).

The roles for IP are typically specified for both value-creation and value-extraction activities. This chapter focuses on the value-extraction activities of the firm's intellectual property as well as on its management. For firms that expect the role of intellectual property in the corporate strategy to be significant, or for firms with a large number of intellectual properties, an IP strategy can guide employee decision-making on issues and outcomes. IP strategies should outline both the strategic objectives of the firm and its related IP activity and the expected use to which the firm's IP is to be put.

Breadth of IP strategy

A company with a *broadly focused business strategy* focuses on *technologies* that can be developed and patented in anticipation of some future use. The portfolio is created with an expectation that it can contribute to the creation of future market demand. The corollary of this is that a technology may be patented in order to stake out an early claim to future design freedom. Companies with a broadly focused business strategy tend to be strategically opportunistic.

Table 4.1 IP value matrix

Kinds of defensive value available from IP	
Protection	
Design freedom	
Cross-licensing	
Litigation	
Bargaining power	
Kinds of offensive value available from IP	
Value	**IP Type**
Revenue	
From products and services	Sale, licensing, joint ventures, strategic alliances, integration
Directly from the intangible	Sale, licensing, joint ventures, co-branding, strategic alliances, litigation damages
Cost avoidance	Litigation avoidance
	Access to technology of others
Strategic position	Reputation / image
	Name recognition
	Competitive blocking
	Barrier to entry
	Customer loyalty
	Supplier control

A company with a *narrowly focused business strategy* uses targeted R&D to develop technologies that will meet a known or narrowly defined current (or potential) market demand. The time horizon for commercialization typically is short. Companies in this group tend to cull their portfolios routinely to ensure that the portfolio's contents continue to be tightly arranged around the focused business strategy.

In most firms the patent portfolio is an outgrowth of the business strategy and merits the attention from senior managers that one would expect for a valued corporate asset. Such firms usually have a well-articulated set of strategic objectives for their intellectual properties and know how they want to use those properties. Companies create a portfolio of intellectual property that they expect to use in support of their business strategy. As noted earlier, IP may be used either defensively or offensively.

IP in a Defensive Role

When an organization anticipates that others may try to use or practice a proprietary innovation, it can develop defensive positions that will strengthen its ability to ward off that use. By their nature, defensive positions are passive. They can be thought of as a modern-day equivalent of the medieval castle. Castle builders created numerous defensive structures, each intended to protect from a different form of attack. Walls protected castles from arrows, stones and artillery shells. Moats kept attackers away from the castle walls; and so forth.

Likewise, modern companies can erect a defensive structure by several means:

Protection by Excluding Others. Innovations protected by patents allow the patent holder to preclude others from using or practicing the innovation. Mere ownership of patents is a defensive activity.

Protection by Trade Secret. Trade secret protection for an innovation typically requires that the company hold secret all information about the innovation. Companies seeking to use trade secrets as a defensive form of legal protection must be able to demonstrate that they can and do maintain the secret.

Design Freedom. Design freedom is the freedom a company has to conduct on-going research in an area of innovation in which it already holds a patent. In other words, companies may not be denied the ability to continue exploring extensions of an innovation even if related patents are issued to others later.

Litigation Avoidance. Companies with large portfolios of IP and I-Stuff[7] are not good targets for lawsuits by less well endowed companies. Large and high-quality portfolios of IP and I-Stuff may intimidate potential litigants and convince them not to pursue legal action.

Negotiation Bargaining Chip. Both patents and I-Stuff may be used as bargaining chips in business negotiations.

IP in an Offensive Role

A firm that uses its IP 'offensively' does so actively (rather than passively). There are three major groupings of offensive value from intangibles: revenue-generating value, cost-avoidance value and strategic positioning value.

Revenue generation: protected products and services

Sale – Revenue may come from the sale of products and services that were developed, manufactured, distributed, and sold using the company's I-Stuff.

Licensing – Revenue may come from licensing others to manufacture and sell the company's products and services.

Joint ventures – Revenue may come through joint ventures created for the purpose of generating a new kind of value based on the company's products and services that are developed, manufactured, or sold using the company's I-Stuff. Companies enter into joint ventures in order to gain access to complementary business assets they otherwise would not have access to.

Strategic alliances – Revenue may flow from strategic alliances created for the purpose of obtaining access to markets the company would otherwise be denied.

Revenue generation: intangibles

Sale – Revenue flows from the sale of the innovation's legal protection (its patent, copyright or trademark), or from the sale of I-Stuff associated with the company's products and services.

Licensing – Revenue flows from licensing an intangible itself.

Joint ventures – Revenue flows from a joint venture where the company's contribution is its I-Stuff.

Strategic alliances – Revenue flows from a strategic alliance related to the company's I-Stuff.

Integration – The integration of I-Stuff into company operations creates indirect value. Employees with very specific skills – how to operate complicated machinery or how to install and set up a factory, for example – possess I-Stuff whose integration into company operations is valuable.

Cost avoidance

Minimizing litigation costs – High-quality patents may help to minimize a company's litigation costs. Patents that are well written and valid reduce the possibility that their owners will be sued for patent infringement.

Accessing the technology of others – Patents are more often more valuable than cash when a competitor needs access to a patent owned by another firm. Technology competitors may be able to establish licensing agreements when other forms of negotiation fail.

Strategic positioning

Reputation/Image – Some companies use patents and other legally protected innovations to establish a reputation or image in the marketplace. For example, IBM, 3M, and Texas Instruments are examples of companies the reputations of which rest on their ability to provide customers with products based on the latest technology. These companies all have substantial portfolios of high-quality patents. Their portfolios bolster their image of technology leadership, which helps them compete in their respective marketplaces.

Competitive blocking – Patents may be used to block competitors from entering certain technology businesses. Where one company has a commanding patent position, others may decide not to enter the field.
Barrier to competition – Companies already in business and may successfully create a barrier to new entrant companies by virtue of their strong patent and I-Stuff portfolios.
Being a player – In some industries, in order to be considered a 'player', a company to bring chips to the table – in other words, evidence that it belongs in the game. A portfolio of patents or known high-quality I-Stuff is such evidence.

A firm's IP strategy must focus both on activities associated with the firm's intellectual property and on supporting the long-term business strategy, which itself supports the strategic vision. The IP strategy must reflect an *a priori* review of the firm's vision and corporate strategy and the roles of intellectual capital and its subset, intellectual property. Those roles may then be codified into a set of activities and practices that are the foundation of the firm's IP strategy.

SUMMARY

Companies use the fundamental practices and concepts described here to extract the most business value from their intellectual property. Often narrowly viewed as a legal asset to be managed by company attorneys, intellectual property has burgeoned into a major corporate asset and a core driver of value for its owners. Its importance to corporations is frequently strategic and in such cases its management should be a top priority. This chapter provides an overview of the landscape in which firms must now operate with mixed-asset portfolios and diverse strategic options for managing their IP. As intangible assets continue to have a more profound and sweeping impact on firm success and national productivity, IP management will become an important strategic field of research and practice.

NOTES

1. As a measure of the amount of literature that exists on the topic 'intellectual property management', a recent search on Amazon.com for books on the topic produced over 700 results; a Google search on the same topic produced more than 60 million hits.
2. Blair and Wallman (2001).
3. J. Bellow et al., 'A New Paradigm for Managing Shareholder Value', Accenture Institute for High Performance Business. July 2004, online: www.accenture.com.

4. The discussion of the history of intellectual property treaties as well as that concerning the World Intellectual Property Organization (WIPO) is largely taken from WIPO (2004).
5. Stockholm Act, Art. 2(viii), July 14, 1967.
6. The ICM Gathering is an informal group of managers in companies that use sophisticated means to extract value from their intangibles. The Gathering has developed new methods and techniques for value extraction and has been at the forefront of the development of new concepts and practices in intangibles management. For more information on the ICM Gathering, see Sullivan (1998), Sullivan (2000) and Davis and Harrison (2002).
7. The ICM Gathering has defined 'I-Stuff' as 'all of the firm's intangibles other than IP'.

REFERENCES

Blair, S. and S.M.H. Wallman. 2001. *Unseen Wealth: The Value of Corporate Intangible Assets*. Washington, DC: Brookings Institution Press.

Davis, J. and S. Harrison. 2002. *Edison in the Boardroom*. New York: John Wiley and Sons.

Sullivan, P. 1998. *Profiting from Intellectual Capital*. New York: John Wiley and Sons.

Sullivan, P. 2000. *Value-Driven Intellectual Capital*. New York: John Wiley and Sons.

World Intellectual Property Organization. 2004. *WIPO Intellectual Property Handbook: Policy, Law and Us*e. 2nd edition, Geneva: WIPO Publication No. 489 (E).

5. Making a return on R&D: a business perspective

Sharon Oriel

This chapter focuses on the business perspective for making a return on investments in biotechnology. Patents will be used as the primary example of intangibles. Most comments and data specific to patents are germane to the other types of intellectual property: trademarks, trade secrets, domain names, copyrights and semi-conductor masks. A brief summary of patent trends and history is used to set the context for the discussion of a model for managing business intangibles. The corporate intangibles perspective will be presented from two positions: a large multi-national firm and a biotechnology start-up. The hypothesis underdevelopment is that the business perspective provides balance to the current conversations about access to the intangibles created by biotechnology research.

The word patent derives from the Latin *litterae patentes* or open letters. The first modern patent law was introduced in Venice in 1474 to attract skilled merchants to the city-state. Anyone who came up with a technique deemed novel was given a 10-year right to its exclusive use. Infringers were fined 100 ducats. Thomas Jefferson said patents should draw 'a line between the things which are worth to the public the embarrassment of an exclusive patent, and those which are not'. In 1851, *The Economist* wrote, 'The granting [of] patents "inflames cupidity", excites fraud, stimulates men to run after schemes that may enable them to levy a tax on the public, begets disputes and quarrels betwixt inventors, provokes endless lawsuits . . . The principle of the law from which such consequences flow cannot be just.'

Researchers, both academic and industrial, are told that patents are good. The question that is seldom asked is: good for what? And good for whom? The growing importance of genetic manipulation of food sources, medicines and medical treatments has heightened the awareness of the role of patents in facilitating or denying access to these advances. Everyone has an opinion about biotechnology patents. If patents are viewed from only a single perspective, a logical case may be developed to support that singular perspective.

The economic perspective is that intellectual property rights ensure there will be investors of capital, time and creativity to produce new products. The difference between tangible property rights and intellectual property rights makes the 'tragedies of the commons' argument inappropriate.[1] Intellectual property is non-rival, so there is no threat of over-use. Multiple uses of the same property may actually enhance the value of the idea. Through unrestricted access, knowledge goods become public goods. How do companies then make a return on knowledge?

It has been more than 20 years since the first gene patent issued and since the United States Supreme Court in *Diamond v Chakrabarty* found 'anything under the sun that is made by man' to be patentable subject matter. Only in this century are companies beginning to generate a profit from the biotechnology knowledge that has been created. With real dollars and even larger potential dollars, the debate about access to this knowledge and ownership of the knowledge has become more than an academic or legal debate. We have trade organizations, courts, entire countries, universities and individuals all claiming their rights.

PATENT CONTEXT

It is critical to understand the context surrounding patents. 1999 UN data showed that entities from industrialized countries held 97 per cent of all patents globally.[2] Further, more than 80 per cent of patents granted in developing countries were issued to residents of industrialized countries.[3] A brief review of current data sets from WIPO (World Intellectual Property Organization) shows little fluctuation in these numbers.

Contrast these numbers with the rise in biotechnology patents that have climbed 15 per cent per year at the USPTO (United States Patent and Trademark Office) and 10.5 per cent at the EPO (European Patent Office), compared with 5 per cent per year overall for patents.[4] The source of funding for these biotechnology patents is also worth noting. 70 per cent of scientific papers cited in biotechnology patents originated in solely public science institutions compared with 16.6 per cent from private sector. 63 per cent of assignees of US patents are private companies and 18.5 per cent of human genes are covered by US patents.[5] These data would appear to be an example of knowledge diffusion. Through unrestricted access (scientific papers), the knowledge goods have become public goods (patents). Patents provide a temporary monopoly to the owner, but it is worth reminding the reader that the purpose of patents is to disseminate knowledge.

In 2003, the United States Federal Trade Commission report on innovation, competition and patent policy concluded that 'in industries with incremental innovation, questionable patents can increase defensive patenting and licensing complication and moreover that questionable patents are significant competitive concern and can harm innovation'.[6] In 2006, we are seeing business models which use these questionable patents to extract royalty payments from companies selling products for which they thought they had legitimate patent protection. In 2008, the US is struggling to reach agreement about what patent reforms should become law. Businesses are also dealing with a series of US court decisions which potentially impact on how they capture a return on their R&D investment.

WORKING DEFINITIONS FOR INTANGIBLES

Any discussion about intangibles needs to begin with definitions.

According to Webster's[7]

Property: Something owned or possessed; something to which a person has legal title.

Asset: The items on a balance sheet showing the book value of property owned.

Capital: a stock of accumulated goods.

According to Financial Accounting Standards Board[8]

Asset: Assets are probable future economic benefits obtained or controlled by a particular entity as a result of past transactions or events.

Intangible Assets (draft): intangible assets are non-current assets (not including financial instruments) that lack physical substance. The list includes agreements, contracts, R&D, patents, copyrights, trademarks, databases, customer lists and software.

For the purposes of this chapter, we will use the following definitions:

Intellectual Property is the set of rights that accrue to the owner of an innovation. These rights are expressed in the form of patents, trademarks, trade secrets, domain names, copyrights or semiconductor masks.

Intellectual assets include all of the above plus codified knowledge with a clear owner.

Intellectual capital is the most inclusive term. It includes intellectual property and intellectual assets plus tacit knowledge (knowledge and relationships which are not codified and often specific to an individual or organization). Collectively, these terms are called intangibles. Depending on the reader's perspective, one may understand the words from a legal, financial or accounting view. Knowledge management and human resource professionals struggle to understand the difference between intangibles management and knowledge management.

These definitions have been published in Davis & Harrison, 2001 and Harrison & Sullivan, 2005.[9] The definitions were created by the ICM Gathering, a group of managers of intangibles for large international corporations who have been meeting since 1995 to share information about how to obtain value from intangibles. Initially, the group focused on the extraction of value from intellectual property and on managing IP as a business asset. Boeing, Dow Chemical and Procter and Gamble are examples of some of the Gathering companies which know how to show the ROI on R&D.

While the group identified intellectual capital as comprising human capital, intellectual assets and intellectual property, they realized that these artificial groupings were not satisfactory. The groupings utilize accounting, legal, and human resource terms which each carry special meanings. Fundamentally companies manage only two types of intangibles: intellectual property and everything else. Because the set of intangibles that is 'non-IP' includes such a wide range of items (knowledge, ideas, wisdom, relationships, customer lists, brands and work processes) it became clear that simple aggregation and naming were difficult. For the purposes of this chapter, we will use IP, IA, IC and intangibles interchangeably.

ECONOMIC PATENT BACKGROUND

The economic logic for granting patents is straightforward. Without patents, inventions could be easily copied and the already difficult path to commercial success would have another twist in it. Companies, absent protection for their discoveries, would have limited incentive to invest in R&D. Even the solo inventor would have no protection for his/her idea and would be dependent on finding funds for commercialization based solely on speed and secrecy.

In 2001 Leonard I. Nakamura, economic advisor of the US Federal Reserve Bank of Philadelphia, presented a paper showing empirically that US private firms invested at least $1 trillion annually in intangibles.[10] This rate of investment roughly equalled the US gross investment in plant and equipment. Nakamura defined intangibles as assets that are 'intangible and necessary to improvements in products and production'. A recent paper, published in 2005 by economists Robert Shapiro and Kevin Hassett, puts the value of intellectual property for the US economy at $5 trillion.[11]

Technology licensing revenue in the US is estimated to be $45 billion annually.[12] The worldwide estimate is approximately $100 billion and growing.[13]

For tangible property, market capitalism works to allocate scarce resources to maximize profits while creating positive social benefits. Blindly applying the market capital system to intellectual property does not work, mainly because intellectual property is not scarce. An idea can be used by billions of people at the same time, unlike a barrel of oil or a piece of equipment. There is often very little cost involved to copy ideas.

One can trace the argument about who owns knowledge back at least to the Age of Enlightenment. The French mathematician and philosopher Condorcet and the encyclopedist Diderot took opposite sides of the argument. Diderot (understanding ideas as subjective) argued that products of the mind are the property of their creator. Condorcet (understanding knowledge as objective) asserted that ideas are not the creation of a single mind. He could see no value in granting individual claims to ideas. This debate continued as the British colonies transformed themselves into the United States. The Federal Constitution of 1787 tries to strike a balance: 'Congress shall have the power . . . to promote the progress of Science and useful Arts, by securing for limited Times to Authors and Inventors the exclusive Right to their respective Writings and Discoveries'. According to the Oxford English Dictionary, the English phrase 'intellectual property' first appeared in 1845. By this time, a consensus, at least in the West, had emerged that inventors and authors could profit from their works and ideas for a limited time span.

Because the contradictory philosophical arguments represented by Condorcet and Diderot were never resolved, the debate continues, in spite of statutes, laws and case law. The easy sharing and decimation of information and ideas across national and corporate boundaries fuels the debate. Who really owns the ideas? Who should profit from their commercial use? And for how long? Today we call the problem private rights versus public good.

The United States is an example of how the view of intellectual property rights changes as a country develops. The country started as an importer of intellectual property. Some could even say the US was an intellectual property pirate. The publishing houses of New York, Boston and Philadelphia built their fortunes on unauthorized and unremunerated publications of British writers. In 1843, a copy of Dickens's *A Christmas Carol* sold for six cents in the United States. In England, the equivalent cost was two dollars and 50 cents.[14]

As the US moved to being an exporter of intellectual property, legal doctrine began to shift in favor of the rights of the creator (inventor, author) rather than the public good. Court decisions starting in the early

1900s started to reflect strong support of the rights of the individual to benefit directly from their creative work.

In the twenty-first century, we see the United States development story being repeated. In general, developing nations, regardless of their religious, cultural or governmental roots, view intellectual property as social in nature. They believe the state has the right to limit the individual claims of citizens and corporations in the name of public good. The United States and much of Western Europe support the commercialization rights of the inventors and strive to protect these rights through treaties, trade sanctions and access to information. Once the proponents of intellectual exclusivity recognize that a balance of rights will insure renewal of ideas and create sustainable economic gain, a new global balance will be found. The extension of time under TRIPS (Agreement on Trade-Related Aspects of Intellectual Property Rights) to 2016 for the least developed countries to have functioning patent systems is one small example that steps are being taking to rebalance the approach.

Richard Posner, a judge of the US Court of Appeals for the Seventh Circuit, acknowledges that the law has not kept up with new technologies, software and biotechnology being two examples. Posner argues that the race for patents we are currently seeing can draw excessive resources into the creation of intellectual property in the pursuit of economic rents and away from other valuable activities.[15] Information technology, chip manufacturers and life science companies race to get patents in order to have bargaining chips that may be used to bring products to market or applied in litigation.

Fundamentally, intellectual property, when protected, provides incentives for economic gain for a limited time. Post-protection the property enters the public domain and potentially becomes a raw material for the creation of new intellectual property. Even the publication of patents offers material for new ideas to be developed upon the foundation of prior art. In this manner intangibles can contribute to a web of innovation which involves many companies and countries. Furthermore, patents may be understood as a form of currency – they are a business asset allowing for the transfer of knowledge while creating a market for that knowledge. Knowledge is the raw material for economic growth.

A MODEL FOR MANAGING BUSINESS INTANGIBLES

To assist in understanding the view of for-profit business in managing its intangibles, a model which has been tested by large and small companies

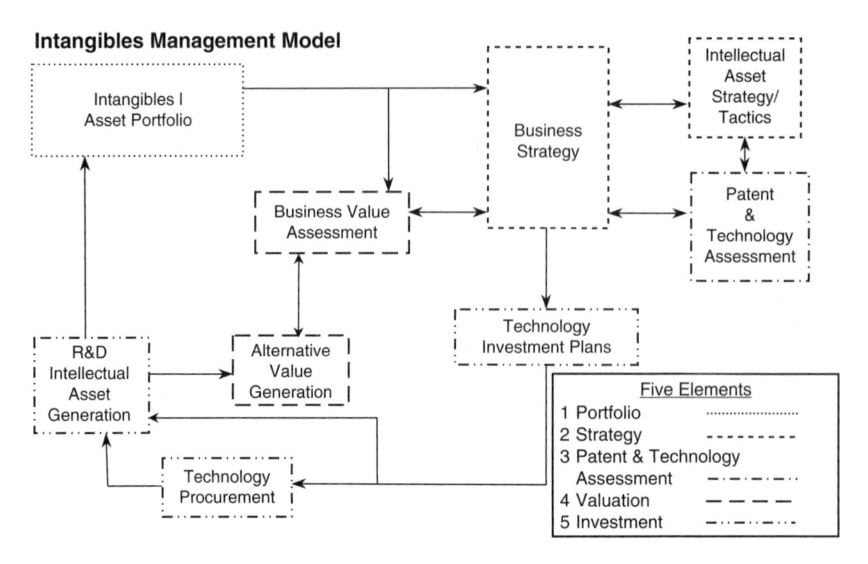

Figure 5.1 Intangibles management model

will be described. Further, this model has been applied to biotechnology companies and to partnerships between large companies and small biotechnology companies. There are five key elements to the model: (1) portfolio, (2) strategy, (3) patent and technology assessment, (4) valuation, (5) investment. They are not called steps because that term implies that the model must be worked sequentially. In fact, a business may start by developing competencies in any of the elements. A start-up may even have a strategy but no portfolio. Eventually they will use all five elements and then an appropriate sequence will develop based on the connections between the components (Figure 5.1).

Portfolio

In 2008, there are still companies which keep their patents, technical agreements and trademarks in physical file drawers. When there is a pending transaction involving intangibles, these firms often rely on patent offices, outside counsel or consultants to tell them what they own and what they can include in the deal. A company cannot begin to manage its intangibles if it does not treat them as inventoried assets. Imagine a company not knowing how many manufacturing plants it owned, or how many pieces of machinery, how many trucks, planes, etc.

As with building a personal financial portfolio, companies must start with an audit of what they have. The audit can identify what intellectual property is being used, how it is being used and who is using it. With this information in hand, quality decisions can be taken about what patents and trademarks to maintain. Royalties will be appropriately paid and collected. New business opportunities may emerge as a result of understanding what intangibles are available. If the results of the intellectual property audit are recorded in a suitable commercial software program, the portfolio management process can begin.

Ideally, the IP software would allow for easy linkage to the corporate financial reporting system with the ability to create reports connecting intellectual property to products, sales and income. By directly linking a patent to sales, a company can examine margins. If patented products have higher margins than non-patented products, this offers solid data that having a patent offers a competitive advantage. If there is no difference in margin, a company should be questioning the value of patents and perhaps even examining the R&D spending on the products in question. A company should also inventory its trade secrets to understand both its investment and return on protecting this valuable knowledge.

Capturing the portfolio requires a commitment to a quality database, dedicated people to input information into the database, resources to maintain the data, and work processes to support the portfolio. There is satisfactory commercial software available for the task, so the firm does not have to spend resources developing its own system. Ideally, a system will be selected which also does docket management for the intellectual property group and is capable of interfacing with the company's financial system.

To reflect the asset dimension of the intangibles portfolio, the patents, trademarks, trade secrets, domain names and agreements involving IP should be aligned to the business unit using the intangibles. A classification system for how the business is using the assets is also useful. Patents can be classified as practicing, licensed, future use, defensive, available for license. A patent may fit in more than one category. Also, a patent may be used by more than one business, and this should be noted in the portfolio. With this information in the portfolio, the business can make informed decisions about maintenance of the properties. The lifecycle of the business may not match the 20-year patent life. Business strategies change and the portfolio may need to be adjusted to reflect this. Patents may become available for licensing or they may be abandoned. All of these decisions should be captured in the portfolio. Trademarks, trade secrets, copyrights and domain names would also be classified and managed in a similar manner.

With an intangibles portfolio in place, reports can be generated for each business unit and rolled up for the entire company. At least yearly,

Table 5.1 US/foreign patent information

	1992	1993	1994	1995	1996	1997	1998	1999	2000
US Patents in Force	4674	4127	3732	3476	3248	2927	2413	2419	2394
US Patents Issued	335	361	280	253	201	181	206	161	195
Foreign Patents in Force	8222	7977	7874	7250	6900	6584	6152	7054	7764
Foreign Patents Issued	1385	1359	1048	974	915	989	1247	1410	1223

a report on the assets aligned with each business should be prepared and reviewed. For start-ups, the intangibles are often the only assets they have in addition to their employees. It is critical that new companies report on their intellectual property to their investors and stockholders.

Table 5.1 is an example of the results of managing a patent portfolio over time. The table shows that a company is filing and maintaining fewer US patents but appears to be getting value from filing outside the US. According to surveys conducted by the Licensing Executive Society and other organizations, the typical company actually uses only about 1/3 of its patent estate.[16] Business specific reports can be generated which show licensing income and expenses, products protected by patents, costs for patents, patents per R&D spending and the list goes on.

Strategy

The strategic element of the intangibles management model is crucial. Once a business knows what is in its intangibles portfolio, it becomes critical to examine if the portfolio supports the business strategy and to determine if the business is leveraging the assets to create maximum value. The businesses can identify where there are gaps in the portfolio and establish a plan to fill the gaps. They may decide to invest more in R&D, either internally or externally, to complete the portfolio. Intellectual asset ('IA') tactics are developed which support the business strategy. Examples of IA tactics include reducing the patent portfolio and using more trade secrets to protect the business; actively seeking a licensing opportunity to fill a technical gap; identifying potential patent infringers and developing the tactics to enforce patents.

By making the intellectual asset decisions business strategic, the quality of decisions on what patents to file, what trade secrets to protect, and what IP licenses to sign improves significantly. Also, the strategic alignment changes the emphasis from counting patents to counting the dollars generated by patents. Some companies may find they actually file fewer patents but that each patent is more strategic. Other companies may file more patents as they realize they are in industries that cross-license patents in order to carry out their business strategy.

Patent and Technology Assessment

When a business knows what is in its intangibles portfolio and has a strategy in place to use the assets, it is then important to understand the competitive intellectual asset landscape. Most firms practice some form of competitive business intelligence. This may be a formal process with personnel assigned to track and report on competitive activity or it may be a collection of individual efforts combined with the occasional use of consultants.

Understanding the external intangibles landscape is also important. When a competitor launches a new product, a more effective response may be planned if a firm knows the following information: (a) if the product has patent protection and where that protection is granted; (b) if it is a licensed product; (c) if it infringes an existing product; and so on. Through the use of patent maps and trees a company can find new opportunities, identify gaps and highlight competitive situations early. A quality patent and technology assessment provides a view of where a company or technology has been and where it is going. Often patent filings provide an early signal of a change in corporate direction, long before products are launched. Monitoring what patents a competitor is maintaining is also useful. Tracking inventors can also be valuable to see if they change research fields or perhaps even change companies. Licensing opportunities may also be identified as a result of tracking competitive patent activity. Linking patent activity to product activity is a good reality check on what a competitor is actually doing.

Business Value of Intangibles

The literature on valuing intangibles is diverse. The value of a piece of real estate or a piece of equipment can be easily determined and agreed upon using standard methods. The value of a patent or other intangible is context dependent. For example: the same patent will have two different values depending on how it is used. One company may require the patent to bring its product to market. Another company may be interested in the patent as part of a licensing business. Unlike real estate, where information on actual transactions is public information, intangible transactions are frequently confidential, which means locating market comparisons is not easy. Moreover, valuation of intellectual assets can apply a variety of methods. For example, the standard methods: market, income and cost, plus methods developed based on the real options model may be used.

Often, it is enough to *evaluate* an intangibles portfolio. Evaluation is qualitative and examines the relative merits of one patent versus another in a business context without assigning a fixed monetary value. When the portfolio, strategy and competitive elements are well understood, a business can qualitatively look at the intangibles available for a transaction and determine which ones it makes strategic sense to offer. The evaluation approach is also useful when determining what to file and where. At the early filing stage one is seeking to determine what filings have the most potential value for the firm, not the actual dollars which could be generated. Intellectual assets are first evaluated and then valued when there is a proposed transaction: such as a joint venture, joint development, license, sale or litigation.

Investment

When a company/business knows what is in its intangibles portfolio, has a business strategy, knows the competitive scene, and has a sense of the value for its portfolio, then it is ready to make technology investment decisions. Do they need to fill a gap? Is this best filled by internal R&D? Or is it best met by acquiring technology externally? Or is a combination of internal and external resources the best approach? Strategic intangibles investing keeps the intangible portfolio renewed and creates value for the business. With the present emphasis on open innovation, intangibles investment decisions are even more critical to a corporate success.

Working the Intangibles Management Model

If we take the point of view that business can successfully achieve a balance between access to knowledge and making money, we need to examine how that can be done. For discussion purposes, we will look at how start-up companies manage intangibles, as well as how an established firm does the job. We will also briefly explore some new business models which seek to address the complexities of access and reward in the biotechnology space.

A start-up biotechnology firm often has a singular focus: a tool, a method, a gene, a drug. The tangible assets of the small firm are limited. They often are renting space and buying time on equipment. Financial resources are limited by the resources of the founders; the size of the investment round. The initial culture of a start-up is science based. Patents provide the credentials for both the firm and the researchers creating them. The patent portfolio is often the company's working capital. The founder/inventors manage the patent portfolio with the assistance of outside counsel. Often, strategic management of intellectual property is replaced by cost management or a survival mentality.[17]

Large firms behave differently. They will have a strategic plan for the biotechnology space. The same company may have an interest in both plant and human biotechnology products. It is seldom focused on working with only one organism, one plant, or with using only one tool. Because of an established track record, the large firm has an easier access to capital. This access to capital creates a culture which is focused on delivering value to stakeholders. While science and technology is part of the company culture, it is not the dominant component.[18]

Larger firms often have their intellectual property managed by a team of professionals. This team can include intellectual property attorneys, licensing managers, intellectual asset managers, or tax specialists. Decision-making can take time because of the need for input from the team. In a start-up, often the founder and chief scientist decide what to patent, what to license, what to abandon. Because of the different context of the large and small firms, intangibles discussions can often be a source of frustration for both parties. Moreover, further differences can emerge based on national culture, country law and ethics.

Both the start-up and the large firm need access to the created knowledge to succeed. Business models which create winners and losers may offer a one-time large monetary reward, but there will not be an ongoing relationship between the two parties. With the base of knowledge creation in the biotechnology space, the ability to create win-win transactions is not only attractive but imperative. Alliances (which include joint development agreements, joint ventures and contract research) are extensively used by biotechnology and pharma companies.[19] Between 1997 and 2002, big pharma formed approximately 1,500 alliances with biotechnology companies.[20] Almost 50 per cent of global pharma development is externally sourced.[21] More than half of the current 20 best selling prescription drugs are co-developed, co-marketed or in-licensed.[22] These data clearly indicate that business is finding a way to have access to knowledge and to financial returns.

In addition to alliances, one of the models being explored in the biotechnology space is that of using a modification of the open source concept. Others are looking at models which blend elements of for profit and not for profit. Cambia is an example of applying concepts from open source into their licensing programs.[23] Perhaps 'open source license' is an oxymoron, but Cambia is doing it. In February 2005, a group of researchers in Australia published a paper in *Nature* describing a way to transfer genes into plants that avoided hundreds of patents.[24] The scientists affiliated with Cambia are making the gene transfer techniques available via a license. The licensee is free to commercialize products based on the techniques. The only obligation imposed on the licensee is to share improvements to the technique itself with all other licensees. This is an important initiative as the gene transfer technique is potentially a way for small companies, or people working on minor crops, to gain access to knowledge without paying either a license fee or ongoing royalties. Current information on Cambia can be found at www.cambia.org.

OneWorld Health is a non-profit pharmaceutical company founded in 2000 by Dr Victoria Hale. OneWorld develops drugs based on donated or royalty-free intellectual property. It is in the final stages of testing a new therapy for visceral leishmaniasis in India.[25] Current information on the company can be found at www.oneworldhealth.org. In the company's own words:

> OneWorld Health directs a worldwide effort to uncover, research, and develop new medicines for neglected infectious diseases.
>
> Our approach is simple. Assemble an experienced and dedicated team of pharmaceutical scientists. Identify the most promising drug and vaccine candidates. Develop them into safe, effective and affordable medicines. Then partner with companies and organizations in the developing world to manufacture and distribute newly approved therapies that will impact the health of millions of people.
>
> We challenge the assumption that pharmaceutical research and development is too expensive to create the new medicines that the developing world desperately needs. By partnering and collaborating with industry and researchers, by securing donated intellectual property, and by utilizing the scientific and manufacturing capacity of the developing world, OneWorld Health can deliver affordable, effective and appropriate new medicines where they are needed most.

The model developed by OneWorld provides access to knowledge which would not be readily applied by large pharma because, while 90 per cent of the world's diseases are found in developing countries, only 3 per cent of pharma budget is directed towards these diseases.[26]

BUSINESS CONCLUSION

Patents are an option for a business or company. The company may choose to exercise the option, let the option expire, trade the property, or use the option as collateral. The biotech patent debate often fails to include the business perspective. Benefits and costs attributed to intellectual property must be examined from multiple perspectives. Economists, lawyers, ethicists, and academicians are the usual parties to the discussions. It has been more than 25 years since *Diamond v Chakrabarty*, yet only now are profitable companies based on biotechnology knowledge and patents emerging. We have the opportunity to combine the rapidly developing biotechnology knowledge base with new intellectual property business models to provide access to knowledge and business profits on a global basis.

NOTES

1. James Boyle, 'Fencing off Ideas: Enclosure & the Disappearance of the Public Domain', (2002) *Daedalus*, Spring, 131: 13–25.
2. World Health Organization, *Genetics, Genomics and the Patenting of DNA. Review of Potential Implications for Health in Developing Countries* (Geneva: WHO, 2005).
3. Ibid.
4. OECD (2004).
5. K. Jensen and F. Murray. 'The Intellectual Property Landscape of the Human Genome' (2005) *Science*, 310: 239–240.
6. Federal Trade Commission Report, *To Promote Innovation: The Proper Balance of Competition and Patent Law and Policy* (Washington, DC: FTC, October, 2003).
7. *Webster's Ninth New Collegiate Dictionary*. Springfield, MA: Merriam-Webster Inc., 1984.
8. Financial Accounting Standards Board, *Proposal for a New Agenda Project: Disclosure of Information about Intangible Assets not Recognized in Financial Statements* (2002), available at http://www.fasb.org/proposals/intangibles.pdf.
9. Davis, J. and S. Harrison (2001) *Edison in the Boardroom*, New York: John Wiley & Sons.
   Harrison, S. and P. Sullivan (2005) *Einstein in the Boardroom*, Hoboken, NJ: John Wiley & Sons.
10. Leonard I. Nakamura, 'What is the US Gross Investment in Intangibles? (At Least) One Trillion Dollars a Year!', Paper presented at the 4th Intangibles Conference, May 17–18, 2001, at Stern School of Business, New York University in New York City, available at http://ideas.repec.org/p/fip/fedpwp/01-15.html.
11. Allan Murray, 'Protecting Ideas is Crucial for US Business', *Wall Street Journal*, November 9, 2005.
12. 'A Market for Ideas', *Economist*, October 22, 2005, at 3.
13. Ibid.
14. Carla Hesse, 'The Rise of Intellectual Property, 700 B.C.–A.D.2000: an Idea in the Balance, (2002) *Daedalus*, Spring, 131, 26–45.
15. Richard A. Posner, 'The Law & Economics of Intellectual Property' (2002) *Daedalus*, Spring, 131, 5–12.
16. Richard Razgaitis, 'US/Canadian Licensing in 2004: Survey Results' (2005) *Les Nouvelles* 40: 4, 145–155.
17. Sharon Oriel, 'Sustainable Value Creation: IP & Ethics at the Core', presented at CIP Forum for the Centre for Intellectual Property, May 24–25, 2005, in Gothenburg, Sweden. Published as 'Towards a more inclusive patent world', (2006) *Intellectual Management*, June/July, 14–17.
18. Ibid.

19. Leslie Gladsone Restaino and Theresa Tackeuchi, 'Gene Patents and Global Competition Issues', *Genetic Engineering News*, January 1, 2006, 10–11.
20. Ibid.
21. Ibid.
22. Ibid.
23. *Economist* (2005).
24. W. Broothaerts et al., 'Gene Transfer to Plants by Diverse Species of Bacteria', (2005) *Nature* 433: 629–633.
25. *Economist* (2005).
26. Data from the OneWorld website.

6. Looking beyond the firm: intellectual asset management and biotechnology

Karen L. Durell and E. Richard Gold

INTRODUCTION

The academic, business and international communities increasingly pay attention to the rights – their nature and number – that companies have over intellectual assets. The single-mindedness of this focus is, we argue, misplaced, as it is not the right itself that is significant but rather the way that companies and communities put these assets to work. Rights in intellectual assets – commonly known as intellectual property rights ('IPRs') – provide their holders with a right to veto the actions of others with respect to the asset (whether that asset be knowledge, patents, copyright, trademarks, and so on). These actions include copying, making, selling, offering for sale or importing goods which incorporate the asset underlying the right. Just because an IPRs holder has the right to block public access to an asset does not mean that he or she will do so. What the right offers is the ability to make choices; how the rights holder chooses to utilize an IPR affects, however, not only him or her, but society at large. We argue that rights holders ought – not only for the public's sake but for their own – make these choices within a framework which takes into account the synergies arising from private and public actors working within a community.

The strategic deployment of IPRs – what rights to keep, license or assign, to whom to give permission to use underlying intellectual assets, which corporate structures to create, and so on – falls into the field of intellectual asset management ('IAM'). In comparison with the weight of literature surrounding IPRs, IAM clearly still is in shadow, having generated significantly less discussion and debate in academic and other spheres. This situation is changing, however, as illustrated by this and the previous two chapters. As IAM rises in interest, we can better define both the opportunities and the challenges it poses. This chapter will examine one such opportunity: our ability to move our level of analysis from a lone firm to a community. This change of focus offers advantages – economic and social – not only to firms, but to the communities in which those firms operate.

The literature currently describes two visions of IAM: firm-centric IAM and firm-networked IAM. Most analysis uses a firm-centric approach, seeing IAM as a strategy which business organizations employ to extract value from both its existing IPRs and its internal capacity to create new IPRs. More recently, analysts have begun taking a firm-networked approach to IAM, under which the firm not only focuses inward, but also seeks to manage its external relationships by sharing IPRs with other firms in order to maximize value from these assets and to support future waves of innovation. The common element of these two IAM strategies is that the firm remains as the nexus of activity and as the center of profit.

In this chapter, we advance a broader – non-firm – concept of IAM: one that takes a systems approach, combining the outward dynamism of firm-networked IAM but applying it to a wider community of public and private institutions. This extension not only increases the options available to firms to manage their intellectual assets, it also introduces and frames public intervention to bolster innovation and dissemination. At the heart of this approach is the recognition that both firms and society can maximize the benefit they obtain through IPRs by mixed private and private–public efforts such as altering business practices on an industry or sector basis and public institution-building.

This chapter will build up to the concept of non-firm IAM as follows. First, we define a basic vocabulary of IAM, given that, at present, there is no clearly accepted defined terms used in the literature. This lack of clarity leads to confusion, hindering understanding between those discussing different IAM strategies. Second, this chapter will examine the two existing forms of IAM – firm-centric and firm-networked – and compare these to the strategy we put forward here: non-firm IAM.[1] Third, the chapter will investigate the benefits of non-firm IAM for both firms and the public. While there are few examples of non-firm IAM in action, we argue that it offers untapped potential to those carrying out IAM in the future.

IAM AND THE ROLE OF LANGUAGE

IAM involves the purposeful use of intellectual assets to achieve the goals of an organization. An entity employing IAM attempts to determine how best to deploy current and future intellectual assets either within the firm (following traditional definitions) or within industry sectors or communities (under our non-firm approach). For example, a firm may seek to increase its revenues or reduce its costs by licensing technology in or out, bundling technology with another firm and so on. Alternatively, a group of researchers, private foundations, industry and governments may combine intellectual assets to develop and distribute a key innovation, such as malaria or HIV/AIDS treatments.

The first step in any IAM strategy is to identify, through an audit, existing intellectual assets within the entity's domain. Only by knowing which intellectual assets it holds can the entity identify gaps, voids or surpluses in its asset portfolio. Second, once the entity has identified existing intellectual assets, it can make several evaluations: it must identify which assets it does not need (and abandon or, better still, license these out to generate revenue); determine where its assets are incomplete (and license in the needed assets); and consider where a novel intellectual asset is required (and provide research and development funds to create that asset). All decisions regarding intellectual assets should conform to the overriding goals of the company.

Within this broad framework, different organizations will seek different goals through the use of IAM. Some will seek simply to increase short-term returns on investment; others will seek to consolidate their expertise in an identified niche; still others will aim to build a new technology standard to open up new business opportunities. Thus, a firm's goals will depend on its overall business strategy and area of business. Moreover, even within a single market or industry different companies will pursue different goals. IAM is therefore not a single strategy that operates across firms; rather, it is an approach taken by different entities in a variety of sectors to achieve their very particular goals.

Due to this diversity of goals and the fact that IAM is a new field, IAM lacks a clear and consistent vocabulary. Different firms describe their IAM strategies by way of a variety of terms and names, making inter-firm comparison and analysis difficult. For example, what this chapter calls 'firm-centric' IAM is called a 'fortress approach'[2] elsewhere, while what we term a 'firm-networked' approach is labeled a 'wider asset view'[3] in other literature. Thus, the same approach has multiple names. Moreover, even where firms use the same idiom, they may mean it to describe different things.[4] We raise these points here not because we want to attempt to solve the problem of vocabulary but to acknowledge that a consistency of language should not be assumed in any dialogue regarding IAM.

With this background in mind, we can begin the process of bringing some conceptual order to the analysis of IAM. The first step in this exercise is to define one of the central terms of IAM and of this chapter: intellectual assets. Intellectual assets refers to all intangibles assets, whether these be codified (that is, recorded in a tangible form – a patent application, written documentation such as books or manuals, software code, etc.[5]) or non-codified (the asset is not stored in a consistent, tangible form, but rather is stored in the minds of one or more individuals as knowledge). On this definition 'intellectual assets' include all IPRs – patents, copyright, trademark, plant breeders' rights, and so on – held by a company, as well as all knowledge of those working within an entity that is not recorded, but is functionally accessible. In sum, intellectual assets include all intangible elements that are available to an entity to deploy for its internal purposes.

FORMS OF IAM

IAM's first foray into practice revolved, naturally enough, around firms. IAM was a response to the recognition that although intellectual assets have, at least since the mid-1990s, constituted three-quarters of corporate assets in the United States,[6] firms had been fairly slow in capitalizing on the value of those assets in comparison with more traditional assets such as land, equipment and inventory. IAM 'now figures as a strategic business issue' involving annual licensing revenues alone worth US$100 billion.[7]

In this section, we describe the initial approach to IAM – the firm-centric approach – and its evolution into firm-networked IAM. A firm-networked approach recognizes not simply the importance of intellectual assets to the firm – as does a firm-centric approach – but the fact that the value of these assets is significantly affected by network effects, thus requiring the firm to develop a strategy that takes into account the various relationships the firm has with its customers and competitors. We argue that the evolution of IAM is not yet complete since there is one other fact that IAM strategies need to embrace: that the value of intellectual assets arises from the economic, social, political and cultural milieu in which they are put to use. We thus outline a non-firm approach to IAM that takes this fact into account.

Firm-centric IAM

A firm-centric approach to IAM focuses on the value that a particular firm can extract from its intellectual assets. The goals that the firm pursues are thus limited to direct financial, or other, benefits to the firm. While a firm-centric approach can, depending on its goals, lead the firm to use its existing intellectual assets to assist it in creating new intellectual assets, other firms will simply seek to obtain revenue by threatening to sue anyone who is actually using the intellectual asset. Several firms make substantial revenues by bringing and then settling such law suits.[8]

Because this form of IAM concentrates so much on the individual firm – regardless of the possibilities of synergy with other firms and organizations – it is, essentially, a protectionist[9] strategy where the rights holder wields his or her rights as if in a fortress.[10] It is protectionist in the sense that the firm sees its intellectual assets as possessions that must be protected from attack by external competition. The firm employs its rights derived from intellectual assets as a fortress, since the firm keeps would-be intruders from making use of the firm's intellectual assets – at least without paying. The firm understands the value of its intellectual assets to reside in the fact that these assets are created by the firm for the firm.

At its base, a firm-centric approach focuses exclusively on extracting value from present or future intellectual assets by using those assets internally and making it difficult for others external to the firm to use them. If others are to be permitted to make use of the intellectual asset, they will have to pay a high price: the firm may pounce on anyone who makes an unauthorized use of the intellectual assets by threatening legal action. Jaffe and Lerner put it this way:

> For numerous large companies – including, notoriously, Digital Equipment, IBM, Texas Instruments, and Wang Laboratories – these types of patent enforcement activities have become a line of business in their own right. These firms have established patent licensing units, which have frequently been successful in extracting license agreements and/or past royalties from smaller rivals. For instance, Texas Instruments has in recent years netted close to $1 billion annually from patent licenses and settlements resulting from its general counsel's aggressive enforcement policy. In some years, revenue from these sources has exceeded net income from product sales.[11]

The management of intangibles is now a 'major determinant of organizational profitability and sustainability'.[12] However, the potential profitability and sustainability that firms may derive from intangibles, in particular intellectual assets, is specifically related to the use to which they put those assets. Choosing to engage solely in firm-centric IAM strategies will by no means deprive a company of future growth or innovation. For example, 3M applied what can be characterized as firm-centric IAM to create masking tape.[13] Specifically, 3M applied its employee know-how and intellectual assets in the areas of adhesives and adhering paper to other surfaces (a process 3M applied in its creation of sandpaper) to the task. 3M engaged in no external synergy to create its new masking tape product and related intellectual assets. Instead, the firm captured the intellectual assets already in its employees' heads and relied on its employees' creativity.

Employing a firm-centric IAM approach does present risks, however. Consider Avon Products as a case in point. Recently, Avon saw the market for its core beauty products plateau.[14] Enthused by the positive buzz surrounding intangible assets, the company bought Foster Medical, a medical products firm and the holder of several patents. Despite Avon's enthusiasm, however, it neither was able effectively to integrate Foster into its existing product structure nor did it develop an IAM strategy to take advantage of synergies between the two firms. As a result Avon ultimately 'ran the medical business into the ground, eventually dismantling it and taking huge write-offs'.[15] Avon thus learned the hard way that merely increasing IPR holdings without a considered management strategy is no guarantee of commercial success.

Three other cases provide positive and negative examples of firm-centric IAM: Xerox, Buckman Laboratories and Monsanto.[16]

Xerox

Xerox Corporation describes itself as a 'technology and services enterprise that helps businesses deploy Smarter Document Management strategies and find better ways to work'.[17] The firm is engaged in selling products as well as servicing business machines. It is the latter activity upon which this case study focuses. It is estimated that Xerox's 19,000 service engineers address 25 million customer service requests annually.[18] To prepare employees to respond to service calls, the company provided traditional step-by-step training to its employees.[19] This involved the use of several intellectual assets, including manuals and lectures, to train its engineers. Xerox decided to look at whether this was the best use of its intellectual assets. Anthropologists hired by the firm discovered that, in fact, the formal training materials and sessions did not fully capture the job of the engineers. Instead of applying a fixed set of knowledge to staid situations, each service call was a trouble-shooting situation which involved the construction of a narrative outlining the steps taken to solve the problem, including false steps and what was learned from these. What the anthropologists discovered was that narrative development is the true learning zone of the engineers.[20]

As a result of this discovery Xerox changed its training strategy so that employees now 'shared their learning through storytelling' in a 'closed system where management was not allowed to listen in to conversations since trust was a key issue'.[21] Moreover, the company discovered that the best intellectual asset to assist in this training was a firm-wide intranet – an electronic water cooler of sorts. Xerox developed software to facilitate the sharing of practice tips and trouble-shooting narratives amongst its employees.[22] The sharing itself had significant value for Xerox engineers, including a 300 per cent learning curve improvement, and a decrease in the length of customer service calls, service time and parts used.[23] The obvious result of these changes was increased customer service at a lower cost across the company.

Xerox's approach to IAM is firm-centric particularly because development and application of its intellectual assets occured entirely within the firm. Xerox employees not only create the intellectual asset, but distribute it solely to other Xerox employees. In this way, the intellectual assets are safe from competitors. Through this approach Xerox supports its corporate goals of improved employee training, augmented customer service, and lower corporate costs. Thus, its chosen IAM is a success. However, this being said, Xerox is also limited to functioning within the strict boundaries of the firm and therefore does not include elements of external cooperation or participation. This strategy epitomizes a firm-centric IAM approach.

Buckman Laboratories

Buckman Laboratories is a 'leading manufacturer of specialty chemicals for aqueous industrial systems' operating worldwide.[24] In 1987, Bob Buckman created the first system directed at sharing information between its employees. In doing so, he aimed to 'reshape the company by providing every individual the fullest amount of information as rapidly, effectively, and efficiently as possible'.[25] Buckman did so by developing software – known as K'Netix – that facilitates a network of forums within which employees are able to post messages, questions or requests for help.[26] This tool facilitated collaboration between associates who are widely distributed around the world and therefore unable to cooperate through face-to-face conversations. In addition to the benefit of establishing greater conversation and collaboration between distant employees, the initiative enabled Buckman to establish uniform service and business standards globally.

The Buckman example highlights a second form of firm-centric IAM. As with Xerox, Buckman employed its intellectual asset, the K'Netix software, within the virtual walls of the firm and only for the benefit of its employees. Buckman utilized its asset to satisfy its goals of global employee collaboration, higher efficiency and lower costs by permitting employees to build on the existing knowledge and experience of their colleagues.

Monsanto

Monsanto is a company that produces agricultural products using plant biotechnology, genomics and breeding.[27] One of Monsanto's products is *Round-up Ready* canola, a genetically engineered plant that is resistant to glyphosate herbicide (which Monsanto sells under the brand *Round-up*).

Plants grown from these seeds survive spraying of the herbicide, permitting farmers to spray their crops against weeds even after the plants emerge from the ground. Regular canola can only be sprayed, on the other hand, while still underground. This provides an advantage to farmers who use the canola seed over traditional varieties.

Monsanto was obviously interested in protecting its lead against competitors in the ag-biotech market. Prior to 1990, it pursued a dual strategy of lobbying the US government for high levels of regulation over genetically modified plants and of educating the public about the benefits of Monsanto's products. Monsanto knew that it would have an easier time of meeting regulatory requirements than its competitors. Any regulation would therefore act as a significant barrier to market-entry for potential competitors. Nevertheless, Monsanto management abruptly changed its strategy in 1990 away from regulation to enforcement of its IPRs, notably patents.

The manner in which Monsanto deployed its intellectual assets is another illustration of a firm-centric IAM approach. While prior to 1990, Monsanto had engaged both the public and the government in achieving its goals – by sharing copyrighted work and knowledge – it radically altered its course to an inward-looking approach with its change of strategy. Instead of openly sharing intellectual assets, Monsanto pursued a course of strictly enforcing its IPRs not only against competitors but against its farmer clients. It did so in several ways.

First, in order to maximize control of its intellectual assets, Monsanto licensed rather than sold its seed. When someone sells a product that incorporates an intellectual asset, he or she loses control over the use made of that product (and hence intellectual asset) by the purchaser. In a licensing arrangement, however, the licensor (formerly the seller) is able to attach conditions on how the licensee (formerly the buyer) uses the product. Monsanto attached two key conditions to its seed licenses. The first was that Monsanto forbad farmers from collecting seed from one year's crop to use in subsequent years. Instead, farmers would have to repurchase seed each year. The second condition was that farmers had to give Monsanto access to their fields in future years so that the company could ensure compliance with the first undertaking. The end result of its licensing arrangement was that Monsanto was able to maintain significant control over its intellectual assets and keep competitors away. At the same time, Monsanto actively sued farmers who either did not get permission to use the seed in the first place or breached one of the conditions in the license.[28]

Monsanto's policy of closely guarding its intellectual assets – a prime example of a firm-centric IAM strategy – ultimately proved less successful than it had anticipated. Partly due to Monsanto's own earlier campaign calling for the need to regulate genetically engineered plants, the public, particularly in Europe, became deeply concerned over the safety of such plants. Led by public fear, European governments, both at the national and Community levels, imposed moratoria on genetically modified organisms.[29] As a result, Monsanto's share price suffered significantly, in direct contrast to technology markets generally that were rising to unprecedented heights at the time.[30] For Monsanto, the change to a firm-centric IAM strategy proved ultimately detrimental.

Firm-networked

If firm-centric approaches concentrate solely on the internal uses of intellectual assets, firm-networked IAM looks both internally and externally. In particular, firm-networked approaches seek out opportunities to trade, sell or license intellectual assets to external entities for their mutual and greater advantage. As with all IAM approaches, how particular firms actually choose to implement their strategy will differ, but the overall effect of this approach is to apply a combination of internal use and external sharing of intellectual assets to meet a firm's goals.

Some authors refer to a firm-networked strategy as a 'wider-use asset' viewpoint, the central characteristic of which is to have a non-defensive attitude towards intellectual assets. The non-defensive nature of this strategy 'opens up more alternative ways to use the [intellectual asset] portfolio for the firm's benefit and thereby allows firms to create more value-extraction opportunities'.[31] A company seeking applications of its intellectual assets beyond its walls will likely use its assets more assertively.[32] For example, a firm may 'enter into a joint venture with another company, set up a strategic alliance with a partner that can give it access to markets it might otherwise be unable to reach or simply sell a technology in which it no longer has a strategic interest'.[33]

Perhaps the main difference between firm-centric and firm-networked IAM is a shift in the context in which the firm evaluates the value and utility of its intellectual assets. '[C]ompanies able to shift their context or, in other words, their self-view, are companies with the strategic perspective necessary to fully extract value from all of their intellectual assets'.[34] While entities following a firm-networked approach may very well transfer or license out their intellectual assets, it is quite possible that, after careful evaluation, the firm's best use of its intellectual assets is not to do so. What is important is that the firm is at least willing to contemplate opportunities of using its intellectual assets outside the firm's walls.

Once a firm is willing to look at outside opportunities in which to deploy its intellectual assets, many options present themselves. For example, a firm may note that certain of its patents bring it no revenue, yet the firm continues to pay fees to the patent office to maintain its rights. In such a situation, the company would be wise to seek out another enterprise that could make use of the patent right and, eventually, assign or license its rights to that enterprise with, obviously, some type of return. Alternatively, the company may find that it possesses a patent that generates some revenue but that it could significantly increase its returns if it had rights to a competitor's patent. In this case, the company may wish to attempt to cross-license – meaning that each company would license its own patents to the other – the patent with that other firm. Because a firm-networked approach to IAM considers possibilities external to the firm, the firm contemplates a greater set of options in respect of the use of its intellectual assets.

A number of examples of firm-networked IAM are available. We will briefly discuss some of these, once again noting that while the strategies applied are not identical, they each set the boundaries for IAM beyond the walls of the firm.

Rockwell International

Rockwell presents an interesting example as the company employed a firm-networked IAM approach to reduce, rather than augment, its activities and intellectual property holdings. Rockwell transformed itself from a firm encompassing a wide variety of businesses under a single umbrella[35] to one that now focuses exclusively on equipment and software for the factory automation market.[36] It sold off its other businesses.

Rockwell did not, however, undertake this transformation because of a sudden change in consciousness; instead, the firm was reacting to a loss of one billion dollars to costs incurred due to adverse patent litigation.[37] At the time of the judgments against it, Rockwell found itself entirely unprepared to deal with the claims. In the course of finding its way out of this situation, Rockwell made the decision to examine its intellectual asset portfolio. In doing so, it shifted its focus from its tangible assets to the significant revenue value of its intellectual assets.[38] It realized that certain elements of its intellectual assets were more valuable to other companies than to it. Drawing on this realization, Rockwell was able to reorganize its efforts and enter into settlement agreements for a combination of a relatively small amount of cash and a significant effort to create, assign, license and make productive its intellectual assets.

The lesson to Rockwell is that 'potential buyers may have needs that differ markedly from those the technology was originally created for, yet that technology may be precisely useful for their particular needs, and they may value it even more highly than it was valued when it was utilized for the purpose for which it was originally intended'.[39] Rockwell acknowledged that the value of an intellectual asset depends on the set of needs and complementary assets held by a firm. In some cases, firms other than the asset holder may be able to extract greater value from an intellectual asset. Thus firms seeking to maximize the value of their intellectual assets should consider the possibility that their most advantageous option may be to assign or license them to another entity.

General Motors

General Motors ('GM') is an automobile manufacturer operating worldwide.[40] It considers its intellectual assets to have application for the company as well as for the marketplaces and societies in which it operates.[41] One example of this is the company's effort actively to take into account customer desires in the course of its determination of the technological advances that the company can offer.[42] That is, GM solicits customer input when shaping the intellectual assets that the company chooses to develop (new car designs, advanced car components such as fuel-efficient engines, and so on). This does not mean that GM simply attempts to anticipate customer wishes and evaluate the level of willingness to pay: it tries to identify new technologies that will be of more universal benefit, even when these technologies will not constitute an added feature from which the company can directly obtain a financial return.

Consider, for example, GM's strategy of promoting E85 – a gasoline blend using 15 per cent of ethanol, a renewable fuel – to its consumers.[43] The sale of renewable fuels does not lead to direct economic returns to GM but it does meet customer and public needs. Moreover, it is related to a facet of GM's intellectual asset production, namely the development of fuel-efficient vehicles. Thus, GM's overall IAM strategy ensures that knowledge and other intellectual assets are directed towards facilitating the development of vehicles that benefit society, whether that be through lower emissions, which help to protect the environment, or other innovations. Obviously, GM is pursuing these technologies not simply for the sake of the environment, but because it wishes to position itself in what it sees as the market of the future. Nevertheless, GM is taking a long-term and contextual approach to its intellectual assets. This allows it to take into account its own internal needs as well as larger consumer and societal factors so as to better align customer needs with the rate of technological development which are not always aligned with those needs.[44]

GM's IAM approach does not, however, only accept external input from its customers and society; GM looks at exploiting its intellectual assets by assessing the needs of the industry as a whole. Thus, GM does not necessarily hold all of its intellectual assets closely; it is willing to work with other car manufacturers to create industry-wide standards based on those assets that benefit the marketplace as a whole. This use of IAM will be discussed at greater length later in this chapter. The willingness of GM to tailor its IAM strategy to integrate external entities, such as customers and industry standards, is evidence that it applies a firm-networked approach to IAM.

Merck and Inbio

Merck & Co., Inc. is an international research-driven pharmaceutical company that has discovered, developed, manufactured and marketed vaccines and medicines since 1891.[45] The firm also publishes and distributes health information, such as *Your Health Now* magazine, as a not-for-profit service that complements Merck's overall business goals. This is one small example of Merck's firm-networked approach to IAM.

A more significant example of Merck's IAM strategy is exemplified in its relationship with INBio. INBio is an organization created by the Costa Rican government to facilitate access to the country's biological resources. INBio agreed to provide Merck with access to the country's genetic resources in 1991.[46] While the agreement itself is confidential, some of its terms have been reported. Merck agreed to pay an initial fee of US$1 million to allow INBio to purchase research equipment and, in addition, to pay a royalty on each sale of a product that Merck developed as a result of its bio-prospecting in Costa Rica.[47] This agreement offered benefits not only to Merck, by way of a larger range of resources for prospecting, but also to Costa Rica in the form of financial support to be used for environmental protection as well as jobs for those collecting and banking bio-materials.

Through this agreement, Merck gained access to resources that may lead to products that would ultimately represent a source of IPRs for the company. Merck transferred some of the benefits of those IPRs to Costa Rica through its agreement to pay royalties. Merck's willingness to share benefits outside the firm represents a firm-networked IAM strategy. Moreover, Merck was obviously conscious of the social and ethical considerations relevant to the bio-materials it sought in Costa Rica and the technology it would ultimately produce itself. The agreement addressed these issues as well, particularly public concerns regarding the outcome of unchecked bio-prospecting.

Non-firm IPM

The first two IAM strategies we discussed take for granted that the firm is the center of our interest. Recall that firm-centric IAM approaches focus only on the firm itself, whereas firm-networked IAM looks at the interests of others, but only in so far as these relate to the firm's ability to extract greater returns from its intellectual assets. A different approach is possible, however, one that views intellectual assets as giving rise to new markets and addresses shared goals. We call this non-firm IAM to emphasize its more open nature. In particular, non-firm IAM takes account of cross-firm, cross-industry and cross-sector synergies that arise from the sharing of intellectual assets.

While intellectual assets represent an immediate use value to firms (as in firm-centric IAM) and an opportunity to internalize the value of intellectual assets to external actors (as in firm-networked IAM), they also can act as a foundation for new industries, new markets and new partnerships, both within the private sector and between the public and private sectors. Firms do not often have the power to create these markets and other opportunities by themselves: they must act together with other entities in the public and private sectors to reach these goals. So, for example, firms and universities can work together to build training programs in intellectual asset management to help all firms better compete in global markets. Similarly, firms can work with government to identify trusted spokepeople – for example, senior government officials, a new professional organization – to promote a country's products, such as natural health remedies derived from its genetic resources. The central element in the IAM strategy is to use intellectual assets as a foundation upon which firms can create value.

Non-firm IAM has three characteristics that make it a unique and viable approach. First it enlarges the set of entities engaged in IAM strategies beyond corporations. Governments, industry organizations and users also have a stake in managing intellectual assets to promote economic development, open new markets and to ensure access to technology. Second, non-firm IAM encourages managers to consider and possibly deploy a greater set of tools to attain their firm's goals. A firm may find, for example, that its optimal use of an intellectual asset is blocked, not by competing IPRs, but by the absence of complementary technologies or market institutions. Consider software that finds no buyers because of insufficient access to wireless technology. A manager who works with government or other private actors to increase the available amount of wireless technology will not only permit the firm to extract value from its existing intellectual assets, but provide a foundation for further innovation. Third, non-firm IAM permits decision-makers to use intellectual assets to achieve a larger set of goals. A firm may determine that it will better be able to attract employees, managers and investment if the city or region in which it operates has a critical mass of people and firms working in the field. It thus may be in the firm's interest to work with other firms and government to develop an industry cluster.

A non-firm IAM strategy does not necessarily attempt to extract value directly from the IPR. Instead, the preferred strategy may be to convert IPRs into a longer lasting asset, such as customer and supplier loyalty and brand name recognition.[48] That is, instead of keeping a patent solely within a firm, it may make more sense for a firm to share the patent and underlying technology with others so as to increase public acceptance of the innovation.

A non-firm approach to IAM is particularly well suited to biotechnology, given that biotechnology touches on some of our most cherished beliefs and social practices: those relating to health care, food safety and security, environmental integrity as well as individuals' self-image and perception.[49] An IAM that ignores the ethical and social implications inherent in biotechnology can create problems. As the Monsanto example described above illustrates, an ill-considered IAM strategy can lead to marketing and regulatory outcomes that are advantageous neither to the firm in particular nor to society in general.[50] When we add the complexity of external social and ethical concerns raised by such technologies as genetically modified crops, genetic tests, stem cell research, and bioengineered foods, non-firm IAM may be better adapted to managing biotechnology innovation adequately.

Taking a broad and dynamic view of IAM thus increases both the possible means of managing intellectual assets and the goals that one may wish to obtain. We will re-examine two examples previously discussed in this chapter – Xerox and GM – as well as two others to illustrate non-firm IAM.

Xerox

As discussed above, as a result of its recognition that learning is best facilitated through story-telling, Xerox established an Intranet through which its service engineers could tell and exchange service-related stories throughout the world. Xerox's benefit from this technology was improved customer service. Imagine, however, that Xerox had gone one step further and extended its training to service engineers working outside the firm. In such a case, it could have obtained additional benefits in terms of design and service ideas generated by these outside engineers.

The main benefit to Xerox would be the construction of a database that highlights problems and solutions to those problems that Xerox's own engineers had not encountered or developed. Drawing on this database, Xerox's own engineers would more easily and efficiently address service problems. More importantly, however, Xerox would be able to use the information in the database to improve the design of its machines using the accumulated knowledge of many more service engineers. There is a second advantage to this approach. By making service stories more available to engineers, customers would have greater servicing options, thus making the Xerox brand synonymous with customer support and care. The benefit here would be in the possibility of increased sales. Of course, Xerox would need to weigh these benefits against any harm it may suffer from permitting competitors to access service information about its machines.

General Motors

As discussed earlier, GM currently takes into account customers' desires in its decision-making about which technologies to develop. GM management states, however, that it has a more ambitious vision, one that moves beyond the boundaries of the firm. Vince Barabba, GM's General Manager of corporate strategy, commented in 2003 that, 'if we can get the companies in related industries to agree on standards and protocols, the industry itself gets bigger and – although we might not increase our share – we open ourselves to new products and services that allow us to better serve the customer overall'.[51] This represents an important step toward implementing non-firm IAM since the strategy seeks to construct and share intellectual assets across the sector for the greater good of all actors.

Similarly, GM's program to promote gasoline–ethanol blends may be characterized as looking beyond GM's own interests to those of the automobile industry and society. As automobile emissions contribute to global-warming,[52] GM's strategy creates goodwill for itself among potential customers but, more importantly, helps the environment.

DVD patent pool

Sony Corporation of Japan ('Sony'), Pioneer Electronic Corporation of Japan ('Pioneer') and Koninklijke Phillips Electronics NV ('Phillips') initiated a unique approach to IAM by creating a DVD patent pooling arrangement in 1998. Essentially, Sony and Pioneer agreed to grant Phillips a non-exclusive license to all essential patents required to achieve compliance with DVD Standard Specification (a world-wide standard for the use and enjoyment of DVDs created by an industry consortium).[53] Phillips took on the task of licensing these essential patents to 'all interested parties . . . to manufacture, have made, have manufactured components of, use and sell or otherwise dispose of'' DVD discs and players that conform to the Standard Specification.[54] All three licensors retained the right to license their essential patents independently of the agreement, but they also agreed that licenses granted under the patent pool would be non-discriminatory and include reasonable terms.

The purpose of this pool was to ensure that all participants in the DVD market could obtain a license to use the patented technologies required to create DVD discs and players in accordance with the sector standard. In the absence of the patent pool, each DVD manufacturer would have been required to seek individual licenses for each patented technology, of which there were several. If a company failed to obtain these rights but nevertheless used the technology, it would find itself vulnerable to a charge of infringement. The patent pool avoids this problem by simplifying the licensing process. By doing so, it supports the establishment of the industry standard, avoiding Sony's earlier difficulties with its Beta technology.

We can learn two important things from the DVD patent pool about non-firm IAM strategies. First, we once again learn that using management tools outside the firm provides advantages to all industry players. Second, the pool itself represents a group of intellectual assets that can be managed in the interests of all actors, not simply a lone firm in a marketplace. In the case of the DVD patent pool, this actor was Phillips. Third, non-firm IAM can significantly reduce transaction costs. By its willingness to license all essential patents to all those wishing to enter into the DVD market, Phillips diminished the risk of blocking patents and the costs of negotiating separate agreements.

PIPRA

The Public Intellectual Property Resource for Agriculture ('PIPRA') is a relatively new patent pooling tool that offers advantages beyond that of the DVD Patent Pool. PIPRA's goal is to facilitate the management of patents collaboratively through the implementation of shared principles 'to ensure the availability of existing and emerging technologies to generate subsistence crops in developing countries and specialty crops in developed countries'.[55] As with the DVD Patent Pool, PIPRA aims to collect patent rights in biotechnology and other plant technologies for the purpose of licensing these patents to PIPRA members, which may be public or private entities. It is hoped that this will increase access to technology in the agricultural research sector. In its present state, PIPRA offers its members an opportunity to employ a firm-networked IAM strategy that involves the management of the licensing process outside the firm.

What makes PIPRA's model particularly interesting is its potential to lead to the creation of new intellectual assets. As PIPRA comes to enlarge its pool and possibly include intellectual assets beyond solely patent rights, such as databases and plant varieties, it could identify gaps in existing technology and encourage its members to address these for the greater good of all. Second, PIPRA could lead members to identify new joint projects or business opportunities prime for exploitation with other members. Such developments would move PIPRA from its present state of offering a mere smorgasbord of patent rights to a proactive entity that fosters exchange and growth among actors.

POTENTIAL BENEFITS OF ADOPTING A NON-FIRM APPROACH

There are several benefits of adopting a non-firm IAM strategy. Firms and their intellectual assets exist within larger systems of innovation, distribution and sales. Existing IAM strategies do little to address this wider environment. For example, while a firm-networked approach may recognize the interrelationship between a firm and other actors the contemplated transactions are licensing or selling its intellectual assets, it does not consider opportunities to uncover new markets. Essentially, only non-firm IAM truly captures the potential depth of the interconnectedness of actors that can be acheived.

Alexandre Styhre and Mats Sundgren review the importance of the interconnectedness of assets in their book *Managing Creativity in Organizations*, in which they apply Deleuze's model of the rhizome to intellectual assets. A rhizome structure sees intellectual assets working horizontally across an industry and even across industries, rather than vertically within a particular firm.[56] According to Styhre and Sundgren, the true advantage of the rhizome model is that 'it underscores that what we are seeing as creative solutions and creative ideas are always produced through association across various entities and events'.[57] Non-firm IAM facilitates creative thinking as to how to use intellectual assets in a manner that offers the greatest advantage not only for a particular actor, but across an entire sector or society. It is through the understanding of the relationships that exist or may be built between assets that a non-firm IAM strategy takes shape. In particular, it encourages us to look at partnerships and tools that we had not previously considered.

One of the more interesting results of adopting a non-firm IAM strategy is the possibility of crossing the public–private divide. For example, we can imagine a future role for governmental research funding agencies in encouraging pooling and licensing of intellectual assets between universities and companies. On a larger scale, countries within a region could band together through international agreements to cooperate on the development and sharing of intellectual assets. This could be particularly interesting for countries with different kinds of competitive advantages – such as genetic resources or research facilities. Indeed the possible configurations of IAM strategies are wide indeed.

CONCLUSION

Within the nascent realm of IAM, the pool of potential strategies to be implemented is fathoms deep. Despite narrow beginnings, IAM is poised to develop into a tool which envelopes a wider context, branching out from its firm-based foundation to embrace industry, government and public relationships for intellectual assets. Of course, in order for IAM to be effective in this new sphere, it must overcome certain hurdles. In particular, we need to develop clear language and concepts so that parties applying or discussing IAM attach the same meaning to the same terms and phrases. Otherwise, IAM will continue to be murky and not lead to mutual understanding or certainty.

As the examples we provided show, IAM has many forms. There is still much to be learned as to which strategies will or will not be successful for the multitude of companies applying IAM. For example, merely amassing intellectual assets will not necessarily meet the goals of a company hoping to increase its profits. A managed strategy is critical, and the new wave of technology companies appear to be approaching IAM in a manner that can easily be augmented to realize a non-firm approach. The DVD standards and PIPRA initiatives in particular show how the involvement of a wider grouping of interests, beyond those that are merely inter-firm, can add depth to IAM.

We have only begun to explore the potential for IAM strategies, especially in the biotechnology sector. While both firm-centric and firm-networked approaches offer short-term benefits to firms, only non-firm IAM offers the flexibility to address the development of new markets and new synergies to which biotechnology gives rise. Public–private partnerships, patent pooling, regional cooperation and novel methods of licensing all become available within the wider vision of non-firm IAM. Enlarging our vision and our tools helps both firms and society to attain their goals.

NOTES

1. The terms, 'firm-centric', 'firm-networked' and 'non-firm', have been coined by the authors to describe IAM strategies and are not taken from the existing IAM lexicon.
2. E. Richard Gold, 'Merging Business and Ethics: New Models for Using Biotechnology Intellectual Property' in Michael Ruse and David Castle, eds., *Genetically Modified Foods: Debating Biotechnology* (Amherst, NY: Prometheus Books, 2002), at 164–65.
3. Patrick H. Sullivan, *Value-Driven Intellectual Capital: How to convert Intangible Corporate Assets Into Market Value* (New York: John Wiley & Sons, Inc., 2000) at 130.
4. Ibid., at 17.
5. Ibid., at 5 and 3.

6. 'A market for ideas' 377 *The Economist*, Oct 22, 2005, 4.
7. *Supra* note 3 at 5 and 3.
8. 'The arms race' 377 *The Economist*, Oct 22, 2005, 8.
9. *Supra* note 3 at 130.
10. *Supra* note 2 at 164–65.
11. Adam B. Jaffe and Josh Lerner, *Innovation and Its Discontents: How Our Broken Patent System is Endangering Innovation and Progress, and What to Do About It* (Princeton, NJ: Princeton University Press, 2004) at 14–15.
12. Ken Standfield, *Intangible management: Tools for Solving the Accounting and Management Crisis* (Amsterdam: Academic Press, 2002) at 119.
13. Jay Chatzkel, *Knowledge Capital: How Knowledge-Based Enterprises Really get Built* (Oxford: Oxford University Press, 2003) at 25.
14. Karl Albrecht, *The Power of Minds at Work: Organizational Intelligence in Action* (New York: American Management Association, 2003) at 202.
15. Ibid., at 202.
16. It should be noted that the examples of IAM in practice that are included in this chapter will not provide detailed information regarding the specific application of intellectual assets. This is due to the fact that details of IAM are retained as proprietary corporate information. However, even without specific details regarding the use of intellectual assets, each example has been included to instruct the reader regarding IAM strategies generally.
17. Xerox, 'About Xerox' online: http://www.xerox.com/go/xrx/template/009.jsp?view=About%20Xerox&metrics=hp_030103_about&Xcntry=USA&Xlang=en_US.
18. *Supra* note 13, at 48.
19. Ibid.
20. Ibid.
21. Ibid., at 48–49.
22. Ibid. at 49.
23. Ibid. at 50.
24. Buckman Laboratories, 'Welcome to Buckman' online: Buckman < http://www.buckman.com/>.
25. *Supra* note 13 at 63.
26. Ibid.
27. Monsanto, 'About Us' online: http://www.monsanto.com/monsanto/layout/about_us/default.asp.
28. See, for example, *Monsanto Canada Inc. v. Schmeiser* [2004] 1 S.C.R. 902, 2004 SCC 34.
29. *Supra* note 2, at 168.
30. During a six-month period ending on August 24, 1999, Monsanto's shares lost 11 per cent of their value: Brown, P. and J. Vidal, 'GM Investors told to sell their shares', The Guardian, 25 August 1999, http:www.guardian.co.uk/science/1999/aug/25/gm.food.
31. *Supra* note 3, at 131.
32. Ibid., at 132.
33. Ibid.
34. Ibid.
35. *Supra* note 13, at 99.
36. Ibid., at 100.
37. Ibid.
38. Ibid., at 101.
39. Ibid., at 104.
40. General Motors, online: http://www.gm.com/.
41. *Supra* note 13, at 130.
42. Ibid., 133.
43. General Motors, Press Release, May 2, 2006, online: http://media.gm.com/servlet/GatewayServlet?target=http://image.emerald.gm.com/gmnews/viewmonthlyreleasede-tail.do?domain=74&docid=25439.

44. *Supra* note 13, at 134.
45. Merck & Co., 'About Us' online: http://www.merck.com/about/.
46. Barbara L. Kagedan, *The biodiversity Convention, Intellectual Property Rights, and the Ownership of Genetic Resources, International Developments* (Ottawa: Intellectual Property Policy Directorate of Industry Canada, 1996), at 70.
47. Ibid.
48. Most IPRs are granted for a specific term – patents for 20 years from the filing date, copyright for 50 years after the death of the author. However, customer loyalty and brand recognition are benefits that might be created because of the exclusive rights of an IPR, but are effects that will last after the expiry date of the IPR itself.
49. *Supra* note 2, at 172.
50. Ibid.
51. *Supra* note 13, at 134.
52. A. Lashinsky 'How to beat the high cost of gasoline. Forever', (2006) *Fortune Magazine* online: http://money.cnn.com/magazines/fortune_archive/2006/02/06/8367959/index.htm.
53. The consortia include: Hitachi Ltd., Matsushita Electric Industrial Co., Mitsubishi Electric Corporation, Pioneer Electronic Corporation, Royal Philips Electronics N.V., Sony Corporation, Thomson, Time Warner Inc., Toshiba Corporation, and Victor Company of Japan. This group set standards for format of the optical disk storage media known as the digital versatile disc (DVD) and captured their decisions in a 1995 study.
54. Jeanne Clark et al., 'Patent Pools: A Solution to the Problem of Access in Biotechnology Patents?' [2000] *USPTO* 14.
55. R.C. Atkinson et al. 'Public sector collaboration for agricultural IP managament', *Science*, 301, 174–75.
56. Alexandre Styhre and Mats Sundgren, *Managing Creativity in Organizations: Critique and Practices* (Basingstoke: Palgrave MacMillan, 2005) at 57–58.
57. Ibid., at 58.

PART III

Intellectual property rights in relation to other measures of innovation

Introduction

L. Martin Cloutier and David Castle

The contribution of innovation to a firm, region or state's competitiveness provides a strong incentive to understand how to stimulate creative activity, to create and capture value, and strategically to protect investments made in an innovation system. Economic performance, often measured as *per capita* gross domestic product (pcGDP), is the most common measure that is used to compare a sector's contribution to an economy, changes to an economy over time, or state-to-state comparisons. Although pcGDP is faulted for being an unreliable measure, it prevails because it is one of the few simple measures available. It is valued for being a somewhat comprehensive endpoint measure, in the sense that innovative, value-creating activity has to happen before an impact on pcGDP is appreciable. Firms and governments, often feeling the sting of late-stage performance appraisals, seek measures of innovation further upstream in value chains to give them tools actively to manage innovation.

The quest to measure innovation to manage it better generates problems for firms and governments. As many economies around the world make the transition from a production base to a service base, one of the hallmarks of knowledge-based economies, they encounter the challenge of evaluating the assets of firms. The market value of real estate and equipment held by firms dedicated to manufacturing was once enough; now the knowledge base of the firm is its chief asset. Measuring intangibles, an activity that almost seems like an oxymoron, has been one of the management sciences' greatest challenges in the past decade. Does one, for example, measure the products that arise from knowledge intensive processes, or are the processes themselves what need to be measured? Does one put R&D costs on the revenues side of the balance sheet because they generate knowledge, or does one put them on the side of revenues? More broadly, does one start from the premise that innovation is something, like an endpoint that one can measure directly, or an elusive property of a system that defies direct measurement?

This part examines the issue of how one measures innovation. Operating in the background is an assumption that measuring patenting activity – both patent applications and patents granted – is not a particularly enlightening approach if taken more or less as a sole indicator of innovative activy. For reasons outlined elsewhere in this book, there are conceptual issues, as well as practical or empirical limitations, on using patenting activity as a measure of innovation. As the deficiencies of patents as measures of innovation have become widely acknowledged, other measures of innovation have been sought. It is this activity of looking for these other measures which is the subject matter of this section, since searching for these measures and then using them to manage innovation is not straightforward. Among the general conceptual problems encountered is the issue of knowing what counts as the right level or unit of analysis if it is not a patent. Some prefer to focus on the role of inputs to innovation such as education, whereas others prefer to focus on outputs of innovative activities like new ideas or services. Furthermore, holding the belief that one can measure implies that one also believes that innovation, whether a product, a process, an organization, or a market, can be quantified with standardized units of measurement.

These conceptual and methodological points, general though they may be, are frustrating to the manager or policy maker who simply wants to get on with the business of doing business. After all, why complicate the already difficult task of staying competitive with what seem like endless loops of theory? In response, the overall point of this section is that theory really matters. Particularly in an economy which is based upon intangible assets, this is no coy statement. The management of intellectual property, and innovation in the processes that lead to products, is not just the invention of new tangible products. Consequently, the chapters in this part take different approaches to the same question about how one goes about the difficult task of first setting the criteria and then implementing the standards for the measurement of innovation. Given what has been said elsewhere in this book, the objective is to do this without relying on bland measures of patenting activity, while at the same time keeping an eye on patents. The chapters that follow are united by the idea that there really is a measurement problem that needs to be addressed head-on. So, for example, in the first of the chapters the conceptualization of the valuation of product innovation is addressed from a purely philosophical perspective. The chapter following examines how organizations address technological challenges and attempt to measure innovation by keeping track of the contribution of intangible assets to dynamic innovative processes. The final chapter is a case study of a firm which has effectively measured and managed innovation by carefully exploited patent value.

In a chapter which is avowedly theoretical, Clinton Francis provides a model of innovation valuation. Francis argues that there are distortions in the way the valuation of innovation is normally conducted. Indeed, this distortion comes from the lack of interpretative complexity that arises when innovation is interpreted in light of its natural and historical contexts. These, Francis argues, severely limit the ability to value an innovation in ways that are appropriately and sufficiently sophisticated that the task of evaluating innovation is on the same level of complexity as one finds it in current market-based contexts and organizational settings. Francis offers three main lines of arguments to call for a re-examination of the way we value innovation. He proposes that valuations should be divided and analysed according to three main groups: IPR-based evaluations, market-based evaluations, and value-chain-based evaluation. IPR-based valuations, which attempt to attach a monetary value to IPRs held by an enterprise, run into several problems. These include the fact that today's technologies cannot be valued in the same way as previous technologies (as each generation acts quite differently), making historical comparisons of little value. Second, IPRs protect only a portion of the value of a new technology. To address the valuation of innovation issue, and the potential for hierarchical confusion in the evaluation of the innovation, Francis presented a model called Language System 3.0, which is an application of a semiotics analysis of complexity. The goal of suggesting a method of innovation valuation that is hierarchically orthogonal to the subject matter, as much as is possible, is to avoid the inversion of dependent hierarchies that could limit the competitive advantage bias against emerging innovations.

The chapter by L. Martin Cloutier and Susanne Sirois borrows from the existing literature to assert that, from a scientific conduct and organizational standpoint, the measurement of innovation in the biopharmaceutical sector is very challenging. In part, this is because of the rapidly evolving science and technology, which put constantly shifting value in the intellectual property that is generated in R&D. Only with difficulty can the value of IP be approximated at any given time using traditional measures by which the value of patents is evaluated. The chapter relies on the distinction between tangible and intangible assets, which makes it very much out of step with the normal approach to asset management which often treats IP as if it were practically tangible. Of course, most innovation measures are based on tangible assets or outputs therefrom, and ignore more generally the contribution of intangible assets. Perhaps more illuminating, it must be pointed out that intangible asset creating and exploitation is not normally associated with the evaluation of market trade-offs in firm management. The chapter uses the example of emerging new technology associated with high throughput screening (HTS) for lead and drug discovery and the *in silico* revolution to show how the technological complexity embedded in the innovation management process only helps to complicate matters. In addition, the failure of most measurement methods to account for feedback loops associated with the management of the organization ignores the contribution of intangible resources in that process. Thus, existing measures of innovation management are likely to suffer from existing biases due to the data and to limitation for establishing benchmarks in that process.

The chapter by Marc Ingham, Cecile Ayerbe, Emmanuel Métais and Liliana Mitkova provides a substantial case study on the role of patents in the leverage of value. The chapter examines the situation at Air Liquide, a French multinational. The chapter is concerned about how a firm decides why and when to change organizational arrangements at each stage of the patenting process, and what values characterizes these management decisions. This case provides insights on how large companies can increase value from patents through dedicated organizational arrangements that contribute not only to improve the efficiency and effectiveness of the process itself but also to support knowledge sharing and creation. These issues have been discussed widely both in the literature on the management of patents and in the literature on knowledge creation and organizational learning, but it is useful to see what they can mean in practice. The two phases that lead to visible organizational arrangements have different impacts on the social interactions which subsequently take place among the members and groups who were involved in the process. This process is primarily informal, and does not appear in traditional measures of innovation. The new organizational platforms have characteristics that are similar to those that have been identified by Nonaka and Takeuchi (1995) in their researches on knowledge creation. The involvement of more multi-functional team members with different backgrounds at the very early stage of the process enrich and nurture social interactions that are propitious to knowledge sharing and creation.

REFERENCE

Nonaka I. and H. Takeuchi. 1995. *The Knowledge-Creating Company. How Japanese Companies Create the Dynamics of Innovation.* Oxford: Oxford University Press.

7. Increasing internal value from patents: the role of organizational arrangements

Marc Ingham, Cecile Ayerbe, Emmanuel Métais and Liliana Mitkova

INTRODUCTION

This chapter explores how organizational contexts are conducive to internally enhancing and exploiting the value of patents. Internal exploitation is adopted for patents that are part of the 'strategic businesses'; the main objective is to keep under control both the invention and the markets. This chapter addresses the following questions: why and when are these organizational arrangements adopted at each stage of the patenting process and what are their characteristics? To investigate these questions, the literature on organizational configurations dedicated to patenting is reviewed. The methodological approach consists in a case study of Air Liquide, a large French industrial group.[1]

Two main characteristics of the organizational context are of paramount importance to increase the internal value drawn from the patent portfolio, the involvement of actors from different functions in the patenting process, and the use of information technology. These 'platforms' enable one to create a context that is conducive to knowledge sharing and creation. The literature on organizational learning and knowledge creation is then used to interpret the case presented in this chapter.

BACKGROUND

The flourishing literature on patents in the fields of economics, law and management rarely investigates how companies develop and use organizational arrangements to increase intra-organizational value from their portfolio of patents.[2] Nevertheless, organizational contexts are of paramount importance to improve the efficiency in the management of patents (knowing why, where and for how long patents should be maintained). This issue is particularly important in the case of large firms that own a large number of patents they exploit internally or externally.

This inquiry begins by investigating two streams of the literature. The first stream deals with the organizational dimensions of intellectual property. This literature argues for an evolution from the traditional approach that considers patents from a legal standpoint towards a more integrative approach that includes organizational and managerial aspects.

Patents can be viewed as knowledge content that has been formalized (and articulated). The key pieces of the literature on organizational contexts that are conducive to organizational learning and knowledge creation are examined to enrich the first stream of literature.

There have been few writings on the organizational dimension of intellectual property. A noticeable exception is Granstrand's (1999) study of the organization and management of intellectual property in large Japanese corporations. According to Granstrand (1999, p. 261):

> Traditionally in Western Companies Intellectual Property matters have not attracted a great deal of resources and attention concerning the organization. A traditional large western corporation has typically added some kind of patent department attached to R&D or to legal department at corporate level with some liaison engineers decentralized.

In his in-depth analysis of organizational structure and management of patent activities Granstrand (1999, p. 260) first describes the various options for intellectual property organization in general. He distinguishes five options:

- centralized at corporate headquarters;
- decentralized to business units and subsidiaries, domestic and foreign;
- decentralized to one business division with corporate-wide responsibility;
- organized as an independent unit; and
- externalized to a supplier organization.

The choice between these various options is linked to the firm's size and its international dimension. The large international companies have a 'centralized IP department with corporate wide responsibilities for patent coordination headed by one central corporate patent manager' (Grandstrand, 1990, p. 262).

The main concerns of *work operations* in intellectual property are the following (Granstrand 1999, 262): (1) obtaining and maintaining patents, (2) opposing patent applications by others, (3) handling infringements,

(4) licensing, (5) dealing with litigation and foreign patent work, and (6) analyzing patent portfolios. Several works have studied the implication of the various functional specialists in these operations. Granstrand mostly focuses on the integration of the intellectual property and R&D, showing that a strong advantage of Japanese companies lies in the involvement of people in charge of patenting in the early stages of R&D. This leads patent management to operate 'proactively' rather than 'reactively' in meeting requests from business and R&D. Further works have shown that not only intellectual property, R&D but also production, finance, marketing V-Ps and CEOs, are mostly involved in the decision-making related to patent activities (Napper and Irvine, 2002). Intellectual Property Department, R&D and CEOs are involved in the early legal stage of registering patents. At this stage, the relationship between marketing and R&D is also important to ensure the success of an invention becoming an innovation. Following this first stage, decisions concerning the external exploitation (licensing) of patents mostly involved R&D and marketing (Marquer, 1985). Production is more involved with internal operations. Whatever these distinctions, the various studies emphasize the importance of multifunctional interactions in the management of patent activities (Sproule, 1999; Grindley and Teece, 1997). That is the reason why, as a typical staff service function, the organization for intellectual property department is a matrix achieving coordination through committees and liaison groups. The increasing resources and financial patent dimension are linked directly to these organizational aspects. Granstrand (1999) explains that the intellectual property department is often a cost center with costs-sharing arrangements.

Finally one of the most important organizational dimensions outlined by Granstrand is the *patent culture* which describes a general orientation concerning patenting. Several elements have to be taken into account to build and reinforce this specific culture, for example: top management involvement in patenting, patenting as a common concern for all engineers, patents strategies and clear objectives integrated in business plans and clear incentives for R&D personnel and organizational units (Granstrand, 1999, p. 265).

Most of these dimensions have also been stressed in the literature on organizational learning and knowledge creation. Amongst the different constructs and processes that have been identified in the literature on organizational learning (Huber, 1991) this section focuses on information distribution and organizational memory. Both are of key importance in the patenting process that primarily deals with explicit and articulated knowledge. According to Huber, (1991, p. 100), 'information distribution is a determinant of both the occurrence and breadth of organizational learning'. Organizational units and actors can develop new information by combining the information they obtain from other units and actors. On the other hand, information synergies and new understanding can be created when information is widely distributed in the organization and when more and more varied sources exist. Organizational memories have to do with the way organizations store information. 'As a result of specialization, differentiation, and departmentalization, organizations frequently do not know what they know. The potential for reducing this problem by including computers as part of organization's memory is considerable' (Huber, 1991, p. 106).

According to Nonaka and Takeuchi (1995), knowledge transfer, sharing and creation are sustained by organizational arrangements such as multifunctional and pluri-disciplinary teams, which nurture and fuel the knowledge conversion process from socialization (tacit to tacit), to externalization (tacit to explicit), to combination (explicit to explicit) and to internalization (explicit to tacit) (Nonaka and Takeuchi, 1995). It may be argued that the patenting process is focused on knowledge creation through externalization and combination in its early stages while its (internal) exploitation is primarily based on the transfer of explicit knowledge and internalization. Finally, knowledge creation encompasses dimensions of organizational structures and cultures that support the processes. When they are well balanced, these technological, organizational, managerial and cultural 'platforms' can leverage knowledge sharing and creation.

RESEARCH METHOD

The qualitative approach appears particularly important because of the exploratory nature of the research question (Silverman, 1993). Very few writings have investigated the theme of organizational arrangements dedicated to patenting. Within qualitative methods the case study approach has been chosen because it enables one 'to understand the dynamics present within single settings' (Eisenhardt, 1989). Moreover it attempts to examine a contemporary phenomenon in its real-life context (Yin, 1994).

The selection of cases is of a great importance when using this research strategy (Eisenhardt, 1989). A single case study is analysed and the firm is Air Liquide. This approach refers to the 'relevant case' (Yin, 1994). In the present research the case has been identified to be typical: it examines in detail a single situation which is considered to be particularly relevant for other comparable firms. Various selection criteria have been used to ensure this relevance: innovation budget, number of researchers, patent activities (number of patents per year, number of protected inventions, patent strategy), organizational structure dedicated to patent activities (status of the intellectual property department, number and activities of patent engineers, relationships with other departments, etc.).

Yin (1994) outlines several sources of evidence when conducting a case study. Two of them are used here: interviews and secondary data. Interviews were conducted with key informants involved in strategic development and intellectual property management. Secondary data enabled the completion of primary data. Two main periods must be distinguished in data collection. The first one mostly focused on the analysis of organizational change that took place in the 1990s (Métais, 1997). The second period specifically analyses the evolution of organizational arrangements put in place to increase the internal value from patents. (Ayerbe and Mitkova, 2005).

THE CASE OF AIR LIQUIDE

Founded in 1902, Air Liquide is the world leader in industrial and medical gases and related services. The company offers integrated solutions to its customers in terms of gas, associated products and high value-added services. With 36,000 employees, 130 subsidiaries in more than 70 countries, and total sales of €9.4 billion in 2004 (80 per cent international), as well as more than one million clients, Air Liquide enjoys good financial performances (Table 7.1). In January 2004, Air Liquide reported a 6.2 per cent increase in its sales for the year 2003. During the last 30 years the Group has achieved solid and sustained earnings with average growth of 11.3 per cent per year. The average rate of dividend distribution has been 40 per cent over the last 10 years.

The company has developed its activities around its core business, offering products and services in two main domains: gas and services (87 per cent of the total consolidated turnover in 2002). Air Liquide has also developed a set of related activities (13 per cent), including engineering.

Gas and services are organized along four main business lines:

- *Large industries* (chemicals, refining, metals etc.), with users of large volumes of industrial gases and energy solutions and long-term contracts (15 years). Customers are generally supplied through pipeline networks linked to plants designed, constructed and operated by the Group (engineering division) (27 per cent of sales in 2004).
- *Other industrial customers* using gases in small or medium quantities and operating in very diverse sectors. The contracts are of short and medium terms (three to five years) and gases are delivered in cylinders

Table 7.1 Key financial data (in millions of Euro %)

	2000	2001	2002	2003	2004
Sales	8,099	8,328	7,900	8,394	9,376
Operating income/sales	14.2 %	14.6 %	15.0 %	14.2%	14.1%
Net earnings/sales	8.0%	8.4%	8.9%	8.6%	8.3%
Market capitalisation	14,528	14,295	12,673	13,998	14,849
ROE	12.8%	13.2%	13.4%	14.1%	14.9%
ROCE (after taxes)	10.5%	10.7%	10.8%	11.6%	11.3%
Net debt/Shareholders equity	40.4%	45.5%	37.1%	31.2%	66.3%

Source: Air Liquide, Annual Reports 2002 and 2004.

or produced in units installed on customers' sites. (46 per cent of sales in 2004).

- *Electronics* covering semiconductor manufacturers (very high-tech sectors). Carrier gases are often supplied through units at customers' sites with long-term contracts (10 years on average), and specialty gases and associated services (three to five year contracts), as well as equipment and installations for gas and chemical products (11 per cent of sales in 2004).
- *Healthcare* which covers the supply of medical gas, services and equipment to hospitals and homecare (16 per cent of sales in 2004).

Air Liquide has also developed expertise in related activities (equipment, products and services) in the fields of welding and cutting, diving, chemicals, space, engineering and construction. This division designs and constructs gas production plants for both the Air Liquide Group and external customers (turnkey units/plants) (12 per cent of sales in 2004). Air Liquide's growth is built on both specialization and related diversification based on innovations resulting from its core competencies and expertise.

The Organizational Context

The organizational arrangements that took place in the early 2000s to increase the value of patents can be interpreted in the light of the organizational context that has been implemented over the previous decade.

In the 1990s, Air Liquide examined the possible evolutions in the economic and international environment, and its position as the leader in the industry. Compared to other companies, Air Liquide conducted its proactive reflections and in-depth analysis while its leadership and performances were still highly satisfactory. A group of 'experts' was established to help define the company vision and to nurture collective reflections. A large number of people with different skills and knowledge backgrounds were involved in this process:

> We should not forget that there was prior reflection, what we called a vision, which involved meeting and brainstorming, where some experts would say 'what I've been dreaming for a long time, is. . .', this was the part of dream, what could be the perfection, but what would be its outcome.

The dreamlike dimension of the vision aimed at defining a 'shared desire for a future state'. This was later translated into a more practical strategic intent (Hamel and Prahalad, 1989).

Up to then, the vision had focused on technological variables. Now its centre of gravity evolved, with technology considered as a necessary but not sufficient condition for long-term success. The focus was on increasing value from technology and establishing closer relationships with the customers–users (external or internal) and translating the vision into a strategic intent:

> We came out (of our preliminary reflections) with a vision that was summarized by the simple idea that 'we should be in closer contact with our clients' and, in order to achieve this we had to create organizations that were close to the field, more flexible and reactive. We had to eliminate hierarchical ladders. The intention at the Group level was also to create a 'shock'. We used the words 'flat structure' to reduce the hierarchical levels. I think we were deliberately a little bit brusque in doing so. [Head of the Gas division (large industries), former head of the engineering division]

This flat structure was considered as the best way to increase value from R&D and innovation and to establish closer relationships with the customers.

Toward a Knowledge and Competence-based Organization

The transformations developed in the divisions covered several steps. The first stages consisted of rethinking the product-services and designing an 'ideal value chain' together with the key processes. The second stage led to the identification of 'core competencies' and their underlying knowledge bases. This in turn led to the creation of a 'platform', followed by the creation of 'competence centers'. The company then developed a new organizational structure based on three organizational entities to facilitate the development, deployment and exploitation of knowledge and competencies to support innovations; competence centers, product teams and project teams.

Rethinking product-services

Implementing the vision and the strategic intent was a major challenge for the divisions and departments. The transformation process was guided by a totally new philosophy of which the essential factor consisted of rethinking and reengineering the interfaces with customers or internal users to discover new sources of value creation.

> We spent a lot of time reflecting on our traditional approach to the business. We thought that if we were to continue to grow out of the pressure of the crazy competition on price . . . because our molecules are commodities, what would make the difference in the future would be the service we could offer around these molecules. We therefore became increasingly interested in the ways our clients use the molecules and how we could add value for them, in their own processes. Was it possible for us to expand the range of our product-services? Were there other competencies we had developed at Air Liquide that we could offer to improve their effectiveness?

This led to a transformational process covering three main stages. The first consisted of 'imagining a new value chain', the second of identifying the key knowledge bases and core competencies (the capacity to group together individual knowledge) required to fulfill the strategic intent and the main objectives, while the third was to build a new strategic and organizational architecture for the divisions.

Creating a new value chain based on knowledge and core competencies

The second stage in the identification of key knowledge bases and core competencies was drawn directly from both the vision and the strategic intent. It consisted of imagining the 'ideal value chain' which could maximize competitive advantage in the future. The value chain was broken up into main processes involving core competencies and their underlying knowledge bases. It is worth noting that the new value chain was designed 'from scratch' through a collective reflection process, disregarding the processes and routines already in place in the organization. The main objectives were to implement the strategic intent and strengthen market knowledge while improving efficiency in the value creation processes. The determination of possible ways to exploit existing (accumulated) knowledge and create new knowledge was at the very center of the process.

Identifying knowledge bases and 'core' competencies

Based on the shared awareness that sources of value creation associated with added services surrounding the 'technical product' (gas), and that the 'ideal value chain' could help achieve the ambitions of the company, the divisions engaged in collective reflection to identify core competencies and their underlying knowledge bases. The first stage consisted of identifying the knowledge bases and competencies involved in each process of the value chain, then isolating the key knowledge bases that support a set of 'core' competencies. A reflection began with the definition of a set of criteria to label knowledge bases and competencies as 'key' and 'core'. First, their *critical dimension* for the central business of the divisions was identified. What key knowledge and know-how should the division master to be the leader in the field (product-service and process)? Second, their specificity to Air Liquide, that is to say, their uniqueness and non-availability on markets. Third, their 'generality', meaning the number of processes in the value chain(s) involved. Finally, their 'range', or the number of product-services involved. This led to the identification of sets of key competencies relating, for example, to technologies, processes, operations, market knowledge, project development and project management.

STRATEGIC ARCHITECTURE: A STEP TOWARDS A LEARNING ORGANIZATION

An important contribution to the new architecture consists of clearly separating the processes in the value chain from their underlying knowledge bases and managing them separately. As a consequence, innovation and change in product-services through the modification processes are made easier. The organization can easily and profoundly modify the configuration of its processes without changing its structure or abandoning its central competencies. The new organizational structure not only allows the creation, accumulation and exploitation of knowledge, but also, and more importantly, the possibility to broaden the variety of learning styles as well as the scope of strategic formulation and implementation. Competence centers are above all justified by the need to isolate, make explicit, formalize and disseminate knowledge. This organization leads to the accumulation of experience in the direction of organizational capabilities that are the common denominators of the processes in the value chain. This reflection, which is led upstream of the value chain processes, focuses on the actual sources of competitive advantages and on the key knowledge bases that are and must be genuinely shared in the organization. This also enables a process to be transformed or changed, leading to more rapid and more radical learning. Due to this separation, the processes in the value chain are the 'intermediaries' between the platform and the market, corresponding to an *ad hoc* and temporary combination of competencies. As a consequence, the improvement to a process results from the enrichment of the knowledge bases and/or in their combination that takes place in the competence centers.

A second contribution to the new architecture lies in the ability of competence centers to develop the organization in a complex and unstable environment. It enables the product-services and the processes in the value chain to be easily modified so as to meet specific customer needs. Competence centers are stable entities accumulating and upgrading knowledge and competencies from which the company can develop a relatively broad, diversified and complex product range. The classical approach of strategy and organization, supporting generic, competitive strategies, founded on value chains and business units, supposes an industrial logic that is stable and clearly identified. Changing a traditional organization, built around processes and a stable value chain, necessitates profoundly and regularly modifying the whole organization, thereby generating high costs and complexity. The more unstable the environment becomes, the more it is difficult to manage this complexity. The new architecture implemented at Air Liquide, with its competence centers separated from the value chain, provides a means to overcome these difficulties. Most of the subsequent modifications and transformations can be easily designed and implemented. Such was the case for the patenting activities.

Innovation and the Role of Patents at Air Liquide

> Meeting today's needs and anticipating tomorrow. For one hundred years Air Liquide's research and engineering teams have been meeting this challenge, continuously imagining and inventing new solutions; gas-production technologies, new applications and high value-added services based on information technologies. The objective is always the same; to fulfill customers' needs while enhancing the Group's expertise and competitiveness. Today is no different from 1902: Air Liquide's commitment to innovation continues to make the difference [Company website, February 23, 2004]

With a budget for research of €150 million, and over 550 researchers from more than 25 nationalities in eight centers in 2004, the company patents about 200 inventions a year (236 in 2003 and 225 in 2004) and owns a portfolio of more than 6,500 patents and 2,600 protected inventions.

> Our eight research centers around the world operate as a network . . . Being present on three continents brings us closer to our customers, but also to the best technological partners in the world.

Technological innovation is one of Air Liquide's great strengths and the firm keeps on developing new solutions to improve its customers' industrial processes. Almost 40 per cent of the innovations of the group deal with process innovations and 60 per cent are product ones.

> Our commitment to technology and innovation has kept Air Liquide at the forefront of the high-tech business of industrial gases for more than 100 years. Our objective is, and always has been, to fulfill customers' needs while enhancing the Group's expertise and competitiveness. The result is relentless pursuit of innovative solutions that improve customers' industrial performance and protect the environment [Company website, February 23, 2004].

The Role of Patents in Strategy

Patents play a central role in the firm's strategy. It may be argued that Air Liquide is particularly active in patenting its inventions. The patenting strategy is extensive at the early stage, (stimulating a great number of announcements) but highly selective in the choices of patents to be registered and/or maintained. The company's strategy in the field is both defensive and offensive. Patents are used both to cover and defend positions held by the company as a world leader but also to reinforce its position through innovations. A central objective is to increase value from patents, externally and internally.

Organizational transformations in patenting activities

A common decision made in the 1990s was to separate 'Intellectual Property' (IP) from the department of legal affairs. This major change can be viewed as a part of the new organizational context that has been created at the corporate level. The main objective of this transformation was to increase value from patents. The sub-objectives were: (1) to have a more strategic and less law-based management of patents; and (2) to involve in the process more multifunctional team members with different backgrounds.

Building new organizational platforms

The management of the activities regarding patents became part of IP Direction, acting as an autonomous unit located at the Headquarters and reporting directly to the Directorate General of the company. The IP unit (about 40 persons, including engineers, legal experts, assistants) is in charge of managing information regarding patents and of supervising and coordinating sub-units located in different countries (US, Japan, etc.). Patent agents,[3] who are dedicated to specialized fields of application or markets, play a central role in the new organizational structure. They are in charge of the coordination of the entire process (from announcement by inventors to the legal responsibility of registration and finally

Figure 7.1 Managerial process and organizational arrangements at Air Liquide

the exploitation of the patent). They manage the interfaces with the different R&D departments but also (because about 40 per cent of patents are used internally) with the engineering and manufacturing departments. Managerial controllers supervise financial aspects regarding the patents' portfolio. Internal legal experts are in charge of writing and monitoring contracts.

Figure 7.1 presents the different stages in the management of patents and the corresponding organizational dimensions. Upstream departments are involved at the very early stages of the invention (idea generation and concept). The R&D axis and the inventors take the initiative to announce the invention to be at the origin of the patent. Patents are increasingly involved at this stage to encourage inventors effectively to declare their intent. An 'Innovation Committee' has been created to complement the role played by patent agents at this stage. A dedicated body, the 'Invention Review Committee', is in charge of reviewing and managing the declaration. The inventor and the patent agents specialized in the field play a central role. One of the key concerns is to support and assist the inventor in her/his effort briefly and roughly to articulate information and knowledge. This entity is rather informal. Meetings dates and deadlines are not clearly defined and the declaration is handmade using a standardized document. This stage is not very important from a strategic point of view. A great number, that is to say, almost every idea that could be patented, is usually declared and registered. Legal deadlines require that a decision has to be made regarding the patent's extension after one year. It is at this stage that fundamental strategic choices have to be made with financial, competitive and commercial key issues.

An *ad hoc* structure has been implemented to deal with this important stage: the 'Patents Committee'. The complexity of choices involves persons from different departments; patent agents, marketing via correspondents who are specialized 'by market' or application for each strategic domain or division who are potentially or actually interested and representatives of a corresponding R&D axis. The process is highly formal at this stage. The committee organizes a monthly meeting and is in charge of assessing each patent using standardized criteria and of writing highly documented reports to be sent to the IP unit. Patent committees not only involve individuals from different departments and divisions. They are a central place where information about the different patents that are in the 'pipeline' are exchanged, discussed and assessed. The patent committee is also in charge of preparing the decision externally or internally to exploit patents.

The next stage takes places once the patent extension has been decided and is effective. The key concern is here to manage the portfolio and to make choices regarding the future of each patent: (1) exploitation, (2) stand by, or (3) abandonment. The 'patents portfolio reviews' have been implemented as an organizational arrangement to facilitate the strategic decision-making process at this stage (see Figure 7.2). Meetings which are less formal than Patent Committee meetings are held every six months. They bring together members of the Patents Committee and experts from different departments, and the discussions and the assessment are based on the scientific, technical, marketing knowledge of its members and the potential offered by patents in terms of their monopolistic value. Choices or recommendations are mainly oriented towards the decisions of abandonment. Management controllers also play an important role in triggering meetings and in the discussions.

The decision to maintain, externally or internally to exploit patents depends on their characteristics, their importance and their contribution to the firm's strategy. Some patents will support a defensive strategy;

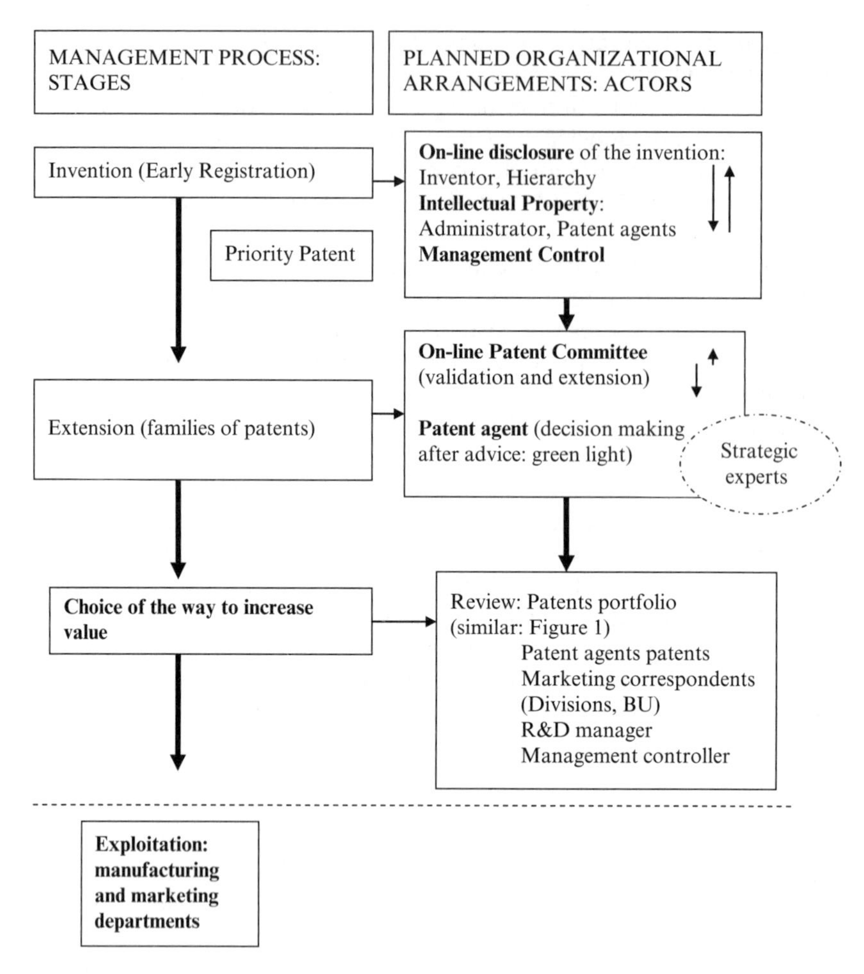

Figure 7.2 New organizational arrangements

others, more specifically those that will be exploited internally, an offensive strategy. Finally, the internal exploitation involves members of the manufacturing and marketing departments within divisions.

A new wave of organizational arrangements: the role of IT platforms

The complexity and the limits of the process led the Head of the IP department to look for new organizational arrangements to improve the process at two stages: the declaration of the invention and the functioning of the 'patents committees'. Difficulties are due to the fact that declarations are 'hand-written', documents are not properly filled in and thus cannot be exploited and used by the IP department, leading to a quantitative or qualitative loss of potentially interesting and valuable inventions, lack of feedback to the inventor on the future of her or his declaration. Difficulties also arose in the functioning and management of the 'Patents Committee'. for example, difficulty to organize meetings bringing together members from different divisions, departments and countries, lack of marketing information regarding patents, lack of formal criteria to support the decision making process. A decision has been made to solve these problems and to 'optimize' the process using information technology.

The declarations of inventions will be performed 'on line' by the inventor, using the standardized format on the Intranet that is available at any time. 'Warnings' are sent to the inventor if the document is not properly filled in or is incomplete. Keywords relating to technology and potential markets are introduced in the file. Once the document has been validated on line by the person in charge of the R&D axis, it is electronically transmitted to the IP department. The IP department appoints the patent agents to the project. Both the patent agents and the management controller have the right to validate on-line declarations. The final document can be consulted by all the individuals involved or with a stake in the patent.

IT solutions are also to be implemented to support key activities led by the 'patents committee' relating to the extension of the protection. Monthly 'on line' committees provide members with the possibility to interact 'on line' during the preparatory stage leading to recommendations. But the system also enables to involve more 'strategic experts' (who are not directly involved in the project) to participate in the discussions. The patent agents write the final report which is sent to the members of the committee for ratification. Face-to-face meetings are organized in case of disagreement on the final decision to be made. All the decisions concerning the extensions are stored in the 'on line' database.

The new technological platform is aimed at 'optimizing the managerial process' but also at strongly supporting knowledge codification. One of the main concerns was to improve the process of knowledge codification, the traceability and the 'on line' availability of information produced and stored. These issues have been extensively discussed in the literature on knowledge management (see Table 7.2 for a summary).

DISCUSSION AND CONCLUSION

The case of Air Liquide provides insights into how large companies can increase value from patents through dedicated organizational arrangements that contribute not only to improving the efficiency and

Table 7.2 Increasing value from patents: tasks, actors, organizational arrangements and synergies

	Options	Internal Exploitation
Key tasks	Analysis: Patents portfolio Decision (Internal, External, Storage)	Industrial development and launch Internal Communication on patented products
Main Actors involved	Unit in charge of Patents Managers: R&D, Marketing, Manufacturing, Innovation	Marketing Manufacturing
Organizational arrangements and synergies between actors	Analysis: Meetings between the unit in charge of patents and R&D and vice versa. Choice: Meetings (R&D axis, unit in charge of patents, marketing and / manufacturing)	Development program led by marketing in close cooperation with Manufacturing Relationship with the unit in charge of Patents in case of infringement

effectiveness of the process itself but also to support knowledge sharing and creation. These issues have been largely discussed both in the literature on the management of patents and in the literature on knowledge creation and organizational learning presented in the introduction. The two phases that led to organizational arrangements have different impacts on the social interactions that took place among the members and groups involved in the process. The new organizational platforms have characteristics that are similar to those that have been identified by Nonaka and Takeuchi (1995) in their research on knowledge creation. The involvement of more multi-functional team members with different backgrounds at the very early stage of the process enriches and nurtures social interactions that are propitious to knowledge sharing and creation. The role of patent agents in charge of specific patents is central in triggering the declaration by R&D team members, in assisting and supporting the efforts made by the inventors to articulate or codify knowledge, and in facilitating the flow of information. They are at the interface between inventors and the Patent Review Committee, and, with other departments such as R&D (and for the patents that are exploited internally), the engineering and manufacturing departments. It may be argued that these organizational arrangements are similar, at the early stages of the process (declaration and elaboration), to those that have been associated with 'rugby style' and at the last stages to the 'American football style' by Takeuchi and Nonaka in their study of innovation and new product development processes (1986, 1995). The second wave of organizational arrangements also illustrates central issues that are discussed in this literature. IT solutions are not only used to facilitate and fuel the flow of information between the different actors involved in the patenting process, but they also enable one better to articulate, codify, share, distribute and store information in 'organizational memories' (Huber, 1991) and, as a consequence, support knowledge sharing.

Finally, it may be argued that these organizational arrangements could also be interpreted in the broader organizational context of the company that constitutes, to some extent, an example of transition towards an 'ambidextrous' (O'Reilly III and Tushman, 2004) or 'hypertext' (Nonaka and Takeuchi, 1995) organization. The first move towards an 'ambidextrous organization' can be found in the separation of the Unit in Charge, or IP, and the fact that the early stages of the process involve multi-disciplinary teams and overlapping that are propitious to the exploration of new knowledge, while the last stages are more oriented towards the exploitation of existing knowledge. In the broader organizational context of Air Liquide, the clear separation of the processes in the value chain from their underlying knowledge bases and the separation in their management as well as the 'competence centers' that have been created also suggest such an evolution towards an ambidextrous organization.

According to Nonaka and Takeuchi (1995, 169–171), a 'hypertext' organizational structure is made up of three interconnected contexts or layers: the business system, the project team, and the knowledge base. The 'business system' is the central layer reflecting the 'classical' hierarchical and organizational design and structure. Air Liquide is organized around divisional structures such as gas and engineering, with clearly identified units and hierarchical levels. This is the context in which the central activities and routines in the value chain are carried out. It may be argued that the newly created units, the 'competence centers' as well as the 'central and standardized components and routines' developed by 'product teams' are part of the 'business system' in so far as they integrate knowledge created in other layers. The IP department that has been created and which is directly related to the directorate general is part of this layer.

The 'upper layer' in the hypertext organization is made up of 'project teams' bringing together members from different units who are assigned to a project team until the project is completed. At Air Liquide, most team members, apart from a very small number, do not see the project through until termination but return to the competence center until he or she is asked to return to the project later or is assigned to a new project. As we pointed out earlier, teams in charge of developing central and standard products can be considered as more stable organizational entities. Project teams and specific product teams play a central role in knowledge creation and exploitation. But it is the competence centers that constitute the key contexts, which fuel the knowledge base and the platform.

The teams that have been created in the new organizational arrangements in the patenting process can be viewed as elements of this 'upper layer'. The Patent Committee is a specific and *ad hoc* unit with permanent and non-permanent members in social interactions, and which fuel learning and knowledge creation. The 'foundation' or 'lower level' in the hypertext organization is made up of the knowledge base (Nonaka and Takeuchi, 1995, 169–170) that is an abstract entity. It is there that the knowledge created in the 'upper levels' is interpreted, stored in organizational memories and made available to the organization. It also includes dimensions such as corporate vision and the organizational culture supporting interpretation and evaluation of the knowledge that is created and exploited at the 'upper' levels.

Similarly, it may be argued that, at Air Liquide, knowledge bases are part of the platform on which the new organization has been built. New knowledge created by product and project teams is accumulated in the knowledge base, justified according to the company vision and the strategic intent of the divisions, interpreted and combined in the competence centers, and made available and exploited by the units forming the business system which support the central activities, routines and processes in the value chains.

In the same vein, the second wave of organizational arrangements and the use of information technologies that took place in the patenting activities can be viewed as a contribution to this knowledge base. Codified and articulated knowledge that is produced is justified and evaluated during the patenting process and the decisions regarding their abandonment, maintenance, or exploitation are made in the light of their contribution to the strategy (offensive or defensive). Articulated knowledge is stored in the content of the patent itself and in organizational memories thanks to the use of information and communication technologies.

Finally, the case of Air Liquide illustrates the need for a good balance between two interdependent dimensions: organizational and technological. This balance could contribute to enhancing knowledge creation through social interactions and stimulate the adoption of knowledge sharing behaviors that could reflect dimensions that are associated to corporate culture.

NOTES

1. The authors would like to thank all the members of Air Liquide for their availability and participation in the interview process for this research.
2. For the purpose of this chapter information drawn from previous works is revisited and integrated. The main source is Miktova (1999). The analysis also draws on Métais (1997) to detail the organizational context of Air Liquide in the 1990s (other references include Ayerbe and Miktova (2005)).
3. Patent agents are in most cases engineers acting as 'attorneys'.

REFERENCES

Ayerbe, C. and L. Mitkova. 2005. 'Quelle Organisation pour la valorisation des brevets d'invention? Le cas d'Air Liquide'. *Revue Française de Gestion* 155 (March-April): 191–206.

Eisenhardt, K. 1989, 'Building Theories from Case Study Research'. *Academy of Management Review* 14: 532–550.

Grindley, D. and D. Teece. 1997. 'Managing Intellectual Capital: Licensing and Cross Licensing in Semiconductors and Electronics'. *California Management Review* 39(2): 8–41.

Hamel, G. and C.K. Prahalad. 1989. 'Strategic Intent'. *Harvard Business Review* 67(3): 63–77.

Huber, G. 1991. 'Organizational Learning: The Contributing Processes and the Literatures'. *Organization Science* 2(1): 88–115.

Marquer, F. 1985. *Innovation et Management des Brevets*. Paris: Les Editions d'Organisation.

Métais, E. 1997. *Intention Stratégique et Stratégies de Rupture*. PhD Thesis, Marseilles: IAE Aix.

Miles, M.B. and A.M. Huberman. 1991. *Analyse des données qualitatives: Recueil de nouvelles méthodes*. Brussels: De Boeck Université.

Mitkova, L. 1999. *Le Brevet d'Invention: Un Nouveau Domaine d'Application du Marketing*. PhD Thesis, Nice: Université de Nice-Sophia Antipolis.

Napper, B. and S. Irvine. 2002. 'Managing Intellectual Assets for Shareholder Value'. *Les nouvelles* 4: 148–154.

Takeuchi, H. and I. Nonaka. 1986. 'The New New Product Development Game'. *Harvard Business Review* 64(1): 137–146.

Nonaka, I. and H. Takeuchi. 1995. *The Knowledge Creating Company*. New York: Oxford University Press.

O'Reilly III, C. and M. Tushman. 2004. 'The Ambidextrous Organization'. *Harvard Business Review* 82: 74–81.

Smith G. and R. Parr. 1989. *Valuation of Intellectual Property and Intangible Assets*. Hoboken, NJ: John Wiley & Sons.

Sproule, R. 1999. 'Case History: Integrate IP Management'. *Les Nouvelles* 2: 70–77.

Yin, R.L. 1994. *Case Study Research, Design and Methods*, 2nd Edition, London: Sage Publications, Applied Social Research Methods Series, Vol. 5.

8. Language system (LS) 3.0: an agenda for a model of innovation valuation

Clinton W. Francis

INTRODUCTION

The task of 'measuring' innovation presents us with the uneasy sense of confronting a paradox – the valuation of modes of valuation and the seemingly infinite regression that this suggests. The increasing rate and complexity of change in our information society highlights the need for business and society to tackle this paradox and adequately value competing sources of innovation. How we value innovation influences the incentives we provide for creativity and thereby affects the things we produce, the world that produces us, and ultimately affects the quality of our lives. Yet we persist in our obedience to older methods of valuing innovation that are less creative than the underlying innovation they measure and that potentially produce a series of paradoxical inversions of dependent hierarchies.

A good part of the task of valuing innovation in our society is relegated to reliance on data concerning intellectual property rights (IPRs), such as density of IPRs and IPR citations levels, supplemented by market-based analyses. While these are useful sources of information, it is important for us to assess the risks associated with their deployment. Our system of property relations and the mechanisms of the marketplace have performed, and will continue to perform, the vital function of social regulation. But this chapter argues that their referential systems increasingly fail to match the complexity of the high-tech and bio-tech 'ecologies', which they frame. Under these conditions the hierarchy of logical types used in legal valuations can potentially attribute a higher (or similar) value to less complex innovations and a lower value to more complex ones. This tends to subvert natural evolution by incentivizing less complex innovation. When such paradoxical constructions go unchecked the valuation of innovation becomes potentially distorted and our ability adequately to identify innovation and the incentives we provide for creativity become skewed. With less complex innovation incentivized ahead of more complex innovation the inability of our enterprises successfully to adapt to their competitive environment is a likely outcome. Loss of competitive advantage in the marketplace (as well as damage to our environment and us) then becomes a greater risk.

The distortions in our valuation of innovation and the incident loss of competitive advantage are associated with a loss of interpretative complexity which erodes natural and historical context. We live in the context of a dependent hierarchy whereby the world, with its attendant biological, ecological and social systems, developmentally precedes our current rationally directed human action. However, humans continually risk inverting this hierarchy and artificially elevating their activities and the systems upon which they are based, producing interpretations that are less complex than their subject matter. This propensity to invert dependent hierarchies is implicit in human utility, which is structured by language classifications that order the world according to logical types in the service of efficient decision-making. By classifying and acting on specific features of the world in order to harness its inherent capabilities, we sometimes position our classificatory systems above that which is being classified. The danger is that the objects of our utility, as shaped and defined within language-based, classificatory systems, are often in actuality embedded within complex systems that we fail to take into account via logical typing, systematic categorization and language.

Paradox ensues when there is an inversion of natural and historical dependencies because the member (the lower logical type – the human) contradicts its membership in its class (the higher logical type – nature and history). 'This statement is untrue' is a simple illustration of how a paradox can be produced by a logical inversion of a natural, dependent hierarchy. An inversion occurs, and paradox results, when something is, or announces itself, at one level as a member of a class upon which it depends, but at another level defies the class definition and the rules of construction of the class. 'This statement is untrue' is an illustration of the power of language-based classification to invert the natural, dependent hierarchy that is traced by the history of language. The statement is self-contradictory in that it contradicts its implicit truth claims as a meaningful utterance. This results because the statement defies the very social conventions of language that support and developmentally precede its truth claims – at one level its speaker uses conventional syntax and word meaning, which were received by the speaker out of the historical development of his language, to produce a self-referential statement the meaning of which, at another level, undermines these very conventions and their history. Language paradoxes are thus the symptoms of a process of logical typing that, when applied to classifications of the world, potentially subvert the dependence of the human subject on a hierarchy laid down by a developmental sequence – a dependent hierarchy in which the lower levels (nature) developmentally precede the higher levels (human culture). The potential for inversion is realized when the dependent hierarchy becomes represented in a classification that constructs a hierarchy of logical types that runs in the opposite direction to the dependent hierarchy. The criticisms leveled against the controversial theory of intelligent design, for example, seem to be implicitly based on the argument that the theory performs such an inversion – the natural world is explained by the theory as the product of human-like intelligence and this, the critics argue, contradicts the Darwinian-based idea that natural world evolutionarily precedes human culture.

This chapter proposes a model – Language System 3.0 (LS 3.0) – that applies a semiotics analysis to complexity in an attempt to restore a creative balance between the method of valuing innovation and its subject matter and thereby to avoid inversions of dependent hierarchies (and associated subversions of evolutionary development) that can impair competitive advantage. The model adopts a semiotic perspective for the following reasons. First, interpretive paradoxes involve language-based constructions, and in order to minimize their adverse affects on the incentives for creativity we need to understand better the language-based formation of logical types underlying the paradox. Second, beginning in biology and more recently in cyber technology we are recognizing that innovation is structured by levels of language, albeit electro-chemical, genetic, spoken, written or digital language. Third, we recognize that language evolves in a dependent hierarchy – digital founded on written, which is founded on oral, which is founded on genetic, which is founded on electro-chemical. For these reasons, the chapter argues that a semiotic-based theory of complexity allows us to understand the structure of innovation and thereby better to achieve competitive advantage, better to diagnose the limits of our methods of innovation valuation (paradox being the symptom of these limits), and better to avoid precarious outcomes by guiding logical typing of our analyses in such a way as to bring them into a balanced relationship with natural, dependent hierarchies.

LS 3.0 (see Figure 8.1) consists of four levels of analysis and interprets innovation in terms of its potential to conduct exchanges across the boundaries marked by these levels. The four boundary points – *signification, valuation, specificity* and *subjectivity* – correspond, respectively, to distinct levels of analysis ordered in increasing interpretive complexity: the subject matter of valuation (*what*), the unit of measurement (*how*), the

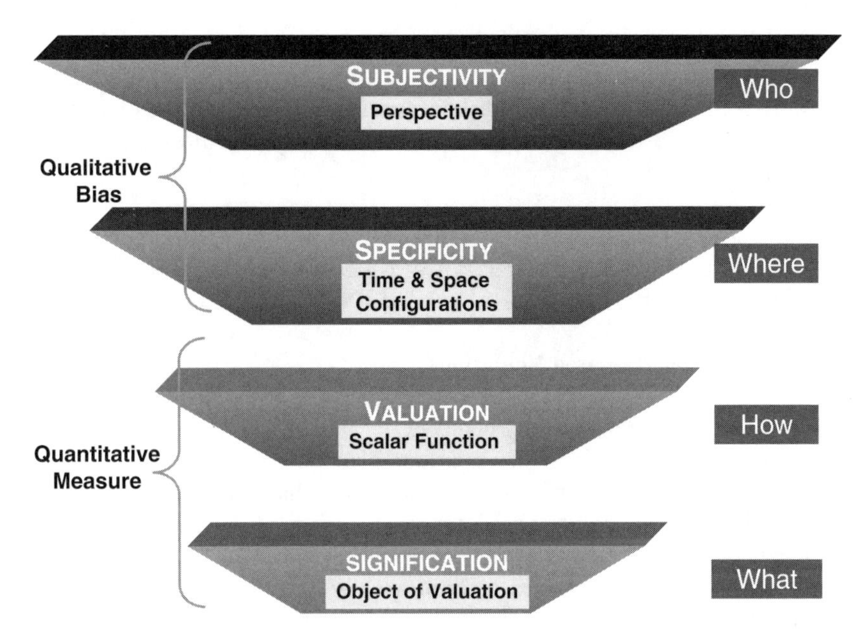

Figure 8.1 LS 3.0

temporal and spatial specificity (*where*), and the perspective (*who*). The first two levels allow for the modeling of quantitative measures, while the second pair allow for the modeling of the qualitative factors that set the bias on the quantitative measures. The goal of LS 3.0 is to suggest a heuristic that will permit innovation to be better identified and measured, and to promote more awareness and control of the practical and epistemological biases that are placed on the resulting valuation.

Section 1 of this chapter evaluates current IPR and market-based methods of valuing innovation. It argues that they are historically and technologically bounded by the past and possess characteristics that distort valuations. IPRs are analysed as characterized by a low level of complexity and exhibiting paradoxes that structure incentives in ways that favor less complex innovation. Market-based methods, because they frequently build on IPRs and outmoded economic assumptions, tend to exacerbate the failures of IPR-based valuations. Section 2 describes LS 3.0 and offers it as the agenda for an alternative method of identifying and valuing innovation – a method that embodies a semiosis of complexity as a metric for measuring innovation and semiosis of economics to help assess its commercial viability. Finally, it depicts the deficiencies of the IPR and market-based valuations as the product of biased principles of selection and economics traceable to the epistemology of a culture the prototypic communication of which is hardcopy. It arises in a culture that is structured by dualities, which make it susceptible to paradox, and that stands in marked contrast to the monism of earlier oral cultures and the triadic formations of later cyber cultures that historically bracket it.

INTELLECTUAL PROPERTY RIGHT AND MARKETS AS MEASURES OF INNOVATION

Intellectual Property-based Valuations

Historically and technologically bounded

Empirical studies of patent prosecutions provide a useful index of technological attractiveness and relative technology share based on such data as the rate of patent grants, the international scope of patent applications and patent citation ratios. These empirical measures operate best when the inquiry is historically and technologically bounded, for example by determination of the current, leading technologies. Use of this information enables identification of better versions of a known technology, such as a smaller version of the microchip, as future sources of innovation. However, because the empirical data are bounded by already known technology and they provide no explanation of the principles that inform the innovation they are accordingly of limited use in identifying the trajectory leading to an envelop-expanding innovation – 'the next internet'.

Equivocal and low-threshold values

Even within this historically and technologically bounded context, IPR-based measures prove of limited practical usefulness. The grant of a patent is an equivocal sign of value. Although a patent grant is preceded by a Patent Office examination and typically an attorney's opinion letter, this is no guarantee of patent validity because nearly half of all patent grants are invalidated in the event of their subsequent judicial review. Infringement of patents is also difficult to monitor and prove, and extremely expensive to litigate. In the area of pharmaceuticals, for example, copying of chemical compounds is hard to establish and the end user often circumvents new-use applications undetected by the patent holder. Also the IPR value is largely independent of the inherent value of the innovation; economic endowment of the patent holder and strategic intellectual capital management ('ICM') are often a principal source of a patent's value. Moreover, the issue of a patent is no guarantee the idea has market potential.

Constructed values

IPR-based assessments of innovation are also controversial because they are based on artificial IPR values that are only a minimal guarantee of underlying innovation. In paradoxical fashion the value of innovation is increased once it qualifies as an IPR, not because this value is inherent in the innovation but because IPRs confer monopoly value on it. This is a teleology of constructed value. Once an embodied innovation passes the threshold test of 'non-obviousness', patent law, for example, elevates the innovation to a monopoly value. Furthermore, the US constitutional guarantee afforded to patents and copyrights turns this teleology of constructed value into irony: patents and copyright have been awarded a perpetual monopoly on their own primacy – the grant of a monopoly on their ability to confer monopoly value.

Distorted values

IPRs further distort valuation because they disenfranchise innovation that is deemed to be in the public domain. In order to ensure that 'top-level' innovation remains open for public use intellectual property law excludes it from being the subject of private property interests. As a result monopoly incentives are reserved for 'lower-level' innovation; top-level innovation is disqualified from monopoly ownership and therefore less commercially attractive. Thus a path-breaking discovery about the etiology of cancer would not receive patent protection, whereas a drug or surgical device used to treat cancer may be protected even though the treatment it provides comes late in the disease process and its therapeutic value may be small and with draconian side effects. Similarly, 'sweat of the brow' factual archives are not protected by copyright, whereas the most menial of 'creative' expressions of this archive are protected. To interpret these distortions in valuation as a case for extending IP protection and thereby shrinking the public domain is to miss the point; the logically prior issue is the social justification for continuing to use IPRs in the first instance.

Low complexity

IPRs also encode a relationship of society to nature in which society is paradoxically placed above nature in a manner that inverts the relations of dependence that society has on nature. This presents us with a situation where IPR-based technologies impact on the world we live in, yet are not subject to forces of natural selection within an ecological context. IPRs (and property generally) are based on a system of centralized judicial and legislative control, whereas the 'innovations' of ecology are radically decentralized. The legal foundations of IPRs exist as *a priori* codes that are subject to weak feedback modification. In marked contrast, the ecological 'value' of a new organism, for example, is tested in a complex of subtle, multi-layered feedback relationships. For an organism to have survival value self protection must be incorporated in every level of its functioning, yet at the same time it must open itself up to, and adjust in response to, the interdependencies of its environment. Given that we have conferred our legal system with the power to award monopoly status IPRs on innovations that manipulate natural processes and which can impact on our bodies and our ecology, this presents us with a potentially dangerous situation. Since IPRs are less complex than the processes they regulate and the monopoly status they confer is more powerful than the inventions they confer this status on would have in their own right, IPRs hold the potential for accelerating the unintended consequences of patented technologies and copyright-assisted systems of product distribution.

In summary, IPRs are binary constructions, embody low threshold qualifications, amplify values to monopoly levels, produce public/private value distortions, are potentially less complex than their subject matter, and incorporate weak feedback modification. These characteristics contribute to paradoxical valuations. IPRs paradoxically confuse their own monopoly values with the inherent value of the innovation they protect, and they incentivize lower-level innovations with private monopoly grants while they provide no incentive to pursue high-level innovations which are left in the public domain. Constitutional reification of IPRs further compounds the paradox by elevating them above the class of innovations of which they are members (albeit a legal innovation about innovations). Accompanying this confusion of logical types and inversion of dependent levels is the IPRs' low level of complexity. Their fundamentally binary form endows them with limited capability to value innovation, and their weak feedback mechanism is symptomatic of an absence of developed meta-levels that contextually modify the operation of IPRs and selectively substitute other regulatory regimes.

Market-based Valuations

Efforts to integrate IPR measures with market growth indices or to augment IPRs with value chain analyses boost the effectiveness of innovation measures by allowing composite measures that produce empirically richer analyses for charting possible interdependencies between technologies and marketable products. Market analyses factor the practical reality of the economics of supply and demand into valuations, something which receives little or no consideration in the legal determination to confer IPRs. Furthermore, whereas IPRs embody substantive principles of selection and confer monopoly-like protection that can distort the valuations on which they are based, the market component of any valuation is subject-matter and value agnostic – market 'feedback' operates regardless of the nature of the innovation or of any pre-coded values.

Historically and technologically bounded

But, while they do not carry any threshold qualifications or any monopoly-value guarantee, market values by definition only record the valuation history of past innovations – a criticism that we have also seen can be leveled against IPR-based valuations. This history can help identify trajectories of successful innovation, and by examining these innovations we can deduce the qualities that may define future successful innovation. But because these predictions are bounded by history and existing technology they are conservative by nature. Accordingly, to use market values as a principal basis for determining the path of future innovation is to make future innovation dependent on the past. To the extent that market empiricism is used in this way, whether in conjunction with IPR-based valuations or not, it engenders a further confusion of logical types.

Content and value agnostic

The agnosticism of market measures means that they also fail to provide guidance as to the core source of innovation – the innovation 'drivers'. While the market success of a product can be measured by its sales, sales figures provide no insight into whether the source of the success was the initial invention, the process of its production, its marketing, the larger branding of the producer, and so on. A more meaningful analysis of the source of innovation requires a principle of selection about which gross sales are silent. Only if some aspect of the product's value chain were to be licensed or sold would the market be able to provide more information, but that presupposes exercise of some principle of selection to separate that facet of the enterprise. In other words, market valuation provides no guidance in determining the relationship of the part to the whole of an enterprise as the source of value. The effects of the problem of determining the locus of value become clear when we look at the proliferation of potential quantitative analyses generally, sales-based analyses being just one such measure. Furthermore, to the extent that market-based analyses are integrated with IPR-based valuations or are built on IPR market values they reproduce the value distortions associated with IPRs. In conclusion, we according see that, contrary to the claim that they more transparently reflect the value of innovation by being content agnostic (market-based) or imposing a low threshold qualification (IPRs), IPR and market-based valuations have the potential to distort values and make the past the model for the future.

Value Chain-based Valuations

Analysis of the economic value-added and strategic value-added components of a product value chain provide help with the task of identifying the locus of innovation within an enterprise. Such analyses employ qualitative principles of selection to help aid with the identification of valuable aspects of an enterprise. Qualitative factors used by analysts include: customer importance; technology clock speed; competitive position; capable suppliers; product architecture. Another evaluative tool is the so-called 'five forces' analysis, which is useful for assessing a product value chain as a whole compared with other potential product value chains. The qualitative factors used in this analysis include: supplier power, buyer power, barriers to entry, and threat of substitutes. While these two analyses make a useful contribution to the task of locating value, their principal purpose is not valuation of innovation. They are directed more to issues such as competitive advantage, outsourcing and strategic evaluation. Accordingly, the qualitative factors they employ lack the explanatory structure needed for a deeper understanding of innovation and its valuation. Although the factors are suggestive of such a structure, their linear presentation in lists means that they lack the complexity needed for the task.

LS 3.0 – A SEMIOTIC-SYSTEMS MODEL

Any valuation model has to have a number of attributes: (1) a scalar dimension that can be used to compare the magnitude of two items of interest; (2) a temporal dimension, because value is always forward looking, whereas returns are always backward looking; and (3) clear identification of, and distinctions between, what is being measured. An adequate model for valuing prospective sources of innovation requires self-referential, feedback linkages between the different levels of its structure to minimize the paradoxes that mark the inversion of dependent hierarchies. This ability to self-monitor defines the capacity of 'recursiveness', which is an important element of a complex structure.

A model's complexity helps correct the asymmetry between the model and its subject matter which, if not remedied, impairs the resulting valuations and can generate incentives for innovation that potentially lead to dangerous, unintended consequences. Remedying this asymmetry and achieving balance is based on the assumption that innovation is structured by complexity and that an adequate model for valuation of innovation must also recapitulate complexity in its own form. The self-reference involved in this recapitulation means that a complex model is subject to

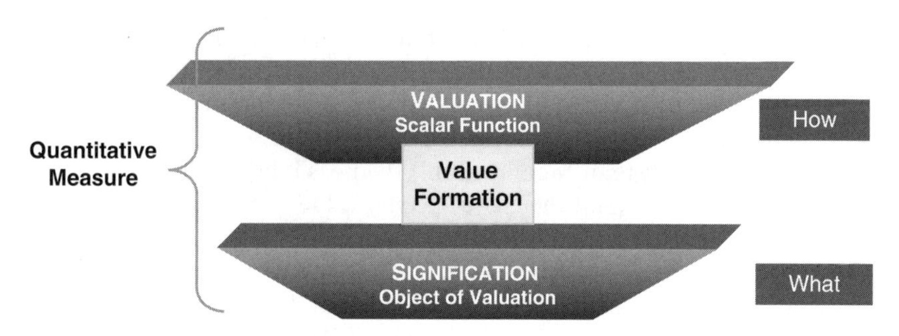

Figure 8.2 Value formation

that charge that it is yet another paradox and tautology – the notion of complexity values itself and as a result has the propensity to be self-biased in its principles of selection and valuation. But unlike the paradoxes and tautologies of current IPR and market-based analyses, the complex, recursive model is self-conscious that paradoxes and tautologies are the warning signs that a model is at risk of inverting the natural, dependent hierarchy and attempts to check this tendency.

Determining Boundaries

The model proposed by this chapter – LS 3.0 – is informed by semiotics and systems theory to help achieve these goals. LS 3.0 interprets innovation in terms of 'boundaries'. Boundaries are conceptualized in terms of four non-linear levels – *signification, valuation, specificity* and *subjectivity* – which, as outlined above, correspond, respectively, to: the subject matter of valuation (what), unit of measurement (how), time and space specificity (where), and perspective (who). The feedback relationship which exists between the upper and lower levels works to achieve a balancing of the quantitative and qualitative valuation potentials that the model offers. Understanding that valuation involves these different levels, helps separate them and avoid the paradox that characterizes IPR and market-based analyses, as well as avoid the linearity of current business analyses of enterprise value.

Innovation in Language – Signification and Valuation

The 'signification' and 'valuation' boundary levels constitute the core, indispensable elements of any language system. When the units of measure formed in a differentiation come to stand for a given subject matter, meaning and therefore value are possible (see Figure 8.2.). The units of measure could be words or numbers, or for that matter any set of symbols the relationships among which create a set of distinctions that can be used to 'map' that which they measure. Once social conventions establish themselves regarding the structure of symbolic differences (what is referred to in semiotics as the structure of 'valuation') and what they stand for (the semiotic structure of 'signification') the possibility of discourse arises producing meanings (and therefore specific values which semiotics refers to as 'signifieds'). The technical functioning of this aspect of language is well understood and uncontroversial. Each of the devices critiqued above necessarily contains this core semiotic function: market analyses do this by having monetary units stand for goods; IPRs do this, for example, by having the differentiation between patent claims stand for innovation; value chain analyses do this by having the strategic and economic segments stand for the enterprise.

This semiotic capacity does not dictate, however, the boundaries of signification and valuation that it employs. From the perspective of the semiotic system the differentiations formed by the boundaries are 'arbitrary'. It is the context of use that they help structure which informs the specific distinctions and the relationship among them – the system 'bias'. We can analyse the distinctions which this bias determines as the 'principles of selection' and the relationships among these distinctions as the 'principles of economics'. Together these principles set the bias on the time and space specificity of valuation. We have seen that current modes of valuing innovation have distinct biases the limitations of which are marked by paradox. LS 3.0 proposes a more complex system bias with the goal of increasing the efficacy of innovation valuation.

Innovation within Systems – Specificity

The third level of LS 3.0 – 'specificity' – frames the two lower semiotic levels – 'signification' and 'valuation' – by helping to locate the 'innovation drivers' and in turn the 'value drivers', and therefore define the site of valuation (see Figure 8.3). By placing valuation in time and space, it helps understand 'where' the 'what' and the 'how' of the semiotic should be applied. In practical terms it helps assess what aspects of the potential innovation value chain are realizable (value chain realization) and what structures of value sharing among enterprises (the participants' 'footprint' in the value chain) will produce maximum realization of this potential.

The selection principle – 'semiosis of complexity'

In place of reliance on IPRs, historically and technologically bounded empiricism, and the various linear factors used in value chain analyses, LS 3.0 employs a concept of semiotic complexity as a principle of selection to

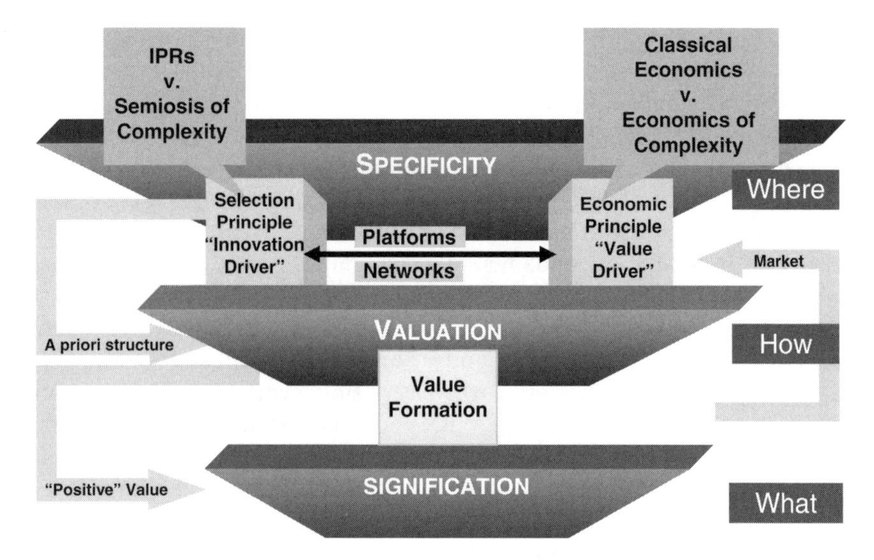

Figure 8.3 The site of valuation

guide valuation. As will be explained in our discussion of the fourth level (subjectivity) of the model, a semiosis of complexity (together with a semiosis of economics which will be introduced shortly) derives from an emerging cyber-based epistemology. The concept is based on the theory that all innovation is describable in terms of the language structure that it employs and that the more powerful the semiotic complexity of the innovation the higher the innovative potential. The semiosis of complexity involves application of the semiotic capacity of the signification and valuation levels of LS 3.0 to measure the potential complexity of an innovation. The result is used as a basis for identification of innovation 'drivers' involved in an enterprise.

When a semiotic complex is articulated 'statements' are formed. If the statements make a valuable difference by crossing new boundaries the statements are innovations. The higher the level of the statement within the structure of semiotics the greater is its innovative potential. For example: the formation of a new language of valuation is potentially more innovative than is the subsequent articulation of the new language to make a specific valuation statement; the invention of money is more innovative than is the purchase of goods using money. In other words, when a statement is a function that is dedicated to the production of new statements its meta-statement capacity endows it with a greater innovative potential. That the invention is not an overtly language phenomenon, as in the previous examples, does not alter the analysis. Innovation of course also occurs with more concrete functions. The conjoining of two metal blades to make scissors is an example. The invention of the scissors is a statement at a higher semiotic level, and therefore more innovative, than the statement involving the subsequent use of the scissors. The scissors are a 'meta-statement'. The innovation could also be in relation to marketing or another aspect of enterprise operations, for example, leveraging brand equity (that is, the use of an established brand name to enter a new product class), co-branding, or fundamentally changing the brand image (that is, what the brand represents or to whom it appeals).

For the purposes of the present model, a statement that creates a new meta level crosses a 'vertical' boundary and produces increased complexity (the invention of scissors), whereas a statement that occurs within an existing level crosses a 'horizontal' boundary (finding a new use for scissors) and linearly extends the functional value of existing complexity. In this manner, we can begin to form an innovation valuation metric, with statements that cross vertical boundaries being exponentially more valuable than those that simply cross horizontal boundaries. An invention that crosses multiple vertical boundaries increases the power of the exponential. For example, the invention of electric scissors crosses two vertical boundaries – first the conjoining of the metal blades and second the conjoining of the electric motor with the scissors. This has an innovation valuation to the power of three: the invention of the scissors, which is about cutting (order two), and the invention of the electric scissors, which is about increasing the magnitude of scissors (order three). Compare this with the development of electric scissors to cut metal instead of just cloth, which is a first order statement that has less innovative value.

Innovations can also be assessed relative to their capacity to spawn innovations that are related to but discontinuous with them. For example, the use of electric scissors to create a business model that leverages the scaling possibilities of electric scissors by creating new production and distribution levels in the garment industry is a vertical statement that increases complexity and that can give rise to further innovative potential. The invention of scissors is therefore the foundation of one vertical 'complex' of statements that has the potential of launching other vertical complexes. Although the second complex (the new business model) is discontinuous with the first (the invention of scissors), the innovation value of the second is potentially additive to the first. Innovations that have the potential of launching new complexes are therefore proportionately more important than those without this potential. Inventions that use complexes as the building blocks of new complexity (rather than single statements as in our scissors example) can produce increases in innovation value that have the highest exponential values.

Within the resulting system of complexes, dependent hierarchies can develop so that later relatively discontinuous complexes can be dependent on earlier complexes and feedback on the earlier complexes operation. For example: the invention of computer software – a complex (software) that stands vertically to another complex (computer hardware and operating system) creating a dependent hierarchy; stem cell therapy – complex (patient cell structure) that stands for another complex (stem cell DNA) that stands for another complex (patient cell structure). From one perspective the later complex seems to stand vertically above the earlier one since it is about (meta) it. However, because the later complex is dependent on the earlier one the temporal relationship cannot be expressed as a simple flow down the vertical relationship (such as occurs in the relationship of the invention of scissors to their new use). That would be an inversion of the dependent hierarchy. Instead the two complexes (the software and the computer; the cultured stem cell and the human body) stand in a feedback relationship where the old complex becomes developmentally nested within the new complex and where temporality is more than notions of causality can accommodate. Identifying dependent hierarchies and understanding the role of feedback relationships in the formation of complexity is an important insight that a semiosis of complexity can contribute to the valuation of innovation. In particular it can inform a principle of selection that can better locate innovation and can help avoid the unintended consequences of inversions of natural, dependent hierarchies and related paradoxes that we have seen can distort valuation and have potentially dangerous outcomes.

The economic principle – 'semiosis of economics'

The use of the semiotic complexity metric can be conjoined with a semiotic perspective on economics to enhance the valuation by facilitating greater time-space specificity in our valuations of innovation. In particular it can help determine what 'innovation drivers' are likely to be 'value drivers' and from that point to calculate its future economic value using, for example, a discounted cash-flow model. While the semiosis of complexity helps model the potential of an innovation, it does not address its economics. Unless inputs, outputs, scarcity and incentive assumptions are factored into the model, boundary complexity alone will not predict what aspects of the innovation are likely to reach end users (innovation 'actualization').

The assumption that innovation requires incentive to produce new knowledge is accepted by the current model as a strongly held social value that currently sets an economic bias on the specificity level – though it should always be kept in mind that curiosity may precede the behaviorism which incentive analysis assumes. This 'rational economic actor' bias is reflected in the notion that while ideas must be replicable to become valuable knowledge ('intellectual capital'), if others can imitate this knowledge there will be no incentive for its production. The apparent paradox that this involves is sometimes described as a tension between 'replicability' and 'inimitability' – for an innovation to be valuable it must be replicable, but if the innovator is unable to control the innovation and competitors can imitate it there is no incentive to commercially pursue the idea.

Notions of 'networks' and 'platforms' that are increasingly entering the dialogue of intellectual capital management suggest that the potential paradox of replicability and inimitability may be addressed using semiotic complexity. The innovation drivers can be located, and the site of application of an economic metric can be thereby determined, when we look at an innovation in terms of its vertical and horizontal levels of language complexity. It is by this means that we can predict that in order for the innovation to be successfully actualized the innovator needs to select specified boundaries to control (the site of inimitability), the other boundaries being left for the next innovator to control (so long as he pays the 'price' for crossing the boundaries the first innovator controls). Such a semiotic perspective on economics, which roots value in structures of language complexity, stands in contrast to more classical economic theory, which is based on physical interpretations of resources and supply chains, and that accordingly fix their limits using positive, finite, unilateral boundaries.

Platforms A platform stands to the application it facilitates across a boundary of complexity (for example, a computer operating system stands as platform to software as application). The function of a platform (a computer's operating system) is to support a separate application (Word software) that in turn supports a function (writing documents). The platform constitutes the installed base of a new dependent hierarchy and facilitates the possibility of new, platform-compatible applications (PowerPoint software). As the range of applications grows the complexes they form can support potentially diverse horizontal applications (which may spawn new innovation, for example, ecommerce). A platform can be 'interoperable' with other platforms or an application can be cross-platform compatible so as to allow it to operate independent of the specifications of a specific type of platform.

To conceive of a platform as a language the successful application of which depends on shared use, helps conceptualize the issue of control – a platform is a complex that operates as a language that supports the application as a statement. The application can in turn be a language complex if it generates another level of statements. A successful strategy of appropriation requires that some level of the resulting complexity needs to be controlled (closed) – the platform, the application(s), and/or the resource inputs for operation of the platform or the application(s). The question of what levels to close and what to make freely available (open) is a strategic decision that directly bears on innovation actualization and its market success. An innovation can be technologically closed in that its design concept does not contemplate and accommodate further complexification, that is, it is not horizontally interoperative or vertically platform-independent. An innovation can also be closed because the innovator limits the end user's ability to apply it or to extend or nest it in a new innovation complex.

Viewed in this way we can better assess how innovators need to structure openness and closure in a product value chain so as to maximize commercial value. From this semiotic perspective a strategy that pursues maximizing control over end users can be seen as potentially limiting value. Whereas one that pursues minimum control, exercising only selective closure and allowing end users the maximum freedom to develop new related innovations, may enhance the value of the original innovation. The business world is replete with examples of better and worse management of the economic semiosis that deserve closer examination and theorizing beyond the bounds of what this chapter permits. Sony retained too much control by refusing to license its now defunct Betamax technology in ways that would permit new innovation conversation by other electronic manufacturers. Atari on the other hand retained too little control as to who could develop games for its early videogame platform, with the result that a flood of low quality games depreciated its value. Apple Corporation has struck a successful balance in relation to its place in the music supply chain by controlling use of its iPod platform in conjunction with another platform, the MPEG-4/AAC protected software, which is freely available for download through its proprietary iTunes website, but also allowing consumers freely to use the iPod in conjunction with other, non-proprietary formats. MySQL and Red Hat (through its Fedora project) allow end users free access to their software but charge commercial developers. Mandrake charges for its support of Linux open-source operating systems. IBM makes freely available the Apache software used by the hardware it sells. Covalent Technologies sells layers of proprietary software that operate over open-source software.

Networks The issue of replicability and inimitability in relation to networks involves economic considerations which, while they are different from what occur in relation to platforms, are similarly susceptible to semiotic interpretation. A new network technology (for example, an automated telephone switchboard) is a complex statement about communications that changes time-space specificity by increasing the size of the community of communicants, the amount of information that they can transmit, and their memory capacity. In a network, strictly defined, the communications are not applications since their content is not overtly about the network with the goal of facilitating a new function in the way that an application 'talks' about its platform; network communicants make no investment of innovation in the network beyond using it as a pipeline to facilitate their communications.

Cyber network effects have been demonstrated to operate exponentially with the addition of each new user, and the economic significance of this is clear. The booms and busts of the dotcom era bear witness to the need for a careful understanding of network complexity as a means to successfully allowing an innovator to capitalize on the potential economic accelerator effect it offers. Many B2B and C2C networks failed because their value proposition was weak or their value sharing between network controller and end users was too asymmetrical. Other businesses pursued more successful strategies. Microsoft and AOL bundle free networking tools – Internet Explorer and Instant Messenger – together with for-pay products. Google subsidizes free internet search technology with for-profit advertising. eBay funds its open marketplace and rating system by charging auction fees.

The strategies that businesses pursue in relation to new network and platform-based innovations can be interpreted in terms of regulation of the structure of vertical and horizontal boundaries. How an innovator regulates the boundaries impacts on value chain realization and defines the 'footprint' of those who participate in the value chain. The greater the barriers an innovator places on any boundary and the lower the level of complexity of that boundary, the more it will limit the value chain, that is, limit the adoption of the innovation by end users and limit the ability and incentive by end users to develop other dependent innovations. For example, the music recording industry's attempt to control music distribution by confining it to CDs as opposed to only controlling music production and allowing others to innovate as to the mode of distribution exemplifies a low-innovative business model that is unlikely to remain profitable. To control a lower-level boundary in a dependent level may yield immediate, high economic gain at that location, but it is likely to restrict the initial (and successive) innovator's economic gain at different locations in the future. Where to control, the method of control, and the time of control are therefore vitally important determinations. Integration of the third level of analysis, specificity, into the current model facilitates the modeling of an innovator's footprint and economic analysis of how boundary control moves supply and demand curves and influences the number and nature of potential successive innovations that are embraced in the innovator's aggregate economic calculus.

To confine the valuation model to three levels means that it does not critically evaluate the issue of *whose* perspective to adopt when deciding on the principles of selections and economics that control the method of innovation valuation. These principles define how time and space relationships are structured (LS 3.0 level three, specificity) and set the bias that controls *what* is measured and *how* (LS 3.0 levels one and two, signification and valuation). Should the current IPR and market-based valuations continue to set the bias on the system, or should we move to a more semiotic approach? Related to this is whether to adopt a private or a public perspective on the matter, and whether the decision should be centralized or decentralized: a choice which has practical ramifications. For example, should the music recording industry be allowed to close the channels of distribution that do not comport with their current business model using centralized legal coercion or should we prefer a hands-off approach that encourages an open supply chain and that favors the convenience of the end users and supply chain intermediaries?

Recursiveness Within Systems – Subjectivity

Unless some acceptable meta-perspective can be provided to resolve whose perspective to adopt, hence what bias to place on measurement, any attempt to answer the above questions becomes mired in the paradoxical loop with which we started this chapter – the valuation of valuation. LS 3.0 recognizes that evaluation cannot stand above paradox; paradox is inescapable because it is inherent in the language that we use to construct any evaluation. Instead it works from within paradox, but attempts a critical perspective on paradox. It evaluates valuation models according to their relative self-awareness of their propensity for paradox and for the inversions of natural hierarchies their paradoxes can engender.

The fourth level of LS 3.0 integrates the notion of perspective into the what, how and where of valuation (see Figure 8.4). It does this by using the concept of 'subjectivity', to frame the other three levels of the model. A defining quality of subjectivity is its language capacity for self-awareness, inherent in which is a capacity for self-criticism. Level four defines different paradigms of subjectivity and places them in a dependent hierarchy of social history. The development of social history for present purposes is crudely divided into three successive, dependent paradigms that correspond with arguably the three most significant social innovations – oral speech, writing, and cyber communication. With the occurrence of each stage a radical innovation of a new language system occurs, each successive system being a complex that nests the earlier system. The resulting change in language complexity initiates a quantum change in the way in

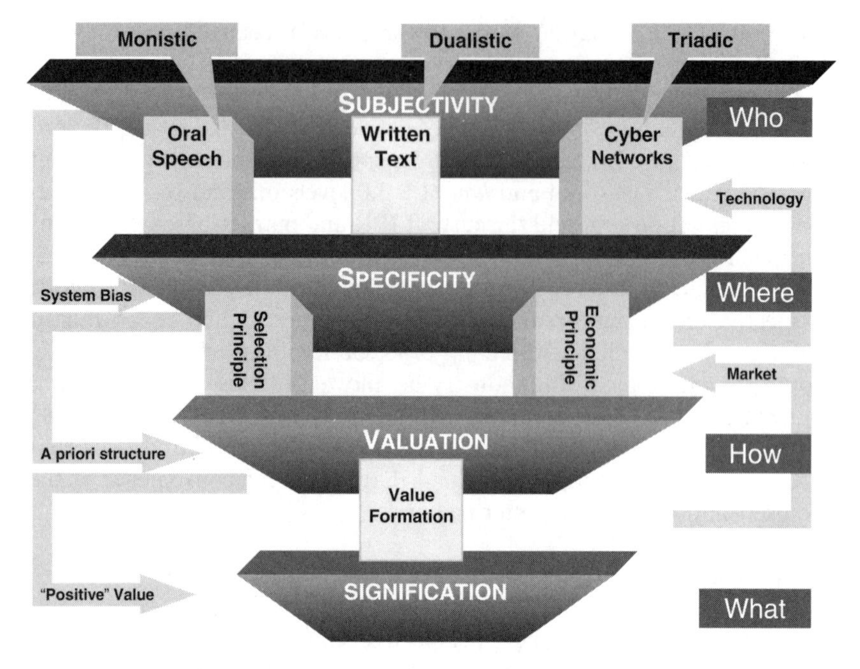

Figure 8.4 Valuation

which temporal and spatial relationships are constructed and interpreted, and therefore in the envelope of possible innovation. Each successive language innovation brings about *functional* changes in the ability of humans to communicate and model their world for themselves and for one another. Not only does the size of the possible community of communicants increase, but also the amount of information that they can transmit, and their memory capacity exponentially grows. New markets and new technologies become possible.

But more than functional changes, the way that meaning is constructed changes and leads to radically new ways of interpreting and configuring the time and space relationships that structure our world – a new *epistemology*. At each step in the evolution of the dependent hierarchy the epistemology that controls the paradigm of social thought changes, producing a change in subjectivity. This fundamental change in the nature of subjectivity produces a discontinuity that privileges a new perspective and changes *whose* principles of selection and economics are adopted. The resulting change in bias in turn alters the *what, how* and *where* of valuation. Within this system the new epistemology and functionality coexist in a feedback relationship, and their interaction lays the foundation for future language innovation, and along with this the possibility of new epistemologies, new paradigms, new levels of subjectivity, and therefore new capacity for critical thought.

Hardcopy Text Paradigm – A Dyadic Epistemology

IPRs and the valuations that are based on them are, I argue, symptomatic of their historical origins as products of a pre-cyber, hardcopy-text paradigm. While IPRs are a form that have adaptive fit to the hardcopy-text paradigm, their suitability as a basis for valuation in a cyber paradigm is questionable. The epistemology of the hardcopy-text paradigm is characteristically dualistic. Viewed in light of the current analysis, this dualism can be seen as attributable to the dual levels of language complexity – a written code that is historically and conceptually built on an oral code and the new textual concept of reproducible hardcopy. The hardcopy-text paradigm produces a characteristic subject–object paradox, which in the legal realm manifests itself in the metaphoric extension of physical property notions to intellectual production in the form of IPRs. These notions turn the subject into an object (flows of information into buckets), or sometimes the object into subject, as a practical and economic way of dealing with the dualism. In other words, IPRs operate, somewhat ineffectively, within the paradox instead of addressing its resolution. This is because the limited complexity of the hardcopy-text paradigm does not support a sufficiently high level of recursiveness fully to understand the history and construction of the paradoxes it recapitulates. Classical economic analysis has also chosen to perpetuate the paradox and its limitations by traditionally choosing to build itself on property notions such as IPRs.

Determining value by identifying and formalizing IPRs has increasing generated paradoxes of this sort, particularly in the area of cutting-edge innovations based on digital and biotechnologies. Under these conditions innovation, valuation and IPRs begin to work against each other within the framework of an outmoded commercial structure that misrepresents its own effective sequences of innovation and production. In part, this is an escalating problem of logical typing, systematic categorization, and the rigidity of language, which can become self-defeating when IPRs are placed prematurely above the increasingly complex and collective processes of innovation that generate the primary valuations on which IPRs rest.

Cyber Paradigm – A Triadic Epistemology

In contrast to the dualism of the earlier hardcopy-text paradigm, the epistemology of the cyber paradigm, it is explained below, is inherently 'triadic'. Triadic thinking facilitates a complexity perspective, which allows us to model integral systems of innovation, valuation, production and IPRs without neglecting the crucial dependent hierarchies on which the functioning of the system is based. The enhanced understanding of the structuring of the systems we live in which accompanies this perspective allows us to avoid inverting dependent hierarchies in such a way that, in an effort prematurely to maximize utility in the form of an immutable property right, we undermine a highly complex system of productive interactions and communications by treating IPRs as logically and categorically prior to the innovations which generated them. The resulting capacity for meta-communication communications has allowed the potential for subjectivity to be released from the bounds of dualism to articulate more complex triadic models – LS 3.0 is a product of this new subjective and its enhance capacity for self awareness.

Central to the formation of triadic thinking that is characteristic of the cyber paradigm is the invention of computer networks. Writing allows code changes from oral to written and from symbolic code to symbolic code. But the 'scrivener' has limited capacity to stand outside himself to record the context of the code translation at the time that he performs it. The most he can do is manually date and sign the document and, if the document is physically transferred it can be endorsed by its transferor. A computer is structured around the invention of a model of dependent hierarchy starting with the hardware as base and working up through middleware (operating system) to software. The meta structure of this machine allows information about code translations to be simultaneously represented in code and the machine quality of the computer allows both levels (the code translation and the coded representation of that translation) to be automated. When the power of computers is harnessed in networked social communications the metacommunicative capacity of the computer feeds back on users incrementally to change their subjectivity. eBay's community of users, which automated the idea implicit in the function of credit bureaus, is a manifestation of this emergent subjectivity. However, the change in epistemology is much more profound and far-reaching than this simple example. Furthermore, the change is much more complex than a simple causal explanation can explain. Computers were not 'instrumentally' responsible for this change; the feedback relationship between functionality and epistemology saw rapid movements, up and down, in the 'artificial' dependent hierarchy of the computer and of the larger dependent hierarchies of the social and natural world that were affected by this dialectic interaction.

The limitations of IPR-based business models are beginning to be recognized in open source platforms and networks. Innovators are selectively forgoing the monopoly profits of IPRs, instead relying on complexity and time-space specificity as forms of decentralized self-protection. Examples of such non-IPR controls include: supply chain relationships, branding (built on complexity and quality), temporally sensitive products, after-market service, real-time performance, first to market, employee loyalty, and so on. New 'communities of creativity' built around 'innomediaries' and other feed-back-like, producer-consumer relationships also highlight the realization that the boundaries created by another property metaphor, the 'limited liability corporation', are controls on innovation complexity the use of which needs be closely evaluated. Innovation is seen as no longer being confined or even best achieved inside the hierarchical structure of the corporation. Corporations that selectively reduce the size of their footprint in the value chain by reducing boundary controls or altering the controls through use of outsourcing, licensing, strategic agreements, attribution and the like are often more profitable and are better able to sustain innovation than are their behemoth competitors who persist with full ownership of the value chain. Implicit in these control reductions is recognition that to place fewer and more select controls on the boundaries of complexity can increase both value chain realization and the economic value of the innovator's share.

Sensitivity to the paradigmatic discontinuities addressed in the subjectivity level of LS 3.0 also allows for rethinking of the relationship between public and private, nature and nurture, and between pro-growth and ecology in the innovation valuation process. The hardcopy-text paradigm works within these paradoxical dichotomies and IPRs valence the dichotomies in ways that have many unintended and controversial effects. For example: IPRs work to privilege pharmaceutical and surgical-based health care (versus preventative health care), at the same time as they result in monopoly pricing that creates enormous asymmetries in patient access; IPR protection of seeds and plants has contributed to reduction of agricultural diversity and the spread of genetically altered food crops.

A developed cyber paradigm may well see the abandonment of IPRs in favor of self-regulation, with the result that innovation would need always to address its own protection and not rely on externally conferred monopolies in order to ensure the innovator appropriates gain and the economics of production are sustained. The new incentive for an innovation's self-protection always to become the secondary goal of the innovation (a recursive process) would fundamentally change the current economic bias. The move from a property bias to a self-protection bias would arguably promote products and production that more rigorously monitored their boundaries for unauthorized crossing, hence that build in a greater capacity for protection against viruses of all types and runaway growth. The result would be less dangerous, more socially responsible and more ecologically sensitive forms of production.

CONCLUSION

The use of LS 3.0's four boundary levels allows for more comprehensive modeling of valuation of innovation. The lower levels – signification and valuation – self-consciously incorporate a semiotic of complexity into the quantitative measurement of innovation. Specificity – the third level – allows a semiotic of economics to be incorporated to define and financially measure innovation actualization. Subjectivity – the fourth level – incorporates historical self-awareness to provide recursiveness into the feedback operation of LS 3.0 by helping understand that the epistemology of a historical paradigm sets a bias on our semiosis, hence the measurement it facilitates.

The aspiration of LS 3.0 is to provide more accurate quantitative valuation of innovation while facilitating modeling of the affect of different biases on its operation. It is recognized that LS 3.0 is itself paradoxical, in that it values itself as more adequate than other models. However, LS 3.0 is a paradox of a different logical type. It incorporates an awareness of the paradox of self-valuation, which is inescapable to analysis. Furthermore, it attempts to compensate for this by building recursive feedback into its operation. It tries to prevent the inversion of dependent levels and embrace the potential to represent the history of valuation (including its own self-valuation) by bringing distinct quantitative and qualitative elements of valuation to bear on each other in a self-adjusting manner. This means that the analytic elements employed in evaluating innovation must be measured against the functioning of the system of property and social relations as a whole. The hoped-for effect would be to embed or nest traditional forms of property, innovation, and valuation within the broader context of their changing social, economic, and ecological relations – rather than dispensing with them altogether. Traditional articulations of property right and innovation have established their usefulness historically, but if they are to continue serving our needs they will eventually have to be reframed, reinterpreted and realigned within an expanding and increasingly complex system of relations between sites of innovation, production, and consumption bounded by an ecology with clearly defined stress-points.

REFERENCES

Barthes, R. 1988. *The Semiotic Challenge*. Trans. R. Howard. New York: Hill and Wang.
Bateson, G. 1972. *Steps to an Ecology of Mind*. New York: Ballantine Books.
Bateson, G. 1979. *Mind and Nature: A Necessary Unity*. New York: Dutton.

Foucault, M. 1970. *The Order of Things: An Archaeology of the Human Sciences*. London: Tavistock Publications.
Eco, U. 1976. *A Theory of Semiotics*. Bloomington: Indiana University Press.
Eco, U. 1979. *The Role of the Reader: Explorations in the Semiotics of Texts*. Bloomington: Indiana University Press.
Ernst, H. 2003. 'Patent Information for Strategic Technological Management'. *World Patent Information* 25: 233–42.
Ernst, H. and J. Henrik Soll. 2003. 'An Integrated Portfolio Approach to Support Market-Oriented R & D Planning'. *International Journal of Technology Management* 26: 540–60.
Garcia, R. and R. Calantone. 2002. 'A Critical Look at Technological Innovation Typology and Innovativeness Terminology: A Literature Review', *Journal of Product Innovation Management* 19: 110–32.
Griffin, A. and A. Page. 1993. 'An Interim Report on Measuring Product Development Success and Failure', *Journal of Product Innovation Management* 10: 291–308.
Sawhney, M. and E. Prandeli. 2000. 'Communities of Creation: Managing Distributed Innovation in Turbulent Markets'. *California Management Review* 42: 24–55.
Sorescu, A., R. Chandy, and J. Prabhu. 'Sources and Financial Consequences of Radical Innovation: Insights from Pharmaceuticals'. *Journal of Marketing* 67: 82–102.
Wilden, A. 1972. *System and Structure: Essays in Communication and Exchange*. London: Tavistock Publications.
Wilden, A. 1987. *The Rules Are No Game: The Strategy of Communication*. New York: Routledge and K. Paul.
Wilden, A. *Man and Woman, War and Peace: The Strategist's Companion*. New York: Routledge and K. Paul.

9. Measurement of innovation and intellectual property management: challenging processes

L. Martin Cloutier and Susanne Sirois

INTRODUCTION

All measurements of economic and managerial activities are, to some extent, limited and flawed. The measurement of innovation also suffers from a number of important limitations: absence of appropriate data, lack of application of research standards, changes introduced in the regulatory environment, to name but a few (Bloch, 2005; Goedhuys, 2005; Jensen and Webster, 2004; Rogers, 1998; Sloan, 2001). This is partly due to the lack of standards and also due to difficulties associated with the object of the measurement. Studies often are conducted for specific units of analyses (activities, processes, business units, firms, interfirm relationships, markets, regions, countries) and are constrained by available data and information that can be employed to obtain the measurements sought with a variety of research methods and techniques. When a 'control' factor can be introduced into the analysis for some of these dimensions, the results obtained can be employed fruitfully for decision-making. The possibility of establishing benchmarks to allow comparisons across studies will always, however, remain both methodologically qualitative and quantitative challenges (Adams et al., 2006). Processes are dynamic and can evolve over time. Products tend to be more static. Products are at times considered innovative as 'new' and at times as 'imitative', when 'improved'. Thus, the object of measurement is typically transient, can only be captured as a snapshot, and may not be representative of the current situation once time passes from the point of measurement.

Nevertheless, better to understand the contribution of economic and management activities and processes in firms, markets and society, it is important to obtain measurements to improve the quality of economic activities associated with innovation over time (Adams et al., 2006). This is why the measurement of innovation in health and biotechnology is such an important activity of interest for many influential organizations around the world, such as the Organization for Economic Cooperation and Development (OECD) and the World Health Organization (WHO, 2006). The measurements of innovation activities and processes are key factors in management and performance improvement. They are additionally important for economic growth because they provide the empirical basis upon which firms can make informed decisions in the immediate, and plan better for future decisions. Not only is this capacity for making better decisions important for biotechnology and pharmaceutical firms, it is also important in the context of policy decisions where decision-makers seek to introduce research and development (R&D), intellectual property (IP), and investment policies that balance the society's objectives of both investors and consumers (Thomassin and Cloutier, 2001).

In short, there is a need to measure innovation better in products and processes associated with R&D in biotechnology. This is a complex problem because biotechnology firms can be engaged in a wide range of innovative activities that defy single measures (van Dyck and Allen, 2006). Firms use, for example, IP management mechanisms, some of which are formal legal tools like patent law and trade secret law. Others are informal, for example contracts, other business practices embedded in marketing, and R&D activities. Firms can use these formal and informal mechanisms jointly or independently to protect their investments and secure future revenue streams from initial and continuous investment (Cloutier and Gold, 2005). One of the most critical challenges in that sphere, however, is the performance measurement of the innovation process and of the contributions made by IP management practices which frequently play an important role in innovation. Measurement of the performance of patents, trade secret laws, and contracts is difficult in its own right because it is not altogether clear what ought to be evaluated or what counts as the best evidence of good performance. Moreover, how these are to be measured as contributors (or detractors) in the innovation process and IP management processes is always challenging. In the rapidly changing context of innovation, knowing what to measure, how and when, becomes that much more complicated and dynamic.

The aim of this chapter is to outline some of the challenges associated with the measurement of innovation in the complex context of an organization with biopharmaceutical R&D activities. Biopharmaceutical firms offer an interesting context to examine the measurement of innovation because they manage unique processes and challenges. For example, their information, knowledge, financial and technological risks are extremely high. Perhaps more significantly, the innovation process can be characterized by very long time delays from project initiation to product commercialization, a factor which introduces its own hazards to biopharmaceutical firms. The next section outlines the biopharmaceutical organizational context for the measurement issues to be examined, that is, the resource allocation process of the biopharmaceutical firms and markets and their characteristics. Then an overview is presented of measurement challenges for the innovation process, accompanied by a brief discussion on the impact of IP performance.

THE BIOPHARMACEUTICAL ORGANIZATIONAL CONTEXT: RESOURCES AND DECISIONS

The organizational context for innovation in the biopharmaceutical firm can be illustrated using the conceptual framework proposed, in Figure 9.1 (Sanchez and Heene, 1996). This framework traces the influences of the composition and control of feedback loops within an organization. The organization includes both tangible and intangible resources, embedded within the management processes. The management process controls these resources. This follows a management logic that includes flows of decisions, resources, information, knowledge, and incentives. It is important to observe that exogenous to the boundary of the organization are two types of markets: the market for inputs, or resources, and the market for outputs. Thus, two types of competition influence the management of the innovation process. First, the market for resources includes capital, human resources, knowledge assets, IP, infrastructure and competencies. The market for inputs can be ferocious as firms compete, for example, for star scientists and other expertise, and capital to complete clinical phases. Second, the market for outputs includes products, knowledge products and often IP in its various forms. The measurement of innovation for the biopharmaceutical firm must take into account the processes involved, as well as the use of inputs and outputs.

The scope of the biopharmaceutical firm can include one or more activities associated with different phases of the drug development process (see Figure 9.2). The entire process can take somewhere between 10 and 15 years, and, as is well known, requires very high levels of financing to progress through the drug discovery pipeline. Several studies conducted on this aspect present the challenge of estimating these costs with alternative methods (Dickson and Gagnon, 2004; DiMasi and Grabowski, 2007a; DiMasi and Grabowski, 2007b; DiMasi et al., 2003). Costing the input side of this type of operation, one that involves several distinct phases over at least a decade, presents many challenges (Rawlins, 2004).

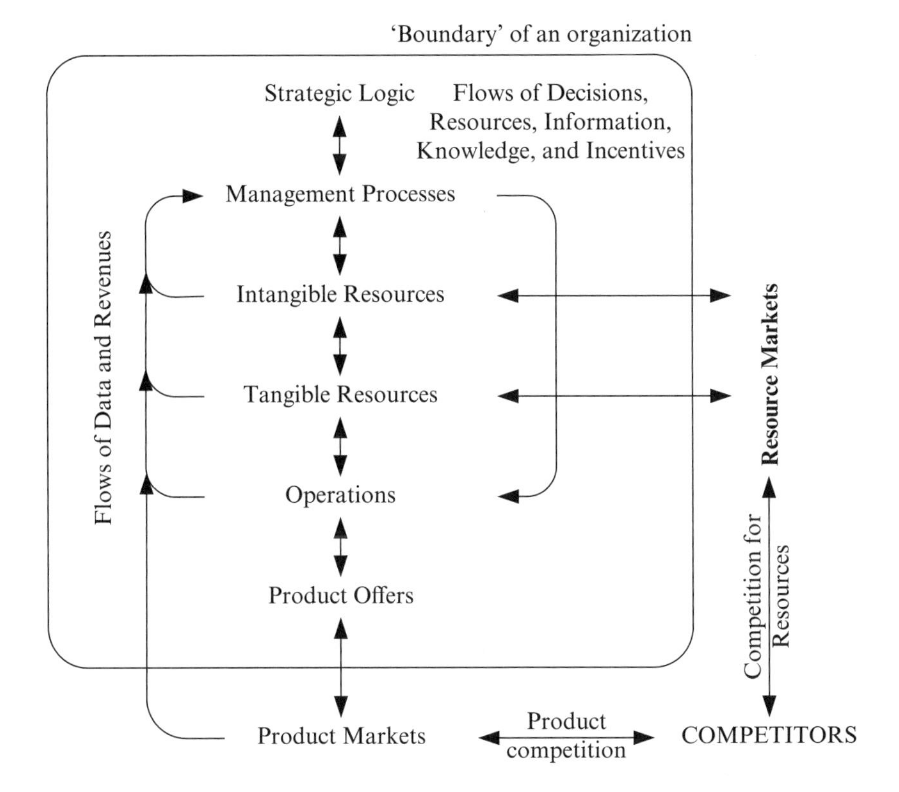

Source: Sanchez et Heene (2002, p. 73).

Figure 9.1 View of an organization as an open system

The development of new innovative drugs for care of patients has become fundamental to improve the quality of life (Drews, 1999). However, in the past few years, the number of new molecular entities (NMEs)[1] for pharmaceutical use has progressively declined,[2] while the cost of pharmaceutical R&D has progressively increased. The so-called productivity gap reported during the past few years concerned the relationship between R&D spending per new molecular entities (NME) (Riggs, 2004) and the total number of NME approved by the FDA. The productivity of pharmaceutical drug discovery is typically indicated by the number of NMEs approved in a given year divided by total R&D investment (Hopkins et al., 2007). This widely used measure, perhaps a poor choice for a productivity indicator (Hopkins et al., 2007), indicates that overall productivity has been falling for the past 30 years (Booth and Zemmel, 2004; FDA, 2004; Hopkins et al., 2007). This indicator

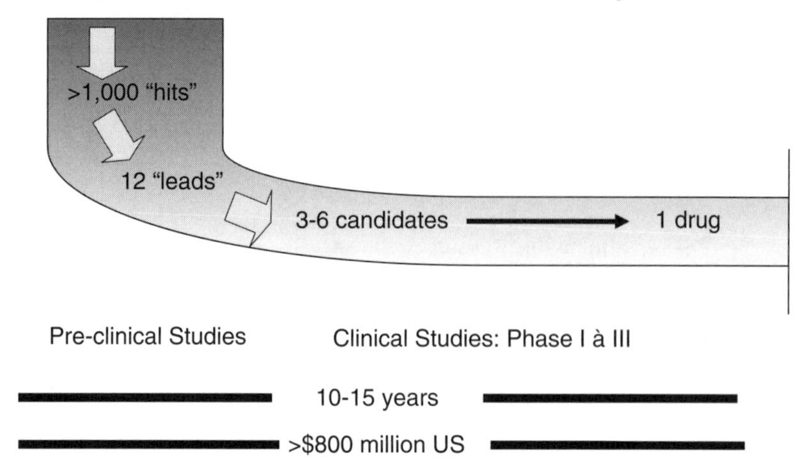

Figure 9.2 Drug discovery timeline

encompasses: 1) important and long time delays between the initial investment and the availability of the product on the market to generate returns (the timeliness, or point of reference period for analysis should be encompassing the 12–15 year period often needed to turn a scientific concept into a new chemical entity (NCE) (Schmid and Smith, 2005)); 2) the insistence on financial returns as a means to appraise R&D performance, rather than multiple indicators; 3) the drug development process; and 4) quantitative changes and substitution in inputs associated with R&D, such as technology and organizational design (Hopkins et al., 2007). Other possibilities include the changing cost structure of R&D, because technology speeds up project execution, which may lead to rising failure rates, coupled with higher standards and organizational design issues (Booth and Zemmel, 2004). The second point of view looking at the same data (Schmid and Smith, 2005; Schmid and Smith, 2006) suggests that although there have been fluctuations in drug launches, there are no major variations in the number of NMEs approved per year by the FDA, except for a major peak in 1996, which cannot be regarded as a benchmark for pharmaceutical innovation capabilities, but instead there has been a steady increase in the number of NMEs launched (Schmid and Smith, 2005; Schmid and Smith, 2006) (see Figures 4 and 5 in Schmid and Smith, 2005 and Figure 1 in Schmid and Smith, 2006).

The different conclusions obtained by various groups of analysts are due to the timeframe considered, and many aspects of this are attributable to differences in the measurement of innovation. The first group of analysts considers the 1996–2004 period, whereas the second group considers a much longer timeframe, that is, the 1945–2004 period. Therefore it is possible that an incomplete picture of the overall trend emerges as being less innovative than previously when looking only at the last decade. Moreover, the analysis made by Schmid and Smith (2005) is based on the split for new medicines into 'priority' and 'standard' (Schmid and Smith, 2005) instead of on NMEs only, as it is conducted by the first group of analysts. Their results show that the current decade is the one which has produced the highest numbers; a total of 117 priority review medicines (priority review refers to FDA-approved drugs that provide a substantial advantage over existing treatments whether or not they are 'first-in-class' for a new disease mechanism). According to Schmid and Smith the sixth decade, from 1995 to 2004, produced the highest total number of drug launches (see Figures 4 and 5 in Schmid and Smith, 2005 and Figure 1 in Schmid and Smith, 2006). A comparison of the total numbers of NMEs produced in each decade can be generated: (i) 127 in the first decade 1945–1954; (ii) 242 in the second decade 1955–1964; (iii) 134 in the third decade 1965–1974; (iv) 190 in the fourth decade 1975–1984; (v) 220 in the fifth decade 1985–1994; and (vi) 307 in the sixth decade 1995–2004 (Schmid and Smith, 2006). Thus, it seems that the peak observed in 1996 is more the result of a short-term indicator in performance measurement than a productivity issue, and it appears to have no connection with the underlying innovation processes and rates of innovation in the pharmaceutical industry (Schmid and Smith, 2005). But, according to John Jenkins, Director of the FDA's Office of New Drugs, the decrease in approvals is a direct result of a reduced number of new drug applications (NDAs) (cited in Owens, 2007). This is because the FDA assesses the probable advantage of a new prospective medicine for the patient at the submission of an NDA. Also, despite the rise in potential drug targets and the increase in R&D expenditure since the 1970s, between 1993 and 2003 the FDA reports an almost continuous decline in submissions for regulatory approval of NMEs, with only 17 in 2002, the lowest for eight years. In 2006, only 18 NMEs, the same as in 2005, were approved (see Figure 1 in Owens, 2007).

The reason there is a gap in the current productivity perception is possibly due to the relationship established between the number of NMEs approved per year and the overall R&D spending per year in the US pharmaceutical industries (that is the inclusion of drug development, and quantitative changes in the inputs to R&D) and not with the R&D cost per type of NME and non-NME. Another indicator for the productivity decline perception is the division of medicines into 'priority' or 'standard'. For instance, the latter indicator shows that over a 60-year timeframe, priority reviewed NME approvals and total NME approvals rose steadily (Schmid and Smith, 2005).

Biopharmaceutical R&D is a sophisticated business, based on complex science and technology. Yet the biological sciences are developing rapidly. As the technology becomes even more complex and the introduction of systems biotechnology emerges, overcoming measurement challenges will become even more important for decision-making at the various steps of the drug discovery process. The diversification of technology creates challenges for firms as they face opportunities to choose among competing, alternative approaches to R&D. This, in turn, will generate more complex and difficult investments puzzles for decision-makers in the industry. For example, consider the impact of having high-throughput screening (HTS) for lead discovery experimental methods which are capable of generating vast quantities of both experimental and *in silico* data. When the generation, management and use of this gigantic data production with integration of data mining and machine learning techniques (Hopkins et al., 2007; Plewczynski et al., 2006; Waller et al., 2007) comes to reality as part of the *in silico* revolution, innovation measurement issues will become even more paramount because the intention is to use these robust data to streamline the R&D process (Bleicher et al., 2003; Cloutier and Sirois, 2008; Davies et al., 2006; Goodnow, 2006; Keseru and Makara, 2006; Myers and Baker, 2001; Sang et al., 2005; Sirois and Cloutier, 2008). Parallel to the integration of HTS technologies in the drug discovery process, major advances in data integration, statistical analysis and hit and leads profiling with cheminformatics methods are becoming standards and a driving force for lowering the DDP productivity gap (Cloutier and Sirois, 2008; Sirois and Cloutier, 2008). Decision-makers will need to understand investment trade-offs and value-creating alternatives using approaches that can help measure the extent of the innovation process (see for example, Figure 9.3). Obviously, IP management, formal, informal mechanisms and business practices, will play an extensive role in that process.

Are tangible resources the only ones that are worth measuring? There are means to evaluate the expected financial value at each stage of the R&D process. For instance, real options theory supports the decision-making in that process, as one moves from one phase to the next. The emphasis is typically placed on the tangible resources used in the process of innovation as they are more readily available for measurement. Other approaches could be useful, as seen in the influence diagram (ID) of Figure 9.4 that depicts a simplified system dynamics (SD) representation of the problem associated with the short-term and long-term allocation of resources. SD is a set of management modeling principles useful to measure performance of business processes including R&D processes of

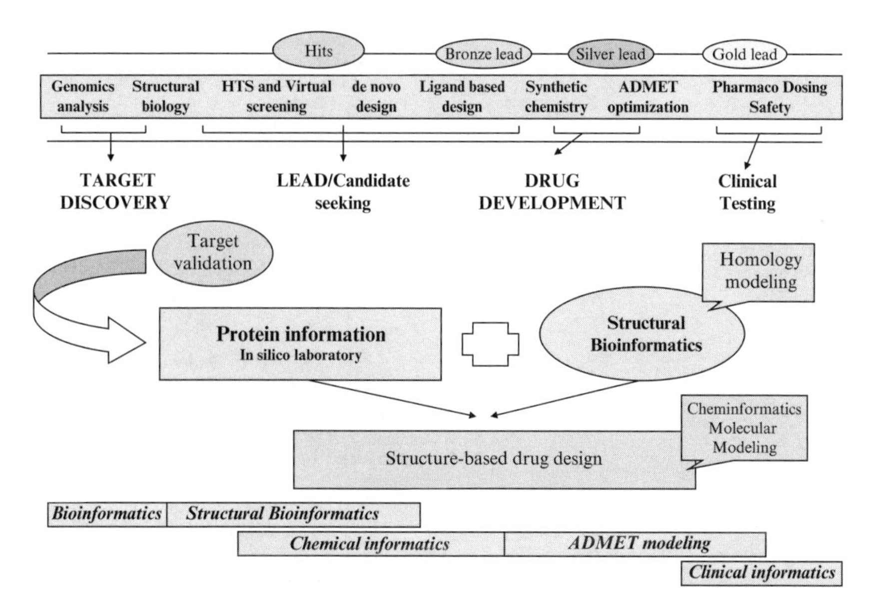

Figure 9.3 From target identification to lead optimization to investigational new drug

innovation (see Sterman, 2000 for more elaborate references on SD principles and references on research papers in that field).

The ID in Figure 9.4 shows the dynamics associated with the use of financial resources (more tangible) and to ignore the formation of intangible assets, these are competencies (more long-term impact intangible resources) (Delorme and Cloutier, 2005).

The ID represents the growth and underinvestment systems archetype, adapted as an example from the organizational learning literature (Senge, 1990). The diagram depicts qualitatively the trade-offs between short-term financial gains at the expense of building long-term competencies. In the ID, there is a reinforcing feedback loop (R) that shows the dynamic growth process. This growth is balanced by the performance of actual competencies in business management processes (balancing feedback loop B1). Competencies need to be built (balancing loop, B2), but they are constrained by the need for short-term, at times immediate financial gains, which are represented by the balancing feedback loop B3. This ID illustrates the situation where intangible resources such as competencies are part of the process, but often are constrained in their acquisition by short-term imperatives. As part of the innovation process, their contribution should be measured, but they often are not explicitly taken into account.

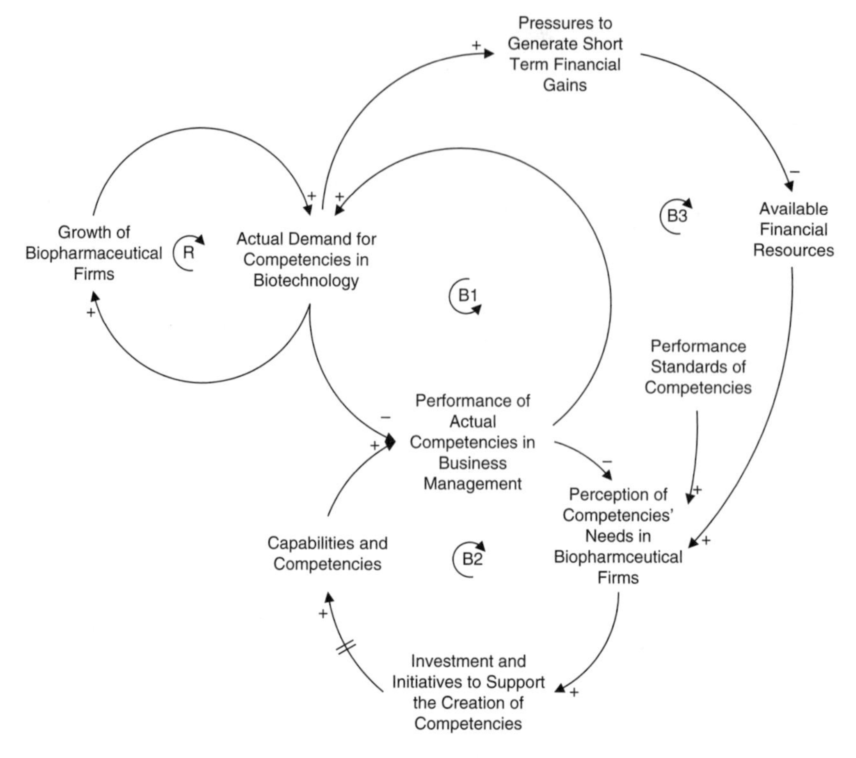

Source: Adapted from Delorme and Cloutier (2005, p. 252).

Figure 9.4 Short-term and long-term measurement of tangible and intangible assets in SD framework

CHALLENGES TO INNOVATION MEASUREMENT

Given what we have said about the complexity of biopharmaceutical R&D processes, the impact of changing science and technology on the innovation process, and the demands for better management tools including the measurement of intangibles, what are some of the key challenges in the measurement of innovation? There are two main considerations. To begin with, one must consider the technical dimensions of the innovation process itself. There is a long list of technical factors, often ignored in management studies, associated with the discovery process. These can interfere with the decision-making process associated with the innovation process as the information becomes available from R&D activities.

Technical Dimensions

There are a number of conceptual issues associated with the measurement of innovation at the firm level. Measures that can be used include such tangible resources such as R&D expenditures, IP filings, and survey data from studies on innovation. One of the most common measures of innovation is tangible resource measures around intellectual property management (for example, the number of patents). There are a number of limitations associated with these approaches. This is not to say that such measures should not be used, but caution should be exercised so as to not draw conclusions larger than the data warrant. Since innovation is a dynamic and evolving process, attempting its measurement is by definition a complex endeavor. The innovation process can only be captured at a point in time if static measurement approaches are employed. As a result, there is a risk of misrepresenting the situation at any given point in time by using data drawn from another point in time. Besides, any analysis will cover a particular time span with a beginning and an end to capture the salient points in an innovation process. The data will be static by definition, which is to say that they are bounded to a temporal context which is the only truly meaningful frame of reference for their interpretation. The number of patent filings which have happened in past R&D activity may not even relate to current R&D activities, for example. When the number of new innovative products is employed, assuming that there is an appropriate coverage of what constitute an innovative product, patent filings and other IP measures, such as trade secrets, perhaps only capture successful R&D outcomes. The product of innovation, however, can be a long and tedious trial and error process that will not be reflected in these measures. Process innovation, in contrast with product innovation, may not be known as such. This is especially the case when many processes are not known or well characterized, or when certain processes begin and end without the possibility of really being able to record their contribution. Innovation is a process that cannot be readily observed because innovative activity is embedded in the functioning of the organization.

There are several other important issues to consider in thinking about how to measure innovation, and IP's role in it. As seen in Figure 9.1, inputs and outputs can be both tangible and intangible resources and the quality and measurability of these can vary substantially. The intensity of R&D efforts and the quality aspects of that effort are intangible and, thus, difficult to capture. Another important aspect to recall is that all patents neither possess the same technological market reach nor have the same potential economic value. Similarly, the cost of inputs and the price for products/processes already in the market can only be used as proxy indicators of their value with respect to the R&D process and the projected value of the end product. But measures of innovation provide no indication of price movements over time. The purchasing cost of a database, for example, is orthogonal to the value generated by the data once put to use. Similarly, the salary paid to scientists or technologists may not reflect the value or intensity of the contribution they make in the innovation process. Finally, there are a few well-known biases that may be part of the problem of innovation measurement. The regulatory environment may also be a source of measurement issues and biases, due to changes in policy that affect the workings of processes, privileging some aspects while downplaying others. Dynamic methods and approaches that take into account changes over time are promising to help attenuate these limitations.

SUMMARY AND CONCLUSION

The objective of this chapter is to provide an overview of measurements issues associated with the innovation process. Far too often, IP, particularly patents, is treated as a reliable data yardstick by which to assess the innovativeness of firms. As seen, the measurement of innovation can be problematic in its own right, and it can be even more problematic according to the nuances and idiosyncracies of a specific industrial context. In the biopharmaceutical sector, the complexity of the innovation process, in terms both of its length and the complexity of the underlying scientific and technological processes, makes it difficult to capture fully the processes that the measure represents. In addition, as we have stressed here, it is important to distinguish between tangible and intangible assets, and to consider the different approaches needed to evaluate the relative contribution of each to innovation. Most innovation measures are based on the evaluation of tangible asset contributions to innovation. As exemplified by the ID in Figure 9.4, intangible assets add an important contribution to the process but are not typically among existing measures. The importance of intangible assets to innovation becomes quite obvious once innovation is portrayed as a dynamic system in an influence diagram. There, the contributions of intangible assets are part of feedback loops embedded in organizational systems. They tend to go unnoticed as resources, and hence escape the decision-makers' data and information pools relied upon in firm decision-making processes. This is a serious omission which needs to be rectified if firms are to make evidence-based decisions, particularly in the cases where the firm's total value may in fact reside more in intangible assets than tangible assets. Ultimately, we conclude that innovation in science and technology is not a process that works according to the 'cascade' model of steps that succeed one another, but, rather, it is iterative and integrated, and includes sets of interlocking complex feedback loops that structure innovative activities. As such, research methods ought to reflect the realities of firm-driven innovation, particularly the important contributions made in the broad category of intangible resources. Future refinement of measures may make the role of intangible assets more apparent in that process. Also, the intrinsic characteristics of a compound, for example in the biopharmaceutical innovation process, have an equal role to play. These characteristics surface and become manifest to scientists and decision-makers during R&D, especially at the modification and optimization phase of the drug development process. They should be modeled as part of the economic and financial appraisal for their changing value over time.

NOTES

1. An NME as defined by the FDA is a medication containing an active substance that has never before been approved for marketing in any form in the United States (FDA 2001). Although there are more new drug products (that is, non-NMEs) on the market than NMEs, their R&D costs are lower than those of NMEs because they constitute incremental improvements on existing drugs. Non-NME encompasses me-too or follow-on drugs that can be subdivided into two categories: 1) they may be innovative products, that is an NME that lost the race to be the first drug on the market in a given therapeutic class (such as anti-depressants, antibiotics, or anti-histamines), or 2) they may be incremental modifications of an existing drug, that is a new drug entity with a similar chemical structure or the same mechanism of action as that of a drug already on the market. That is, a me-too drug is a new entrant to a therapeutic class that has already been defined by a separate drug entity that was the first in the class (sometimes referred to as the breakthrough drug) to obtain regulatory approval for marketing. Me-too drugs have also been characterized in a more value-neutral way as follow-on drugs.
2. Complete reports are available at http://www.fda.gov/oc/initiatives/criticalpath/whitepaper.html, access verified August 2008.

REFERENCES

Adams, R., J. Bessant and R. Phelps. (2006). 'Innovation Management Measurement: A Review'. *International Journal of Management Reviews* 8(1), 21–47.

Bleicher, K. H., H. J. Bohm, K. Muller and A. I. Alanine. (2003). 'Hit and Lead Generation: beyond High-throughput Screening'. *National Review of Drug Discovery* 2(5), 369–378.

Bloch, C. (2005). 'Innovation Measurement: Present and Future Challenges'. In Working Paper from The Danish Centre for Studies in Research Policy, http://www.cfa.au.dk/fileadmin/site_files/filer_forskningsanalyse/dokumenter/working-papers/WP2005_6.pdf.

Booth, B., and R. Zemmel. (2004). 'Prospects for Productivity'. *National Review of Drug Discovery* 3(5), 451–456.

Cloutier, L. M. and E. R. Gold. (2005). 'A Legal Perspective on Intellectual Capital'. In B. E. Marr (ed.), *Perspectives on Intellectual Capital: Multidisciplinary Insights into Management, Measurement, and Reporting*. Boston, MA: Elsevier, pp. 125–136.

Cloutier, L. M. and S. Sirois. (2008). 'Bayesian versus Frequentist Statistical Modeling: A Debate for Hit Selection from HTS Campaigns'. *Drug Discovery Today* 13(11–12), 536–542.

Davies, J. W., M. Glick and J. L. Jenkins. (2006). 'Streamlining Lead Discovery by Aligning in Silico and High-throughput Screening'. *Current Opinion Chemical Biology* 10(4), 343–351.

Delorme, M. and L. M. Cloutier. (2005). 'The Growth of Quebec's Biotechnology Firms and the Implications of Underinvestment in Strategic Competencies'. *International Journal of Technology Management* 31(3–4), 240–255.

Dickson, M. and J. P. Gagnon. (2004). 'Key Factors in the Rising Cost of New Drug Discovery and Development'. *National Review of Drug Discovery* 3(5), 417–429.

DiMasi, J. A. and H. Grabowski. (2007a). 'The Cost of Biopharmaceutical R&D: Is Biotech Different?'. *Managerial and Decision Economics.* 28, 469–479.

DiMasi, J. A. and H. G. Grabowski. (2007b). 'Economics of New Oncology Drug Development'. *Journal of Clinical Oncology* 25(2), 209–216.

DiMasi, J. A., R. W. Hansen and H. G. Grabowski. (2003). 'The Price of Innovation: New Estimates of Drug Development Costs. *Journal of Health Economics* 22(2), 151–185.

Drews, J. (1999). 'Research & Development. Basic Science and Pharmaceutical Innovation'. *Nature Biotechnology* 17(5), 406.

Food and Drug Administration FDA. (2004), *Introduction or Stagnation: Challenge and Opportunity on the Critical Path to New Medical Products*. Washington, DC: FDA, http://www.fda.gov/ScienceResearch/SpecialTopics/CriticalPathInitiative/CriticalPathOpportunitiesReports/ucm077262.htm.

Goedhuys, M. and L.K. Mytelka (2005). 'Measuring innovation: making innovation surveys work for developing countries'. Technology Policy Briefs 4:12, *UNU-INTECH, Maastricht, the Netherlands.*

Goodnow, R. A., Jr. (2006). 'Hit and Lead Identification: Integrated Technology-based Approaches'. *Drug Discovery Today: Technologies* 3(4), 367–375.

Hopkins, M. M., P. A. Martin, P. Nightingale, A. Kraft and S. Mahdi. (2007). 'The Myth of the Biotech Revolution: An Assessment of Technological, Clinical and Organisational Change'. *Research Policy* 36(4), 566–589.

Jensen, P. H. and E. Webster (2004). 'Examining Biases in Measures of Firm Innovation'. Melbourne Institute of Applied Economic and Social Research, Intellectual Property Research Institute of Australia vols. The University of Melbourne, Melbourne Institute Working Paper No. 10/04, http://www.melbourneinstitute.com/wp/wp2004n10.pdf.

Keseru, G. M. and G. M. Makara. (2006). 'Hit Discovery and Hit-to-lead Approaches'. *Drug Discovery Today* 11(15–16), 741–748.

Myers, S., and A. Baker (2001). 'Drug Discovery – an Operating Model for a New Era'. *National Biotechnology* 19(8), 727–730.

Owens, J. (2007). '2006 Drug Approvals: Finding the Niche'. *National Review of Drug Discovery* 6(2), 99–101.

Plewczynski, D., S. A. Spieser and U. Koch. (2006). 'Assessing Different Classification Methods for Virtual Screening'. *Journal of Chemical Information and Modeling* 46(3), 1098–1106.
Rawlins, M. D. (2004). 'Cutting the Cost of Drug Development?' *Nature Reviews Drug Discovery* 3(4), 360–364.
Riggs, T. L. (2004). 'Research and Development Costs for Drugs'. *Lancet* 363(9404), 184.
Rogers, M. (1998). 'The Definition and Measurement of Innovation'. Melbourne Institute Working Paper No. 10/98, http://www.melbourneinstitute.com/wp/wp1998n10.pdf.
Sanchez, R. and A. Heene. (1996). 'A Systems View of the Firm in Competence-based Competition'. In R. Sanchez and A. Henne (eds), *Dynamics of Competence-Based Competition: Theory and Practice of the New Strategic Management*. New York: Pergamon Press, pp. 39–62.
Sang, Y. L., D. Y. Lee and Y. K. Tae. (2005). 'Systems Biotechnology for Strain Improvement'. *Trends in Biotechnology* 23(7), 349–358.
Schmid, E. F. and D. A. Smith. (2005). 'Keynote review: Is Declining Innovation in the Pharmaceutical Industry a Myth?'. *Drug Discovery Today* 10(15), 1031–1039.
Schmid, E. F. and D. A. Smith (2006). 'R&D Technology Investments: Misguided and Expensive or a Better Way to Discover Medicines?'. *Drug Discovery Today* 11(17–18), 775–784.
Senge, P. M. (1990). *The Fifth Discipline: The Art of the Learning Organization*. New York: Doubleday-Currency.
Sirois, S. and L. M. Cloutier (2008). 'Needed: System Dynamics for the Drug Discovery Process'. *Drug Discovery Today* 13(15–16), 708–715.
Sloan, B. (2001). *Developing the Linkage between Policy and Innovation Measurement*, Brussels: European Commission, DG Research.
Sterman, J. D. (2000). *Business Dynamics: Systems Thinking and Modeling for a Complex World*. New York: Irwin/McGraw-Hill.
Thomassin, P. J. and L. M. Cloutier. (2001). 'Informational Requirements and the Regulatory Process of Agricultural Biotechnology'. *Journal of Economic Issues* 35, 323–333.
van Dyck, W. and P. M. Allen. (2006). 'Pharmaceutical Discovery as a Complex System of Decisions: The Case of Front-loaded Experimentation'. *E:CO Emergence: Complexity and Organization* 8(3), 40–56.
Waller, C. L., A. Shah and M. Nolte. (2007). 'Strategies to Support Drug Discovery through Integration of Systems and Data'. *Drug Discovery Today* 12(15–16), 634–639.
World Health Organization Commission on Intellectual Property Rights, Innovation and Public Health (WHO) (2006). Public Health, Innovation and Intellectual Property Rights Geneva: WHO, http://www.who.int/intellectualproperty/report/en/.

PART IV

Beyond patent length

Introduction

Amy J. Glass and Fabricio X. Nunez

Patents are often the first thing that comes to mind when thinking about how to protect intellectual property rights and encourage innovation. But policy options exist for stimulating innovation other than lengthening the duration of patents to generate larger rewards. A broader perspective that explores other means of increasing innovation is needed in order for one to know what options are available. This part is organized to encourage discussion of how to stimulate biotechnology innovation by means *beyond patent length*. Our hope is that these chapters will help encourage thought and study about using less conventional means of boosting innovation.

With much attention focused on harmonizing patent length across countries, gains from other policy options could be overlooked. While promoting innovation is a key goal, other objectives such as improving diffusion of technologies and access to new technologies are also important. Using a wider range of instruments achieves a larger number of goals. With more policy levers, a faster rate of innovation could be achieved for any specified level of another goal. Expanded policy options might achieve a better outcome in terms of multiple policy goals.

Industries have differing characteristics that influence how best to promote innovation. What stimulates innovation in one industry may have little impact in another. Some industries patent frequently; others almost never. Increasing patent length will not succeed in stimulating innovation in industries that do not patent. A wider mix of policy options may also generate a better distribution of innovation across industries.

This part covers a disparate set of approaches: international economics, international political economy, and comparative institutional economics and legal perspectives. Economics, politics, institutions and the law all interact in determining intellectual property (IP) systems and the impact of IP systems on innovation. Thus, we sought viewpoints from each of these perspectives on the common theme of how to stimulate biotechnology innovation.

From an international economics perspective, in the first chapter, Alan Isaac and Walter Park examine whether an open source approach could be used to stimulate innovation in biotechnology. To assess the likely strengths and weaknesses of an open source approach in biotechnology as a means of stimulating innovation, they begin by considering the existing IP system to determine what failures of the current system we might hope to improve upon. Critiques of the current IP policy applied to biotechnology argue that it may actually deter innovation through patent thickets, fragmented rights and transaction costs, given the sequential and cumulative nature of the research and innovation process in the sector. The critiques also point out that the proprietary system places a disproportionate burden on developing countries by denying them access to biotechnologies that can address some of the most pressing issues in the developing world including pharmaceuticals, food security and concerns about biopiracy. However, the alternatives that have been proposed to the current IP policy (like plant breeders' rights and *sui generis* modalities of protection) do not address the fundamental characteristic of the system, namely, a proprietary mechanism based on exclusion. Isaac and Park ask if a non-proprietary approach, such as the open development system, could improve upon the current IP system by fostering innovation and by promoting diffusion of technology to poor countries.

They discuss three varieties of open development possibilities and then explore the relevance of those strategies to modern biotechnology by comparing similarities and differences between the software and the biotechnology industries. They examine some of the underlying principles that could make open source work for biotechnology. They also highlight some concerns associated with the application of those strategies in some areas of biotechnology.

Open development has proven to be quite successful in the software industry, and Isaac and Park argue that there exist multiple similarities between biotechnology and the software sectors. Both sectors are characterized by cumulative and sequential innovations that are mostly research tools. In both sectors, it is quite difficult to distinguish pure from applied R&D, which in some cases has led to the risk that protecting basic research could stifle future innovations.

The authors also warn that there are differences between the two sectors that may render the open development mode (what they call 'open bio') inappropriate for biotechnology. For example, software is protected by copyrights while biotechnology is protected by patents, which are shorter and more expensive than copyrights. The authors warn that there is some reason to believe that the case of software may be particular, and that an open development system could actually reduce innovation in other industries.

The application of open bio in developing countries requires special attention, especially issues relating to neglected diseases and biopiracy.

Park and Isaac differentiate between possibly increased access to data and research tools and (less likely) increased development of targeted biopharmaceutical innovations needed in those countries. The open bio alternative could allow developing countries to obtain access technologies needed to create a body of basic R&D that would make development of commercial innovations like tropical medicines profitable. Moreover, the open bio alternative could address the complaints of biopiracy that developing countries have against developed countries by allowing developing countries to enhance their capabilities for indigenous innovation and research, and to level the playing field against firms from the developed world.

The open development system could allow developing countries to jumpstart their own indigenous innovation efforts. However, Isaac and Park emphasize that an open bio mode is a complement, not a substitute, for government supported research programs. As for innovation incentives, the conclusions are even less certain. It is not that clear whether open bio would be an improvement over the proprietary system in fostering innovation.

From an international political economy perspective, in the second chapter Chris May addresses the implications of increasing protection of intellectual property in the biotechnology industry. He argues that political power, economic interests and philosophical uncertainty make the development of policies of intellectual property more political and less technical than the traditional analysis of innovation versus market power acknowledges. He makes the case that the trend to expand patent scope will continue due to industry lobbies at the national and international levels. He argues that the TRIPs agreement is only the beginning of an effort which intends to harmonize and globalize a patent regime which will benefit private interests over the common good.

May finds this expanded and globalized patent regime particularly troublesome for biotechnology because of the specific characteristics of the industry, namely, the difficulty in separating basic research from applied research. At the international level, a global policy regime would constrain nations, particularly developing countries, to choose policies that may not reflect their own preferences and needs. This, in turn, creates distributional consequences biased against developing countries that must adopt policies that may not be in their own best interest.

May views the patent system as not working to balance the public and the private interests and to protect the public realm. For example, there is a strong tendency to grant numerous patents and to assume that the judicial process is there to repair mistakes, without considering associated costs. He fears that patents are transforming scientific endeavors into technology and commercial industrialization, and that this transformation can create problems like slowing down research and creating uncertainty related to risks of litigation. He explores areas of action to reform global IP such as improving patent quality, reducing uncertainly, controlling costs, setting up a pre-grant process of opposition, and establishing fair-use measures.

May argues that the current policy debate is centered on the technical discussion of balancing innovation and market power, which he believes to be inappropriate. He argues that the proper policy making approach should emphasize the political nature that permeates the debate on intellectual property. He proposes adopting an approach that stresses the multidimensionality of the policy debate, taking into account moral, distributional, technological, environmental issues simultaneously.

This policy making approach, May argues, is a viable alternative for developing countries as it would not force them to support the current patent system. Moreover, he argues that such a multidimensional approach also addresses the needs of developed countries, and, thus, it is an alternative to a global governance of the patent regime that truly looks after the global common good. May suggests that the issues related to biotech innovation and IP be put within the broader context of the organization of science in modern societies, and that the opposition between social forces and the commodification movement be addressed at national and the global levels.

From a comparative institutional economics and legal perspective, in the third chapter Scott Kieff examines a role of IP institutions too often neglected: coordination. He argues that intellectual property could help coordinate among users to ensure the best use of assets (with easier negotiation, improved diversity and socialization) but that it was often misused as an incentive mechanism or as a way to reduce transaction costs and problems related to monopolies. He explains how those latter problems are not always solved by IP but instead can give rise to liability rules and enhanced antitrust regulation. He then compares coordination with other IP goals, some related to externalities, rent dissipation and direct incentive, and also contrasts property rights' coordination effects with those of alternative institutions. He then assesses the coordination role of IP with elements of general property theory. For example, he assesses the compatibility of the coordination function of IP with the commercialization theory, highlighting many similarities, for instance with regard to coordination of complementary and non-competing users. He also suggests some implications of this approach for further research, in particular for research tool innovations, and provides some preliminary thoughts on the implications that this coordination role could have on the development agenda of developing countries and for local institutions.

These chapters raise several important issues, not the least of which is the appropriateness of using open source in the context of biotechnology. There are differences between the biotechnology industry and the software industry. In the software industry, open source is mostly used (and is particularly useful) for disclosing the underlying technology: the source code. The situation is different in the biotechnology sector where disclosure happens early on, at the time of applying for a patent. Open source would therefore mostly be used to increase access to already disclosed technology. This point emphasizes the different purposes of open source in different sectors and the need to distinguish between them.

Another issue is the importance of the public sector in research in relation to the open source movement. Indeed, it might be important and useful to get the public sector more involved in the open source movement. However, a concern can be raised that the growing tendency of public institutions to get involved in commercial endeavours and to license their technology to private enterprises could interfere with this purpose. Public institutions are less likely to advocate open source approaches if they are interested in pursuing payoffs from commercializing discoveries based on public R&D they funded. In this respect, the connection between the public sector and the open source movement should be explored further.

These chapters also raise important issues about the different modes to assess IP in biotechnology and the tension between the public and the private. One can note the many levels and types of information involved, highlighting the complexity of the different structures present. While representing one option amongst many others, property rights should be given proper consideration and assessed taking the other approaches into consideration. Along with the need to keep them as an option, property rights have a complex mix of costs and benefits relative to other structures.

The pattern of research output has a large random component. Often, one person discovers a research technique that turns out to be useful elsewhere. Property rights are needed to reduce the transaction costs. Otherwise, there is excess duplication of effort and insufficient gain from specialization according to talent and ability. If two steps are needed, and someone has already discovered the first step and someone else the second, the two need to be able to get together and contract without cost barriers. It would be great if one person discovered both steps, but it does not always happen that way. Transaction costs are everywhere, even related to status and other rewards to innovation such as publications. It is important to keep markets from getting too concentrated. Thick markets, with many buyers and sellers, typically operate more efficiently than thin markets: a greater number of potentially mutually beneficial exchanges are realized.

Finally, there remains the issue of what consideration should be awarded to wealth distribution when assessing the patent system. Inequitable wealth distribution makes a patent system less appealing as a means of encouraging innovation. It can be pointed out that in reality there is no guarantee that people have enough resources to participate in transactions over patented products and that this fact puts pressure on the political system for alternatives. Concern over wealth distribution is stronger in Europe than in the United States for cultural reasons. However, due to the absence of criminal sanction in the US patent system, patentees often allow a vast amount of infringement, especially when the infringing parties lack the funds to pay or transaction costs are prohibitive. In either case, the transaction would not have occurred anyway so the infringement does not cause direct harm to the patentee's profit.

10. Open development: is the 'open source' analogy relevant to biotechnology?

Alan G. Isaac and Walter G. Park

INTRODUCTION

During the past quarter century, innovation and growth have characterized the biotechnology industries. At the same time, multinational agreements have strengthened and harmonized global intellectual property standards. The link between these two developments is the subject of much controversy. Did economic growth occur because of the growth of IPRs? Did economic growth occur despite the growth of IPRs? Or was the growth of IPRs an institutional manifestation of the economic interests concomitant to economic growth?

Academic research and public policy discussions during the 1990s explored many concerns about the increased scope and global reach of IPRs, including IPRs in biotechnology. A spectre of IPRs is haunting the biotech industry, the critics say. Despite the rapid growth of the biotechnology industries, one common concern is that the proliferation of IPRs may raise the costs of innovation and thereby slow technological progress. A second concern is distributional: the increasing global scope of IPRs appears to disadvantage developing countries, which accede to a regime of global IPR harmonization without possessing the IPR riches of the developed countries. A related concern is that the assertion of IPRs over the genetic resources of developing countries may constitute a kind of 'biopiracy' by developed countries. Such concerns are entangled with a concern that proprietary rights (such as patents) may be inappropriate in the field of biotechnology, where innovations may be mere discoveries and where substantial public-sector research funding can make it difficult to determine the substantial contributions of private agents.

Claims to biotechnological innovations are currently asserted as patent rights, plant breeders' rights, trade secrecy, trademarks, and to a lesser extent copyrights. The present system has many defenders, but it also has vocal critics who call for reform. Proposals for reform include shifting the mix of proprietary instruments (for example, emphasizing plant breeders' rights instead of patents) and creating new specialized forms of protection (for example, '*sui generis*' provisions). Such responses leave unquestioned the traditional proprietary model of innovation. In this chapter, we explore an alternative: we consider a non-proprietary mode of innovation, known as 'open development'. While open development relies on contemporary legal institutions and does assert some limited IPRs, the open development approach to innovation is distinct.

An open development movement recently has emerged in biotechnology. We call this the OpenBio movement. (In some circles, it is known as 'open source biotech'.) Some participants in this movement have consciously emulated the free and open source software (FOSS) movement. This emulation often appears quite natural: both software and biotechnology are emerging fields of study, and there are some important parallels between the two fields.[1] Indeed, the open development movement in some areas of biotechnology, such as computational biology, is largely an extension of the FOSS movement. Our goal in this chapter is to expand interest and encourage further inquiry into the OpenBio approach to biotech research and innovation. We consider the approach broadly, including open licensing schemes, open innovation communities, open standards, and even open business models at the product market end. We address questions about whether these approaches conform with recognized needs in the biotechnology industry and in developing countries and what economic effects of open development on innovation and economic development might be expected. The current state of knowledge renders answers to such questions highly speculative, so we also suggest research strategies that might lead to more definitive answers.

Both the OpenBio and FOSS movements are reactions to the proliferation of IPRs and to concerns that IPRs may restrict research and access to new innovations. These concerns stem from a similar basis: both software and biotechnological innovation are often cumulative and sequential, and innovations in both areas often constitute research tools. Moreover, in both areas some observers have claimed that IPRs are too often granted for inappropriate subject matter: pre-existing art, or pure science, or even pure mathematics. For example, in software, certain innovations appear to be essentially mathematical algorithms, and in biotech certain innovations appear to be essentially scientific discoveries. In both fields, it is sometimes difficult to differentiate between basic and applied research and development (R&D) (Stokes 1997). There are also important differences between the two industries. These differences determine the extent to which lessons from the FOSS movement are applicable to the OpenBio movement. In particular, we are interested in differences that matter for research, innovation and economic development.

This chapter is organized as follows. First we provide a very brief review of the IP framework relevant to biotech research and development. In the process, we briefly introduce some controversies related to IPRs in biotechnology. We then discuss the open development alternative and provide examples of open development in the biotech industries. Along the way, we review some economic principles underlying open development and innovation. Finally, we explore some implications of the OpenBio movement for developing countries.

BIOTECHNOLOGY AND INTELLECTUAL PROPERTY

This chapter uses a broad definition of 'biotechnology' following the UN Convention on Biological Diversity (CBD): 'any technological application that uses biological systems, living organisms, or derivatives thereof, to make or modify products or processes for specific use'. Ambiguities in this definition (for example, in the meaning of 'technological application') will not be important for our purposes. We wish to explore the role of intellectual property institutions in supporting or restraining biotechnological innovation and developing country access to such innovation.

The Green Revolution typifies early agricultural biotechnology innovation in an important way: the public sector and non-governmental organizations were heavily involved in its development. In contrast, industrial biotechnology consistently has been centered in the private sector. This distinction was never absolute and may no longer be tenable: large multinational firms are heavily involved in agricultural biotechnology, and in the fast evolving area of genetically modified organisms there have been complaints that public sector participation is largely missing in some important negotiations (such as the Cartagena Protocol on Biosafety).

The predominance of the developed-world private sector, and also its assertion of private intellectual property rights in innovations based on developing country resources and knowledge, has generated substantial international conflict. Developing countries have been seen as 'gene-rich' in their biodiversity, and even rich in traditional knowledge of potential therapeutic agents, but firms from developed countries possess the technological know-how and financial resources needed to bring innovations to market and to establish contemporary intellectual property rights in these innovations.

Intellectual Property Background

The past century has seen radical shifts in intellectual property institutions, and the creation of new kinds of property rights has been particularly striking in biotechnology. The US led the way to biotech patents with the United States Plant Patent Act of 1930. Another major shift occurred in 1980, when the US Supreme Court narrowly ruled (in *Diamond v. Chakrabarty*) that living organisms can be patented. Patent filings in biotechnology began a steady rise. Europe has been slower to allow patenting life,[2] but since the late 1990s European biotech filings have also risen steadily.

Until 1980, academe had little patent presence in biotechnology despite having a large research presence. Academe contributed to the common pool of knowledge through publication in peer reviewed journals. In the US this was partly, perhaps largely, due to issues surrounding the ownership and control of patents generated by federally funded research: in particular, universities had little ability to offer exclusive licensing of government funded innovations. The Bayh-Dole Act (Public Law 96–517) changed that: signed in 1980 and subsequently amended to encourage university patenting and exclusive licensing, it has led to an increase in university biotechnology patents.[3] Some observers now express concern that journals are being supplanted by patents as vehicles of knowledge dissemination in the area of biotechnology (NRC 2003). The implications for researchers are profound. As an example, consider Cornell University's particle delivery system ('gene gun'), developed in the 1980s and then exclusively licensed to DuPont. Despite its academic origins, it is not a freely available tool of research. American Cyanamid claims that lack of access to this tool substantially delayed their development of herbicide tolerant crops (Pray and Naseem 2005). The swell of patenting in the biotech industry has raised fears that such stories of delayed development will become increasingly frequent.

Proprietary Lifeforms

International agreements and organizations have established the mechanisms by which rights to biotechnological inventions are protected, including trade secrets, plant breeders' rights (PBRs) or plant variety protection (PVP), and patents. We briefly discuss the relationship between PBRs and patents, and then discuss patents in more detail.

The WTO's Agreement on Trade-Related Aspects of Intellectual Property Rights (TRIPS) was negotiated in the 1986–1994 Uruguay Round. TRIPS was a far-reaching agreement on various aspects of IPRs (including copyrights, patents, trademarks, geographical indications, trade secrets, industrial designs, and semi-conductor layouts). It was an effort to harmonize and internationalize intellectual property rules. The TRIPS agreement allows countries to exclude plants and animals from patenting, but then obliges them in that case to offer some other form of protection (essentially PBRs of some kind, possibly country specific). TRIPS did not formalize protections of traditional knowledge and indigenous genetic material: how to ensure these remain available to the country of origin has not been resolved. Currently the core international understanding of PBRs is governed by the International Union For The Protection Of New Varieties Of Plants (UPOV).[4] Like patents under TRIPS, PBRs under the UPOV Convention generate proprietary rights (to exclude others) for at least 20 years (25 years for trees and vines).

Patents are generally more expensive to obtain (to file, translate and litigate) than PBRs. Even so, the cost of obtaining a UPOV authorized Breeders' Right certificate in developing countries is expected by some observers to exclude all but the largest seed companies (Sahai 1999).[5] The most economically important differences between patents and PBRs are the criteria for grant. Patents are granted if inventions are judged to be novel, non-obvious and industrially applicable (that is, the invention has utility). PBRs are granted if the protected organism is distinct (compared to previous varieties) and has never before been commercialized. Note that patents are not supposed to be granted for discoveries of substances found in nature, but PBRs may be granted for discoveries (of things in the wild).

Another crucial distinction is that patent grants (in principle, at least) have an enablement requirement. The patent application must allow someone ordinarily skilled in the art to replicate the invention. Enablement with biotechnology can be tricky: replication may not be ensured (for example, with mutation). It may also be that, lacking a specimen, third parties are not able replicate an invention. Hence, often it is necessary to deposit biological materials with a designated center. The 1977/1980 Budapest Treaty on the International Recognition of the Deposit of Microorganisms for the Purposes of Patent Procedure requires that the deposited materials be open to the public. In contrast, PBRs do not require that a third party be able to 'repeat the invention'. All that is required is that there be uniformity and stability in reproduction. Moreover, there is no public deposit requirement: the seeds or samples are held in confidence.

PBRs may provide for breeders' and farmers' exemption. For example, under the UPOV Convention, a country may elect to leave farmers free to use the saved seeds of a protected plant for their own use, so long as they do not sell them. More generally, a protected variety can be used without permission or royalties as a starting point for breeding other distinct varieties. Under patents, such actions may constitute infringement.[6]

Article 27.3b of TRIPS does allow countries to exclude from patentability 'plants and animals', as well as essentially biological processes other than microorganisms and microbiological processes. Here 'essentially biological' means rooted in processes occurring naturally in nature (not implemented by the scientist). The term 'microorganism' is ambiguous: it could mean any microscopic organism, or only a unicellular organism. While Article 27.3b allows countries the right to exclude plants and animals from patentability, it does not prohibit them from allowing it, as for example the US and Japan have chosen to do.

FUNCTIONS OF PATENTS AND PBRs

We have seen that patents and PBRs share many characteristics. We turn now to some of the anticipated economic effects of such IPRs. Since my use of your invention does not reduce your ability to use it, we say that knowledge is non-rivalrous in consumption. If in addition you cannot exclude me from using your invention, we say that the knowledge is non-excludable. Public goods are characterized by non-rivalry and non-exclusion, so in this sense knowledge can be a public good. But in a society that creates intellectual property rights, knowledge may become excludable. Patents and PBRs are designed to transform a public good into a club good. Club goods are characterized by non-rivalry and exclusion, so in this sense knowledge can become a club good. From a public policy perspective, exclusion may mean that a club good is under-utilized. Ideally, it might be argued, society benefits most when knowledge is non-rivalrous; it is natural to explore the relative merits of creating excludability. The conventional view is that there are trade-offs between the dynamic benefits and static costs.

Fostering Innovation

Biotechnology firms make extensive use of IPRs, especially patents. This suggests that these firms find it profitable to claim IPRs, but it does not imply that this activity is socially beneficial. Public justifications of strong IPRs generally presume that the creation of such rights will promote innovation and commercialization. Ideally, the dynamic public benefit of bringing additional innovation to market will outweigh the static costs of granting rights of excludability. Theoretical and applied work by economists indicates that this ideal need not always be achieved, offering important qualifications to the standard understanding of innovation.

From a public policy perspective, one view of intellectual property regimes is that they attempt to make sure that socially profitable activities are commercially profitable, so that private individuals will pursue them. Trade secrets, patents and PBRs increase the economic reward to innovation by limiting the free access of competitors to the fruits of invention. Patents and PBRs accomplish this by granting temporary rights of exclusion. Since patent applications are public documents, patents may additionally speed knowledge diffusion to the extent that the innovation is truly novel and the patent application truly enables others to understand and implement the innovation. Such benefits clearly depend on the quality of patent institutions.

Polanvyi (1944) famously argues that 'pioneer enterprises should in general be protected against free competition', and certainly many research activities of biotechnology firms qualify as pioneer enterprises. The basic intuition is simple when large sunk costs are required to realize important innovations that might easily be copied *ex post*. Additionally, Polanvyi refers to a 'strong presumption' that patent protection is required to secure profitability adequate to justify both the research investment and venture capital, especially when we consider innovation occurring outside established firms.

In other circumstances, such justifications of IP protection fail. When innovations and their commercialization require small sunk costs to achieve, involve little novelty, and are likely to be independently produced in the absence of IP protection, then IP protection may work against the public purpose. Modern IP regimes, especially patent regimes, have been heavily criticized by some as failing to protect the public interest. Some observers believe that extensive IPRs are being granted for innovations that are not novel or are obvious to anyone skilled in the relevant arts. Such critics have been especially vocal in software manufacturing, where prominent computer scientists (notably Donald Knuth and Richard Stallman) have claimed that patents have been granted on programming techniques comparable to undergraduate homework assignments. Similar criticisms are now being raised concerning biotech patents.

Constraining Innovation

The tale of the mid-19th century dyestuff industry appears as a cautionary tale in the discussion of patents and innovation: vigorous patent enforcement in Britain, home to the initial innovations, appears to have stifled the industry, which in contrast grew explosively and prospered under Germany's looser IP laws (Murmann 2003). Some studies of the contemporary software industry sound a similar cautionary note. In an empirical examination of software manufacturing, Bessen and Hunt (2003) find that software patenting activity can substitute for firm innovation effort. They argue that the predominant use of software patents appears related to strategic 'patent thicket' behavior, rather than being a means to protect R&D investments. Naturally such results call into question the role of software patents in bioinformatics. More generally, these results raise serious questions about the role of strong IPRs in any industry that shares key characteristics with software manufacturing. For example, Bessen and Maskin (2000) show that strong patent protection can reduce innovation in industries where innovation is sequential and complementary, that is, later innovations rely on earlier innovations to be practiced. While Bessen and Maskin focused on IT related industries, sequential and complementary innovation appears characteristic of many biotech research efforts. Indeed, survey evidence finds that many industry researchers in the biological sciences have delayed, changed or abandoned research projects due to complex licensing negotiations over necessary technologies (Hansen, Brewster and Asher 2005).

Patent Thickets

Rai (2004) observes that 'large pharmaceutical firms – once vertically integrated engines of innovation – must now negotiate a complex array of university and small firm proprietary claims on research inputs', some of which are subject to exclusive licenses. Economists refer to the need for such negotiations under the general rubric of 'transactions costs'. When the transactions costs associated with overlapping and dispersed IPRs begin to constrain innovation, we refer to a 'thicket' of IPRs. The literature on such constraints has focused on patent thickets.

More generally, when innovation is cumulative and sequential, multiple parties may hold overlapping and/or fragmented IPRs to the different components necessary to constitute a larger innovation. The resulting transaction costs are potentially innovation reducing (Isaac and Park, 2004). Inventors reduce their level of effort, knowing that in the future they will face these costs. These effects are worse when the extent of the patent thicket is unknown, since an inventor who sinks costs into innovation will be in a weaker negotiating position with the possessors of relevant IPRs. Indeed, 'blocking' may occur if the IP holder refuses to license or demands a high royalty.

Potrykus and Beyer's 'golden rice' is a well known example. A biotech innovation of the mid-1990s, golden rice produces beta-keratin, and thus could potentially mitigate deadly vitamin-A deficiency in millions of children in rice-consuming developing countries. Striving to bring this product to the developing world, Potrykus learned that more than two dozen different biotech companies claimed patents on the technologies used to create golden rice (Piore 2003). In this case the thicket was successfully penetrated, and eventually the primary patent holder agreed to offer golden rice seed freely to small farmers in developing countries. Unfortunately, the mechanism for cutting through the patent thicket may not be replicable: it appears to reflect a pressing need felt by biotechnology firms to garner some favorable press.

Research Tools

IPRs on core research tools may be particularly problematic. In the commercial sector, they may increase transaction costs on a wide range of innovators. Academic researchers may also be affected, as discussed below.

Heller and Eisenberg (1998) explore how the patenting of research tools can be innovation reducing. As we discuss in greater detail below, they consider ESTs (expressed sequence tags) an important tool in genome research. In contrast, Walsh *et al.* (2003) find in survey evidence that an increase in patents on research tools important for drug discovery has generally not slowed innovation in existing projects. (Projects that failed to emerge due to research tool patents are of course not in their sample, potentially biasing their results.) They note an important exception to their core finding: patented genetic diagnostics appear to be constraining university research efforts.

Overall, Walsh *et al.* find that transactions costs have usually been manageable rather than prohibitive. But some cost-reducing factors may decline in effectiveness. For example, some firms went off-shore, where proprietary rights could not be claimed. Increasing international homogenization of IPRs will reduce the effectiveness of this strategy. Additionally, academic researchers sometimes claimed a 'research exemption' and used proprietary tools without seeking licenses. In the US, a successful 2003 patent infringement lawsuit against Duke University (*Madey v. Duke*) suggests that the 'research exemption' strategy will see declining use in the US However European and Japanese patent law is more inclined to protect the unrestricted access of academic researchers to patented tools of research (Eisenberg 2003).

The Anti-commons

If IPR holders are blocking one another, their technologies may be underutilized. This blocking may be intentional, or it may simply represent the high transaction costs of negotiating with diverse IP holders. In either circumstance, useful technology may never be developed, and extant innovations may remain unexploited. The term 'anti-commons' is meant to capture the role of strong property rights in producing an under-use of knowledge when IPRs transform it into a club good by creating rights of exclusion. The term is chosen to contrast with the theory that real property held in common (the commons) will be overused in the presence of inadequate property rights.[7]

Murray and Stern (2005) broke new ground by offering modest empirical support for the anti-commons thesis: they look for citation frequency declines for scientific publications (in one journal, *Nature Biotechnology*) after patents are granted in the innovations described in the publications. They find a decline relative to other papers, whether or not the patented knowledge involves a research tool. Unfortunately citation rates are an extremely indirect measure of knowledge diffusion, and the authors consider only scientific citations, not patent citations. They also did not address lags in scientific publications (for example, due to the refereeing process), which might be rectified by adding working papers to their analysis. Nevertheless, their results are suggestive and should be followed up with more empirical research.

Heller and Eisenberg (1998) wrote what is probably the most widely cited paper on the anti-commons in biotechnology. They discuss the implications of expressed sequence tag (EST) patenting. ESTs are considered an important research tool in gene discovery and sequence identification. In 1991, a group of National Institutes of Health researchers attempted to patent a large collection of ESTs, but they abandoned this attempted after encountering resistance from the United States Trademark and Patent Office (USPTO). Many human ESTs are now contained in the freely accessible 100 gigabase GenBank database of sequence data. Many scientists opposed allowing EST patents (HUGO 2000). Industry however has tended to support EST patents,[8] and the USPTO has taken a series of positions that appear in principle to allow EST patents. Nevertheless in 2005, Monsanto's assertion (making analogy to the microscope) that ESTs should be patentable as a research tool was rejected for failure to identify the specific utility of individual ESTs. If we accept the arguments of Heller and Eisenberg, this decision appears to erect an extremely thin barrier against the anti-commons in this area of biotech research and development.

VARIETIES OF OPENNESS

A recent movement in biotechnology is that of open development, which explicitly borrows ideas from the free and open source software movement. We call this the OpenBio movement. This section considers a variety of open development possibilities, relates them to the free and open source software movement, and explores the relevance to modern biotech enterprises. We consider open standards, open development, and open source.

Open Standards

Standards are pervasive in any complex industry, and biotech is no exception. Commercial and government standards can affect every aspect of biotech research, production and sale. These standards may be regulatory requirements, commercial contractual requirements, or simply industry practices. We illustrate with a few quick examples.

A multitude of government agencies and intergovernmental organizations are heavily involved in standards setting. At the international level, the Codex Alimentarius Commission (CAC, the UN's food standards agency) promotes food safety standards affecting genetically modified (GM) food. The World Organization for Animal Health (OIE) has a biological standards commission which, like the CAC, is involved in antimicrobial resistance testing standards. The International Cooperation on Harmonization of Technical Requirements for Registration of Veterinary Medicinal Products (VICH) provides internationally harmonized guidance for the stability testing of new biotech veterinary medicinal products.

In this chapter, however, we are primarily interested in standards that directly facilitate the coordination of innovation efforts. Technology standards promote interoperability, even in the absence of direct contact between developers. For example, Standard Reference Materials (such as those purveyed by the NIST) support efforts to achieve accurate, reliable DNA devices in pharmaceutical and forensic laboratories.

Corporations and non-profits are involved in standards development, sometimes with government funding. Many biotechnology standards are focused on various kinds of data exchange. For example, the Clinical Data Interchange Standards Consortium (CDISC) develops industry standards to promote medical and biopharmaceutical product development. CDISC has dozens of corporate sponsors. A smaller scale but no less interesting example is the Open Bioinformatics Foundation (OBF), a volunteer organization focused on supporting open source programming projects in bioinformatics. Supported projects include BioMOBY and BioDAS. BioMOBY is a small, grant-funded open source project dedicated to the creation of standards and tools for the registration and exchange of biological data stored on multiple hosts. BioDAS supports the development of an open source distributed annotation system (DAS) for exchanging and collating annotations on genomic sequence data.

Technology standards committed to transparent and reasonable licensing terms can reduce patent thicket problems. Unfortunately, as Shapiro (2000) stresses, *ex ante* price setting may attract the attention of anti-trust enforcement, so standards setting bodies may need accommodation from the anti-trust institutions. Such anti-trust concerns are diminished when standards are free and open.

Whether standards emerge via formal or informal industry processes, the intellectual property implications are immediate: industry adoption of a standard valorizes patents essential to implementation of the standard. A patent holder may profit directly through increased licensing fees or indirectly by refusing to license competitors. It is natural that industry groups in standards development organizations (SDOs) will struggle to reconcile the interests of intellectual property owners with the interests of others who wish to practice the standard. This suggests that SDOs should require participants to disclose any patent interests in the standard. It also seems natural that SDO participants should agree as well to license all patents essential to compliance with the standard on 'fair, reasonable, and non-discriminatory' terms. In the absence of such safeguard, an SDO risks that its standards will be 'captured' by a strategic member.

Free and open standards diminish the risk of such capture. A standard is called 'free and open' when any party is licensed to read and implement it without payment. This is usually understood to imply an open standards setting process, structured to circumscribe the market power of specific vendors or groups. Free and open (FO) standards are clearly non-discriminatory and also clearly set *ex ante* price limitations. Such open standards can benefit all industry participants, but the benefits appear especially great for new entrants with small or non-existent IP portfolios. The incentives of established firms with substantial IP portfolios actively to participate in open standards bodies are less clear, and economic analyses remain incomplete.

From an economist's perspective, SDOs appear to be involved in the private provision of a public good. This raises the concern that standards will be undersupplied. Offsetting this may be a snowball effect of accepted standards: when biotech firms participate in an SDO in order to align their development processes with emerging standards, these standards movements may become dominant as the number of participants reaches a 'critical mass'. In any case, it is clear that open standards at times garner widespread industry support. We consider one example from the software industry, and one from the biotech industry.

The World Wide Web Consortium (W3C) is the preeminent open

standards body for the internet. The focus of these standards is data exchange and display. Hundreds of organizations participate in the development of interoperable technologies (specifications, guidelines, software and tools). The W3C has struggled to remain independent of specific vendors, who face incentives to co-opt an open standard through 'embrace and extend' tactics or to patent standards-related technologies. The W3C has worried publicly about 'the growing challenge that patent claims pose to the development of open standards for the Web'. To ensure the continuing openness of both the standards and their implementation, the W3C adopted a patent policy that its specifications must be implementable on a royalty-free basis. Adoption of W3C standards has been extremely widespread, and this has ensured a remarkable level of interoperability on the internet. This in turn has supported tremendous innovation, manifested as a multitude of competing, standards compliant technologies.

Biotechnology does not have a single SDO that plays a comparable role to the W3C. However there are a number of efforts to promote open standards, especially in the areas of data exchange and interoperability. Consider for example the standards developed by CDISC, which reflect its mission of improved data quality and accelerated innovation in medicine and biopharmaceuticals. CDISC explicitly commits to vendor-neutral, platform independent standards. A variety of pharmaceutical companies, biotech companies and researchers participate in developing these standards. Companies that practice these standards can evade capture by single vendors while increasing their ability to interoperate (for example, through reliable data exchange) with the rest of the industry.

Free and Open Development

A development process where innovations are shared freely rather than fenced off with IP claims is called 'free and open development'. Distribution of modified technology may be restricted to ensure that the modification also remains free and open. Enabling disclosure is public. While IP recognition may be sought and granted, the licensing to use, redistribute and modify the technology is provided gratis by the developer. So while enabled technology in the public domain is obviously free and open, patented or copyrighted technology may be as well.

Free and open (FO) development accompanying rapid innovation has been noted in many industries. This seems to challenge the presumptions of many intellectual property arguments. Rosenberg (1976) documents FO development in the machine tool industry, von Hippel (1988) in the scientific instrument industry, and Allen (1983) in the iron industry. It appears typical of FO development that users of technologies are actively involved in the innovation process. Motivations for participation are diverse. Some consider it essential to preserve core research tools in the public domain. Others are concerned about data exchange and interoperability. Generally there is no intent to preclude commercial development, although some licenses constrain actions that would render proprietary a modified version of an open development tool.

Free and open source software (FOSS) is the most famous contemporary example of FO development. Software development is considered to be free and open only if the source code is readily available and freely redistributable. There are no legal restrictions on the redistribution of the unmodified open source software to others, and in practice FOSS software has generally been available for download without charge. FO development practices in biotechnology have often been deliberately modeled on the FOSS example. This is most evident in bioinformatics, where many FO development efforts *are* FOSS development efforts. For example, computational biologists make heavy use of a public domain sequence search tool (BLAST). In addition, well known FOSS software tools from the Linux operating system to the Python scripting language often play a supportive role in bioinformatics.

Database projects have also been heavily influenced by open development paradigms. An example is the International HAP MAP project, an open collaborative database project among scientists from various countries. This project aims to map common patterns of variations in the genome. The goal is to put the completed data in the public domain. Data access policy imposes minimal constraints: 'users agree not to reduce other's access to the data, and to share the data only with others who have made the same agreement'. Patenting of subsequent discoveries is allowed 'as long as [patentees] do not prevent others from obtaining access to the project's data'.

The SNP Consortium (TSC) makes an interesting case study of the effort to cut through patent thickets. Several large pharmaceutical and technology companies joined with Wellcome Trust and academic researchers to file patent applications on single nucleotide polymorphisms (SNPs) that will be freely accessible to all. (They will not allow the patents to issue.) The SNP Consortium characterizes an SNP map as 'an important, but essentially pre-competitive, research tool'. TSC has already characterized about two million SNPs, so the potential for fragmented patent rights with associated high transactions costs for SNP dependent research was obviously high. The SNP Consortium is a successful example of open development providing tools to advance industry goals. Apparently the prospect of large transactions costs motivated private firms that normally seek proprietary rights to turn instead to free and open development.

The success of the SNP Consortium appears surprising. Each individual firm seems likely to face substantial incentives to 'defect' and patent its SNP discoveries. This will provide bargaining chips if the other firms patent, and it will gain a competitive advantage if the other firms do not patent. Explaining this success is beyond the scope of the current chapter. Our brief analysis suggests, however, that the SNP Consortium would not be a Nash equilibrium outcome in a 'one-shot' game without precommitment, so that to understand its existence we would need to consider the roles of precommitment and of the diachronic relationships among the consortium members.

Incentives

Open development is often considered a puzzle. Why do individuals select an open development model? Is it a viable business model? What are the effects on research and innovation? These are large issues that transcend the scope of this chapter, but we briefly point to a few possibilities: inducing complementary innovation, collective invention, and strategic revelation by users.

A producer may find that open development in one area can stimulate product demand or lower production costs in another area. Open development may raise the rate of innovation in product complementary to a firm's commercial enterprises: it may pay freely to reveal innovations in razors if this stimulates additional razor innovations that increase the demand for blades. Or a firm may freely reveal an innovation in order to stimulate the development of a research tool it needs in a separate commercial endeavor. The issue is one of opportunity cost: such induced innovation may prove the least cost route to the desired innovation.

Under a collective invention process, researchers freely exchange information about new techniques. They share innovations and participate in experimental and developmental projects, working on different modules and/or revising one another's contributions. Famous examples include the development of the Cleveland blast furnace for iron production and Cornish pumping engine for mines (both during the 19th century in the UK). These important innovations were not the result of explicit research and development programs. Rather, innovations emerged gradually as a by-product of capital investment and experimentation (what some economists term 'exogenous innovation'). An engineer who freely revealed his design innovations was more likely to see them implemented and thereby tested in real world conditions. This is economically similar to the case of induced research tool development. Owners may encourage their engineers freely to reveal knowledge that might lower the costs of competitors in anticipation both of their own lower cost and of a capital gain on complementary assets (for example, the value of mines). Additionally, if the industry is oligopolistic, prices may not be viewed as tightly tied to costs of production.

Traditional theories of innovation tend to be manufacturer-centric (von Hippel 2005). Yet users also contribute to innovation. Users may freely reveal information to producers in hopes of having an innovation implemented. Users may know their needs better than manufacturers. Rather than be passive, they may actively participate in innovative activity and develop solutions customized to their needs. Users of research tools often propose improvements in functionality or interface. They may rely on open revelation to encourage the adoption of innovations that might otherwise be ignored.

Open revelation is easy to imagine, and easy to find, when users are skilled professionals desiring innovation in a heavily used research tool. It is more difficult to imagine how consumers of pharmaceutical products might contribute innovations to the pharmaceutical industry. And yet a recent example suggests that even in apparently unlikely circumstances, motivated users may invent new ways to foster innovation. People suffering from orphan diseases (like cystic fibrosis) found no existing drugs or treatments perfectly matched to their needs (because the market is too small). Using the Internet to lower the cost of coordination, groups offered themselves as test subjects (*Economist* 2005). They now participate in creating clinical trials and developing databases, for example, tissue banks.

Concerns

Might open development actually reduce innovation? Consider the impact of OpenBio on innovation in research tools. OpenBio certainly promotes access to certain research tools, and it encourages user innovation that might be absent from a different development process, but it may also reduce the profits from commercial R&D in the area of research tools. The effect on innovation will depend on conflicting influences: an open innovation process may lower the cost of research tool innovation by eliminating the transaction costs of license negotiations, but the potential benefits of inventive activity may no longer include possible profits from licensing the innovation. This may be because the open innovation process uses licensing that restricts a developer's ability to commercialize improvements or modifications of an open research tool, or it may simply be that a useful innovation cannot successfully compete with free research-tool substitutes. Such effects could lower development effort in the production of research tools, which could have a net negative effect on innovation in an entire field of research.

The extremely difficult project of anticipating the impact of OpenBio on private incentives is therefore critical. Recall that it was suggested earlier that the core public policy justification of patents is that they stimulate innovation and diffusion by raising the private return to research, development and commercialization. If open development lowers this private return, growth may suffer. In a simple endogenous growth model, Saint-Paul (2003) shows that philanthropic innovation can reduce long run growth by crowding out proprietary innovation. In his model, the philanthropic innovator offers a substitute for the proprietary good, which reduces the sales and profits of the proprietor, and additionally uses up human capital that could otherwise work in the proprietary sector. However growth reduction is only one possible outcome: it depends on the parameters of the model. If philanthropic innovation lowers the cost of private innovation, private innovation may be 'crowded in' even if the proprietary sector will face a lower price.

IS THE OPEN SOURCE ANALOGY RELEVANT TO BIOTECHNOLOGY?

Attempts to harness the communitarian, licensing and organizational innovations of the free and open source software (FOSS) movement to promote biotech innovation and diffusion have been spreading. For example, the BiOS Initiative explicitly speaks of trying to 'extend the metaphor and concepts of open source software' to biotech innovation.

Analysis of these attempts has been somewhat contentious. Like Maurer *et al.* (2004), some have seen in open development a hope that underserved populations will be more effectively targeted. Like Saint-Paul (2003), others have worried that open source practices will reduce the overall rate of innovation.

Analogies Between Biotech and Software

Popular discussions of FOSS applications can leave the impression that enterprise quality software is being produced as a hobby by amateurs, perhaps even by teenage hackers. This would offer a substantial contrast to some biotech fields, where research can require a team of scientists with advanced degrees, and the credentials of scientists and engineers matter. Useful FOSS software is indeed written by individuals and small groups. However, writing enterprise quality software is generally not a trivial project, and substantial professional and corporate resources are also involved in the development of FOSS software.[9] Biotechnology projects like software projects vary in size and scope: some may need only the resources of a Luther Burbank or a Jonas Salk, while others may require the resources of Monsanto or Pfizer.

Some biotech research shares important characteristics with software development. For example, some research in computational biology focuses on algorithm development. It *is* software development. Other biotech related research focuses on facilitating data exchange, which bears analogy to the W3C efforts to develop standards for information exchange on the web. Lessons drawn from the FOSS movement are most likely to be applicable to such near neighbors. But in biotech considered more generally there are some important differences.

Disanalogies between Biotechnology and Software

Traditionally, software code has been protected by copyright, while biotech innovations have been protected by patents. Obviously this is not a hard and fast distinction, given the large role of software in bioinformatics, and given the growing use of software patents. Copyright grants rights of exclusion for a very long time, for example, life of author plus 70 years but protects only a particular expression of an innovative idea. Patents offer a shorter period of protection, for example, 20 years, or even less if the rights holder chooses not to renew his patent right but protect the innovation and not just a particular expression. Copyright is much cheaper to obtain than a patent. Copyright is automatic, while patents applications involve filing fees, lawyers' fees, translation fees, and depositing of materials, and even then the application may be rejected.

Outside bioinformatics, open development is generally not simply a matter of source code sharing. Hope (2004) notes that technical information often is not enough to convey an innovation. 'Uncodified' knowledge may be needed to understand how to practice an invention. Once we move outside bioinformatics, 'open source' is largely a metaphor for open development: sharing the underlying technological secrets or information and giving access. When a patent holder excludes, knowledge should still be disclosed (through the patent application), but permission to practice the innovation is restricted.

In particular, the 'open source' metaphor is misleading if it is applied to the 'code' in biochemical sequences, including ESTs, SNPs or even genes. If these are patented, the 'code' is fully disclosed. Indeed, the revelation of 'code' in such cases goes far beyond the standard applied in software patents, where software patents often do not have to reveal the underlying source code in order to satisfy the enablement requirement.

Another divergence of biotech from software manufacturing is that certain fields of biotechnology appear inherently capital intensive, requiring large laboratories and other physical capital. In addition, substantial sunk costs may derive from needs to comply with regulations (on health and safety) in order to get to market. Venture capitalists may have no incentive to fund biotech startups who cannot claim adequate proprietary rights in their innovations. This of course is just a contemporary application of the Polanvyi (1944) argument that 'pioneer enterprises' need protection from free competition.

Hope (2004) notes a subtlety in the capital intensity argument. While our first instinct may be that open development in software manufacturing has low capital intensity, this may be false. Of course the incremental project may be able to proceed with a low capital investment. But huge sunk costs support the current software industry: operating systems, fiber optic cables, DSL lines, wireless access points/towers, and modern computer hardware (including modern processors and memory) all reflect tremendous capital investments. Rather than presume that biotech has capital needs beyond those of the software industry, researchers need to explore the extent to which a similar supportive infrastructure may be emerging.

Finally, analogies between software manufacturing and biotechnology are most persuasive when the final products share the key characteristics of knowledge goods: an essential non-rivalry and non-excludability rooted in negligible costs of reproduction. Of course, this non-excludability may be over-ridden by IPR institutions; however the final products of biotech research are often not software or databases but rather physical substances or devices that must be produced and sold, and these products are essentially rivalrous in consumption and excludable.

Innovation and Open Development: Where Can it Work in Biotechnology?

Biomedical research appears increasingly proprietary and secretive, generating fears that future progress may be impeded by access and licensing difficulties. Some propose mitigating this by requiring researchers to offer easy access to certain types of data and research tools. Others propose that open development models, including what Rai (2004) calls 'open and collaborative' science, promise relief from potential patent thickets or problems of hold up. Finally, there is of course the possibility, highlighted by Saint-Paul (2003), that open development may reduce innovation effort by reducing anticipated profits.

Any assessment of these possibilities must be extremely tentative. However at the moment it seems reasonable to say that open development is in use and is working in a number of areas, including bioinformatics, database development and data exchange, and 'pre-competitive' sequencing efforts. Many research tools in bioinformatics are free and open software, including operating systems, scripting languages and sequencing algorithms. Database development has illustrated additional possibilities for free and open development. From an economist's perspective, one of the most fascinating developments in OpenBio is The SNP Consortium, where commercial interests were driven to open development in an effort to reduce future transactions costs.

While user innovation has been important for research tools in bioinformatics and computational biology, it is much less plausible that consumers of biopharmaceuticals will propose improvements, and perhaps scarcely conceivable that even the most sophisticated consumer will be implementing improvements. If OpenBio really has few prospects in biopharmaceutical development, many proponents will be disappointed. Yet end user innovation appears rather unlikely, modularity looks low, and the costs of safety testing and regulatory compliance are high. Prospects for OpenBio look much better as we get closer to basic R&D, projects involving platforms, or enabling technologies.

IS OPENBIO GOOD FOR DEVELOPING COUNTRIES?

Might OpenBio promote economic development more favourably than existing proprietary modes of innovation? Might it at least reduce the costs developing countries face in licensing products and research tools from the developed world? These are extremely general questions for which we highlight some specific considerations.

It is natural to suspect that developed and developing countries anticipate very different effects of an international harmonization of IPRs. Many observers expect a resulting wealth transfer from the IPR-poor South to the IPR-rich North. This transfer may be offset if stronger IP regimes stimulate increased innovation in developing countries, or perhaps even if adequately increased global innovation lowers the relative price of technology and improves the developing world's terms of trade. Currently, neither empirical nor theoretical work supports such hopes. Theoretical work suggests that stronger global IP regimes will reduce welfare in developing countries (Deardorff 1992). Recent empirical work suggests that stronger IP regimes do not stimulate technological innovation or diffusion in developing countries (Park and Lippoldt 2005).[10]

Since the innovation response to IPRs is not detectable in developing country data, these countries should be cautious about establishing stronger IPRs in their present environment. At the same time, it is difficult to pin down causality in such data. For example, perhaps the R&D response is low because IP enforcement is currently very weak. Or perhaps there is a vicious circle: the resources to create and maintain a vigorous IP regime (with an intellectual property office, specialized courts, enforcers, education and training for IP professionals) may be impractical given the level of development – even if necessary for development.

Development economists often characterize developing country innovation systems as relying on imitation and adaption (Evenson and Westphal 1997). In this case, premature enforcement of developed country IP claims could undermine the nation's chance for an indigenous biotech industry. IP enforcement may restrict opportunities to imitate and produce while simultaneously reducing access to IP inputs.

In fields where innovation is characterized by incremental, imitative and adaptive innovation, OpenBio therefore may be a suitable mode of innovation for developing countries. OpenBio may offer developing countries opportunities to imitate, learn and innovate without violating their IP agreements. In such fields, OpenBio effectively lowers the cost of entry into biotech research. Particularly in a world where there is mounting pressure for expansive IPRs asserted by the developed countries to be enforced in developing countries, open development may help foster the human capital accumulation of developing countries and thereby alleviate the problem of their shortage of resources complementary to innovation. These arguments closely parallel arguments that have been offered for developing countries to embrace free and open source software (UNCTAD 2004).

Neglected Diseases

Love (2002) argues that the patent systems in many African countries are 'relics of colonial regimes, and serve the interests of the former colonial powers better than they serve the people who live in Africa'. As an example of this bias, he notes that as of 2002 not a single African country imposed compulsory licensing on any medicine patent, despite tremendous need, especially in the treatment of AIDS. Some researchers have proposed that the OpenBio movement may aid developing countries in getting needed medicines that current institutions have denied them.

Maurer, Rai and Sali (2004) propose that open source software can provide a model for improving innovation in tropical medicine. Major tropical diseases affect large populations, but poverty offsets size in limiting market demand for new medicines. In these circumstances, the market power provided by patents may be inadequate to ensure the recovery of research costs. The result is 'neglected diseases', where treatment innovation is slow despite the large size of the affected populations.

Maurer *et al.* hope that establishing a body of basic research that cannot be patented (at least in developing countries) can lower the costs of development to the point where the commercial sector can profit from R&D on neglected diseases. They seek an alternative to subsidizing third world purchases of first world medicines or attempting to have benevolent nonprofits play a venture capitalist role. Their 'Tropical Diseases Initiative' (TDI) raises the hope that the increasing reliance of biotech on computation implies that the open source model of software innovation can be transplanted into the arena of early-phase drug discovery. Since 'open source' discoveries will not be patented, it is hoped that zero licensing fees and competitive pressures will conspire to keep prices low. Clearly the choice of license will be crucial. Maurer *et al.* wish for a key licensing restriction: 'open source' drugs cannot be patented in developing countries.

Advances in computational biology make it plausible that OpenBio can facilitate some aspects of early-phase drug discovery. Many difficulties remain in actually bringing a drug to market, however. Firms in developed countries face extraordinary regulatory costs in the introduction of new drugs. The rise of a generic drug industry in the largest developing countries may allow them to support late-phase drug discovery and even development and testing efforts. We suspect that, for now at least, large sunk costs in drug development will continue to pose a substantial barrier to OpenBio drug development.

Biopiracy

Some developing countries complain that foreign firms have been 'pirating' their genetic resources and traditional knowledge. They claim that patents are granted for products derived from the genetic resources of the South without consent of the owners of the resources, and even without the knowledge of the owners. They also claim that patents are based on the traditional knowledge of the South, in which case the putative innovation is actually common knowledge (or prior art). This is the charge of 'biopiracy'.

Modern IP regimes do not deal easily with collective and traditional knowledge. Developing countries have been clamoring for international reform in this area, to require Northern firms to obtain consent and to share the distribution of the benefits from patents. This concern is reflected in the UN Convention on Biodiversity (CBD), which however has no enforcement power. It is also reflected in the efforts by India to pre-empt biopiracy by digitally placing traditional knowledge into the public domain. Central to this effort is the Traditional Knowledge Digital Library (TKDL), under the auspices of India's National Institute of Science, Communication and Information Resources (NISCAIR).

It is natural to ask why the South has not simply used the global intellectual property rules to assert IPRs in their 'genetic' resources, which would allow them to charge Northern firms for licenses. Part of the answer lies in the current rules of the game: these can require that traditional knowledge and crops be modified before they can be protected, and developed countries currently have an advantage in producing such innovations.

Stiglitz (2004) explores additional ways in which developed countries are disadvantaged in the IP race. Applying for IP protection is expensive. The process of obtaining international protection requires expertise that is more likely to be available to a large multinational firm. It is expensive to litigate or make invalidity challenges against developed country firms. In addition, the international system is based on first to file, not first to invent. This interacts with the need for IP-system expertise: it takes time and expertise properly to draft an application, giving an advantage to a rich experienced inventor (for example, a multinational pharmaceutical firm). So even if a developing country inventor is first to invent, he may not be first to file. Such considerations raise concerns that the North will continue to amass massive patent portfolios and property claims, leaving the South with reduced access to new technologies and higher costs for technological goods, even those that are substantially derivative from the knowledge and resources of the South.

An OpenBio presence in research into indigenous crops and knowledge has the potential to reduce biopiracy. This is beneficial to developing countries which fear that foreign firms will end up with 'pirated' IP rights that constrain indigenous users. Such an OpenBio effort may also foster local capabilities for innovation. But developing countries are sure to be interested in the revenue stream that may be harvested from local biological diversity and traditional knowledge, and open development is unlikely directly to support this. Whether OpenBio will generate large complementary advantages, as discussed earlier, remains an open question.

CONCLUSION

The biotechnology industries are dynamic and innovative. There is a substantial public interest in keeping them that way. While it seems likely that

IP policies will prove important, we cannot predict whether stronger IPRs will promote biotech innovation or hinder it. Indeed, the role of IPRs in promoting innovation is likely to differ by industry and by manufacturing stage. Research activities that generate large sunk costs in the production of final consumer products appear most likely to be well served by strong IPRs, especially if the products allow easy reverse engineering. Research activity that resembles basic research or is focused on research tools appears least likely to be well served by strong IPRs, which may constrain innovation by creating 'thickets' of IPRs or by raising the costs of research tools (even to academic researchers).

Despite a growing understanding of the problems posed by overlapping and fragmented IP claims to innovation in industries characterized by sequential and cumulative innovation, the public policy momentum of the last quarter century has favored an increasingly expansive understanding of what constitutes appropriate subject matter for IP protection. Today there is little evidence of a public policy commitment to constrain an expansive biotech industry demand for IP protection. Even when industry participants would find a legislative constraint beneficial, we expect that they will often find it difficult to constrain themselves: group agreements not to assert IPRs are undermined by incentives to defect.

In an interesting development, we observe that industry occasionally overcomes this 'prisoner's dilemma'. The SNP Consortium is an outstanding example: important industry participants in gene research chose an open development model and formed a consortium to set SNPs 'off limits' to patenting efforts. While such commercially focused efforts are not yet typical of the OpenBio movement, we find them extremely promising. Still, it is important to note that the successes we observe do not imply that legislative constraints are not needed: basic economic considerations suggest that such consortia will be in 'undersupply'.

The global expansion of IPRs also raises distributional concerns. To many observers, developing countries appear disadvantaged: they are acceding to a regime of global IPR harmonization that will extract substantial payments for developed world IPRs, but they do not possess IP riches of their own. Some observers argue that this disadvantage extends even to the assertion of IPRs in their own biological riches and traditional knowledge: developed countries have been accused of a kind of 'biopiracy' as their firms race to establish IPRs in the genetic riches of the developing world.

Some proponents hope that the OpenBio movement will lessen the burden on developing countries. We agree that it is likely to increase developing country access to the data and the research tools that will be needed for indigenous biotech research efforts. Here we are speaking of lawful access: OpenBio developments are freely available to developing countries, which reduces the pressure on those countries to transgress recently harmonized IPR standards. However we are less hopeful that, on its own, OpenBio will lend much stimulus to the development of the biopharmaceutical innovations so desperately needed by the developing world. Instead, in this arena, we expect OpenBio to work best when complementary to targeted government research support. A detailed exploration of this complementarity is an important area for future research.

NOTES

1. As Dawkins (2004) writes, 'genes are software subroutines that perform cellular operations'.
2. TRIPS has accommodated only part of the European reluctance to patent life. In June 1999, the EPO decided to implement the European Union Biotech Patents Directive (98/44/EC), which allows patents on life, but still forbids plant and animal patents. This decision was controversial in Europe, and some have argued that it conflicts with provision of the European Patent Convention, Article 53, which forbids patents on plant or animal varieties.
3. History suggests an irony in the timing. The 1980 Cohen-Boyer patent on 'gene-splicing', often cited as a fundamental innovation that proved essential for the biotech industries, was licensed non-exclusively (to more than 300 licensees).
4. The UPOV is a intergovernmental organization established by the International Convention for the Protection of New Varieties of Plants (UPOV Convention). This convention restricts farmers' rights, including for example the right to save seed from protected crops for reuse.
5. Stiglitz (2004) argues that developing economies have a big disadvantage in playing the patent game against the patent players of the North (whether it is in claiming priority to a discovery or in challenging the validity of a patent grant).
6. These important exemptions may have been undermined in 1991, when UPOV was modified so that IP holders can simultaneously obtain a patent and PBR.
7. Common goods are rivalrous but non-excludable, whereas club goods are non-rivalrous but excludable.
8. For example, the International Association for the Protection of Intellectual Property (AIPPI) urged that 'ESTs, SNPs and entire genomes must be considered as patentable subject matter'.
9. Consider the heavy involvement of large corporations in the Carrier Grade Linux working group of the Open Source Development Laboratory.
10. The same work, however, does claim to find positive effects on developed country innovation and diffusion.

REFERENCES

Allen, Robert C. (1983), 'Collective Invention', *Journal of Economic Behavior and Organization*, Vol. 4, pp. 1–24.

Bessen, James and Robert M. Hunt (2003), 'An Empirical Look At Software Patents', Federal Reserve Bank of Philadelphia Working Paper No. 03–17 August.

Bessen, James and Eric Maskin (1999), 'Sequential Innovation, Patents, and Imitation', Massachusetts Institute of Technology Working Paper No. 00–01, http://www.researchoninnovation.org/patent.pdf.

Deardorff, Alan (1992), 'Welfare Effects of Global Patent Protection', *Economica*, Vol. 59, pp. 35–51.

Dawkins, Richard (2004), *The Ancestor's Tale*, New York: Houghton-Mifflin Co.

Economist (Technology Quarterly) (2005), 'Just What the Patient Ordered', 17 September, pp. 34–5.

Eisenberg, Rebecca. S. (2003), 'Patent Swords and Shields', *Science* 299, 1018–19.

Evenson, Robert and Larry Westphal, (1997). 'Technological Change and Technology Strategy', in Hollis Chenery and T.N. Srinivasan (eds), *Handbook of Development Economics*, Vol. 3, Amsterdam: North-Holland.

Hansen, Stephen, Amanda Brewster and Jana Asher (2005), 'Intellectual Property in the AAAS Scientific Community: A Descriptive Analysis of the Results of a Pilot Survey on the Effects Of Patenting on Science', American Association for the Advancement of Science, http://sippi.aaas.org/survey/Survey%20Report%20 ExecSumm.pdf.

Heller, Michael. A. and Rebecca. S. Eisenberg. (1998), 'Can Patents Deter Innovation? The Anticommons in Biomedical Research, *Science* 280, 698–701.

Hope, Janet Elizabeth. (2004), 'Open Source Biotechnology', available at http:// rsss.anu.edu.au/~janeth/OpenSourceBiotechnology27July2005.pdf.

The Human Genome Organization. (2000), 'HUGO Statement on Patenting of DNA Sequences', available at http://hugo.hgu.mrc.ac.uk/PDFs/Statement%20 on%20Patenting%20of%20DNA%20Sequences%202000.pdf.

Isaac, Alan G. and Walter G. Park, (2004), 'On Intellectual Property Rights: Patents v. Free and Open Development', Enrico Colombatto, *The Elgar Companion to the Economics of Property Rights*, Cheltenham, UK and Northampton, MA, USA: Edward Elgar.

Maurer, Stephen. M., Arti Rai and Andrej Sali, (2004), 'Finding Cures for Tropical Diseases: Is Open Source an Answer?', *PLoS Medicine* 1, 183–86.

Murmann, Johann Peter (2003), *Knowledge and Competitive Advantage: The Co-evolution of Firms, Technology and National Institutions* Cambridge: Cambridge University Press.

Murray, Fiona and Scott Stern, (2005), 'Do Formal Intellectual Property Rights Hinder The Free Flow Of Scientific Knowledge? An Empirical Test Of The Anti-commons Hypothesis' NBER Working Paper No. 11465, http://www.nber.org/papers/w11465.

National Research Council Committee on Responsibilities of Authorship in the Biological Sciences, (2003), *Sharing Publication-Related Data and Materials: Responsibilities of Authorship in the Life Sciences*, Washington, DC: National Academies Press, 2003.

Park, Walter and Douglas Lippoldt (2005), 'International Licensing and the Strengthening of Intellectual Property Rights in Developing Countries', *OECD Economic Studies*, Vol. 40, No. 1, pp. 7–48.

Piore, Adam. (2003), 'What Green Revolution?', *Newsweek* 01637053, 42.

Polanvyi, Michael (1944), 'Patent Reform', *Review of Economic Studies* 11, 61–76.

Pray, Carl E. and Anwar Naseem, (2005), 'Intellectual Property Rights on Research Tools: Incentives or Barriers to Innovation? Case Studies of Rice Genomics and Plant Transformation Technologies', *AgBioForum*, 8(2&3): 108–17.

Rai, Arti K. (2005), 'Open and Collaborative Research: A New Model for Biomedicine', in R.W. Hahn (ed.), *Intellectual Property Rights in Frontier Industries: Biotech and Software*. Washington, DC: AEI-Brookings Joint Center for Regulatory Studies, pp. 131–58.
Rosenberg, Nathan (1976), *Perspectives on Technology*, Cambridge: Cambridge University Press.
Sahai, Suman (1999), 'Protection of New Plant Varieties: A Developing Country Alternative', *Economic and Political Weekly* 34, http://www.grain.org/bio-ipr/?id=34.
Saint-Paul, Gilles (2003), 'Growth Effects of Nonproprietary Innovation', *Journal of the European Economic Association* 1, 429–39.
Shapiro, Carl (2000), 'Navigating the Patent Thicket: Cross Licenses, Patent Pools, and Standard Setting', *Innovation Policy and the Economy* 1, 119–50.
Stiglitz, Joseph E. (2004), 'Towards A Pro-development And Balanced Intellectual Property Regime', available at http://www2.gsb.columbia.edu/faculty/jstiglitz/download/2004_TOWARDS_A_PRO_DEVELOPMENT.htm.
Stokes, Donald E. (1997), *Pasteur's Quadrant: Basic Science and Technological Innovation*, Washington, DC: Brookings Institution Press.
United Nations Conference on Trade and Development (UNCTAD) (2004), Free And Open Source Software: Policy And Development Implications. Geneva: UNCTAD, http://www.unctad.org/en/docs/c3em21d2_en.pdf.
Von Hippel, Eric (1988), *Sources of Innovation*, Oxford: Oxford University Press.
Von Hippel, Eric (2005), *Democratizing Innovation*, Cambridge MA: MIT Press.
Walsh, John P., Ashish Arora and Wesley M. Cohen (2003), 'Working Through the Patent Problem', *Science* 299, 1021.

11. On the border: biotechnology, the scope of intellectual property and the dissemination of scientific benefits

Christopher May

Many commentators want to discuss the intersection of biotechnology and patents (or intellectual property more generally) as if this was a novel problem, requiring new solutions, and a new politics. However, as the politics of intellectual property has always been about making property from new techniques and knowledge (intellectual property almost by definition is the commodification[1] of innovation), this claim for historical novelty is far from conclusive. In this chapter I will suggest there are two responses that can be made to the argument that biotechnology requires a reformation of the patent system; first we can examine the possibilities for incremental 'problem solving' – reworking and renegotiating how the system deals with biotechnology 'innovations'; second, it can be taken as a question regarding the general scope of intellectual property itself. The latter argument suggests that this is a category, or border, problem; biotechnology is erroneously included in the system of patents, and many of the political, ethical and practical issues might be solved by establishing that biotechnological 'products' and techniques should not be subject to patent at all. Thus, rather than a question about how patents might better support innovation, another way through the current disputes over biotech patents would be firmly to (re)establish patent criteria to exclude biotechnological tools and bio-medical materials/products altogether.

Certainly, in many ways biotechnology is not a revolutionary technology; rather, it builds on centuries of husbandry, the accumulation of (traditional) knowledge about the manner in which the 'building blocks' of nature work together. Much of this knowledge has been consolidated within a regime of extensive public-sector support for science since the middle of the twentieth century. The development of sophisticated methods for DNA cutting and splicing during the 1970s and into the 1980s coincided with a plateau in the chemical development of pharmaceuticals.

Searching for new compounds to commercialise, and new or innovative products, companies working in agriculture, food, medicines and more generally chemicals saw in the 'biotechnology revolution' a new set of resources to commodify, exploit and develop further. While the accelerating biotechnology sector, in the first instance, hardly relied on intellectual property rights (IPRs) (not least because many innovations were produced in public laboratories), industrial interests demanded exclusive control of these new resources to exploit them profitably (Drahos and Braithwaite, 2002: 154). Hence, the increasing recognition of biotechnology's commercial potential touched off a patent 'gold rush', and the inclusion of the biotechnology industry in the pro-patent lobby both in domestic and international politics.

In the rest of this chapter, I briefly examine some questions relating to the interaction of patents and the biotechnology industry. To this end after setting out the general realm of the problem in the first section, I move to examine two specific problems: the question of 'trade-related-ness' and the scope of patents; and the more technical question of the emergence of an anti-commons in innovation and issues of incremental system change that this has raised. I conclude that we can approach the 'patent problem' in biotechnology from a practical (reformist) direction, or we can examine the more fundamental question whether the biotechnology industry needs to be included within a patent regime at all. These two approaches are not mutually exclusive, and the technical (incremental) improvements are valuable even to those seeking a more fundamental reorganisation of the sector's interaction with the (global) intellectual property system.

THE 'PATENT PROBLEM' IN BIOTECHNOLOGY

In the wake of the US Supreme Court's decision in *Diamond v. Chakrabarty* in 1980 (by a five to four majority) that genetically *modified* organisms could be patented, unsurprisingly the rate of patent applications from the emerging biotechnology industry accelerated. Although the US Patent and Trademark Office (USPTO) had originally rejected Chakrabarty's application, the Supreme Court's decision went on to have a global significance. Previously, in 1969, both the Australian and the German Patent Offices had reached a similar conclusion, but it was only the US decision that really established the legal grounds for the emergence of a strongly IPRs-oriented bio-tech sector, not least as most of the key players were based in the US (Drahos and Braithwaite, 2002: 158–159). Although the Supreme Court may have intended this decision to be read narrowly, the outcome has been that since 1980 the scope of patentable materials has been widened considerably to encompass much that might have previously been regarded as natural, and hence outside the realm of intellectual property.

Despite this legal shift, in biotechnology difficulties remain in recognising 'invention' and 'technology', key elements of the underpinning of the intellectual property system. While there are certainly elements of biotechnological advances that might be easily located in the public realm of scientific discoveries and thus not patentable (at least by historically dominant criteria), there are also advances that one could argue are potentially *only* industrial in application and at least in process-of-innovation terms might be parallel to more usual notions of patentable technologies. Distinguishing between these two types of outcomes of research might be regarded as relatively easy in an abstract sense, related to the distinction between pure and applied research. However, not only have powerful corporate interests been deployed in shifting any supposed line of patentability towards the former pole (reducing the scope of un-patentable biotechnological 'discovery'), as one approaches the borderline between these two 'ideal types', any easy distinctions collapse and political power, economic interests and philosophical uncertainty enter the debates to try and establish the favoured line of particular (political) groupings.

It should therefore be no surprise that in the last quarter century there has been a significant expansion of the scope of patentability (what can legally be granted a patent), especially in biotechnology, where new processes (and the refinement of older practices) have allowed new 'products' and technologies to be developed. Certainly this is not unprecedented; across the 500-year history of the institutionalisation of intellectual property there has been a steady (if at times contested) expansion of the criteria for awarding grants of patent rights. From their emergence as monopoly 'privileges', intellectual property generally and patents more specifically have been required to balance the needs and rewards of inventors (alongside other creators and innovators) with a more generalised social need to ensure economic development, and thus the widest dissemination of useful innovations.[2]

Two principal methods have been used by states to balance private rewards with public benefits in the institutionalisation of IPRs: first, limits on the scope of patentability and, second, the period of protection. A shorter period of protection broadly maps on to a heightened recognition of the public side of the balance (material enters a public domain of free access more quickly), while a longer period reflects the privileging of private rewards (control and thus income over an extended period). As regards patentability, however, the balance is a little more complex, and hence issues of patentability have always been a major element in the *politics* of intellectual property.

We might initially expect a similar position to obtain as regards patentabilty as it does for the length or protection, that private interests would be best served by expanding the scope of protection (patents available for a widened range of classes of technologies and innovations), whereas the public good might be best served by closely limiting the sorts of things for which patents could be awarded. However, as many industrial and/or scientific organisations utilise patented technologies or processes in their own research as inputs, it is not necessarily clear that all private interests would want to support widened patentability (Landes and Posner, 2004: 15). However, despite the suggestion that even private corporate interests should actually seek to establish limits to the scope of patentability, the actual history of IPRs has seen those representing the corporate interest push for an ever widened scope of patents. Thus, while certainly corporations have utilised intellectual property as inputs, in most cases the advantages of establishing protection for their outputs has outweighed any interest in seeking cheaper (non-commodified) inputs. This is to say the post-production control issue is more important than the pre-production input-cost issue.

Moreover, if we are to accept the utilitarian justification of intellectual property – that by establishing rewards for innovation and creativity these endeavours are encouraged – then we may also suggest that there is a public interest in awarding these rights to support 'progress' and human creativity. Here, arguments about human motivation and its relation to material rewards have engendered significant (but essentially inconclusive) arguments. It is difficult to ascertain, isolate or generalise what role patent awards may play in the motivation of innovators. Thus, although it is taken on trust that IPRs encourage valuable human activities in the realm of discovery, there is little in the way of firm proof. Nevertheless, there is a clear faith in certain powerful and influential sectors of society that the protection through patents of research outputs is a significant driver of innovation (Samuelson, 1989: 179).

Examining the history of bio-medical science it is difficult to argue that advances were stymied or constrained by a lack of opportunity to patent advances, although, conversely, one can make an easier argument about the development of commercial industrial sectors depending on the protection of intellectual property. For instance, the development of the European and US pharmaceutical industries relied to a considerable extent on the utilisation of patenting (see May and Sell, 2005: chapters 5 and 6). However, for the most part discussion of patenting in biotechnology has focussed on the relatively technical problem of how the patent system might be modified to recognise the particularities of scientific endeavour and the requirements for economic 'efficiency'.

This reflects, as James Boyle has identified, the limited range of legal discussion of intellectual property; for most patent specialists, he points out, 'claims about the effect of commodification on attitudes towards the environment, about distributive justice and the common heritage of mankind or the moral boundaries of the market are simply outside the field' (Boyle, 2003: 8). Indeed, he asks, 'Does being an intellectual property scholar offer a license to be parochial?' (Boyle, 2003: 12). Of course, for a political economist the answer here must be no! There is no sustainable argument that the legal and political arguments can be separated from one another. Again as Boyle puts it:

> It will ill-serve us if we have a double-decker structure of policy debate about [patents]: a lower deck of tumultuous popular and non-legal arguments about everything from the environment to the limits of the market, and a calm upper-deck which is once again fine-tuning the input-output table of the innovation process. We need a fusion of the two; or at least a staircase between them [Boyle, 2003: 13].

The problem is that if we focus merely on how particular forms and/or scopes of patent grants effect a notional model of innovation then the discussion remains moribund and hardly speaks to the real political debates outside legal scholarship. The economic arguments about the pros and cons of various patent arrangements smuggle into the analysis assumptions about the definition of 'efficiency' and the distribution of the rewards and benefits in society. Indeed, Boyle identifies the realm of gene patenting (and biotechnology) as a microcosm of a wider and more serious shortcoming of much scholarship on intellectual property – it is insufficiently embedded in (if it has contact at all with) the social political and economic context of social relations.

This suggests that if we are to understand the question of patents in biotechnology, we should not merely examine the biotechnology industry and how innovation has been enhanced or limited within the sector, but rather should set the problem into the wider context of how science is to be organised in contemporary society, and perhaps most importantly (and more generally) how social forces are arrayed against the movement to commodify more and more. It is in this wider argument that the work of Karl Polanyi (1944 [2001]) is frequently alluded to, and certainly there is a clear need to balance the increasing commodification of the life sciences thorough the increased penetration of biotechnological companies into once publicly controlled research, with an appeal to social embeddedness; to ensure that the market *serves* the public purpose, not that the public interest is merely a residual once all possible private rights have been exercised.

If the modes of privatisation and control of IPRs that have patterned other areas of technology take hold in the agriculture sector, for instance, there is a very real danger that small agriculture communities will find themselves living under the threat of litigation from the new rights holders to particular genetic resources they continue to utilise and share between themselves. Much of the critical commentary as regards agricultural patents (see for instance Shiva, 2001) has taken this position, noting cases such as: RiceTec Inc's patent on Texmati, a variant on the Indian staple basmati rice; Monsanto's soya and maize patents; and W.R. Grace's patent on the neem seed and its linked medical and agricultural uses.

Court challenges of politically problematic patents are distinctly possible, but this is both an expensive and time consuming process. In 1997 for instance the Indian government spent over 200,000 USD fighting the RiceTec patent on basmati ('texmati'), settling the case when a compromise, although not outright cancellation, was mediated between the litigants. However this was merely one rice-related grant among many (in 2000 alone, over 500 patents were granted relating to rice) and the cost of fighting each questionable grant is prohibitively expensive. The Indian government's Ministry of Commerce reportedly told campaigners in 2003 that there was no money available to fight further cases whatever their merits (Ramesh, 2004). Less wealthy developing countries will no doubt come to the same decision, while those campaigners in other areas (from software to pharmaceuticals) who seek to challenge patents on their own account, will need to raise large amounts of money without any real confidence of mounting a successful challenge.

The problem is that patent examination is not as stringent as it should be for balancing a public-regarding interest with the grant of private rights. While this is certainly a problem for patent grants for specific agri-biological products, when patents are granted for tools relating to (and 'built' from) bio-resources, the issues may have a different (although still serious) impact. As Adam Jaffe and Josh Lerner, recently put it, as regards biotechnology, 'if the [patent office] is doing its job, a patented research tool will be one that might have not been available at all, if the researcher who secured the patent had not developed it' (Jaffe and Lerner, 2004: 203). This is to say, if the patent system works as it we might presume it had been intended to work in some ideal moment of regulatory history, then techniques and technologies that otherwise might never have been developed will be stimulated and accelerated. This of course points us towards the mobilisation of interest and the political deliberations behind innovations and changes in policy and/or prevailing legislation.

When, in patents' early history, the question of monopoly was at the forefront of governments' concerns, early forms of examination were developed and deployed to ensure that the grants were justified by measures of specific public-regarding policy ends. However, as the policy focus has increasingly shifted towards the individual, such examination seems to have become less important than enhancing the private rights of applicants. Thus, most patent offices see their job as to issue patents (and in many cases the fees from doing so generate income for their operations); they are not overly concerned with any particular grant's relationship to the public realm. Critics have cited business method patents as particularly egregious examples of ill-conceived and misguided patent grants (Gleick, 2000). As any grant *can* be challenged in court, many patent office personnel seem to have accepted, at least implicitly, that their 'mistakes' can be tidied up through a judicial process without considering that the expense of this process severely limits its effectiveness as a mechanism for protecting the social interest in maintaining any limitations to the scope of patent.

This suggests that the key issue as regards the impact of patenting on the biotechnology sector is centred on (unsurprisingly) the question of patent scope, and how regulation in this area may respond to (global) political developments. The clear dynamic in the (global) governance of intellectual property rights is towards the expansion of patent scope and the increasingly robust inclusion of the biotechnological sector's key resources and tools within the regime of intellectual property protection.[3] In one sense it may be too late to ask the question: should the protection accorded to grants of patent be extended to the resources and tools that are utilised in this new (bio)technological sector? On the other hand, there are significant differences of opinion about how these shifts in patentability will affect scientific endeavour in the bio-medical sciences. Moreover, as Susan Sell and I have demonstrated at some length (May and Sell, 2005) there is nothing inevitable or teleological about the institutionalisation of IPRs; the current settlement is not necessarily final or fixed, and certainly continuing processes of political dispute may well shift and change the future regulatory systems for (bio)technological properties.

TRADE-RELATED-NESS AND SCARCITY

If the current settlement is not some final point of development for intellectual property governance, then it is worth unpacking the underlying issues that influence and affect the role of intellectual property in the biotechnological sector. First it is as well to be clear exactly what IPRs are: in most cases IPRs are 'owned' not by individuals (despite extensive narratives of justification that focus on the individual as worthy recipient of rewards[4]), but rather by companies, because many individuals can only hope to achieve significant rewards by licensing their IPRs to companies that have the resources to invest in the large scale manufacture and reproduction.

These companies may have significant and legitimate rights to profit from their investment in the production of goods and services, but the confusion of commercial rights with human or natural rights leads to an over-emphasis on the protection of IPRs related to others' rights. While human rights are natural rights accorded to persons with no qualification, commercial actors are legal entities subject to legal definition and limitation (Bakan, 2004). Although they find their origins in the products of individual endeavour (with natural persons), IPRs' largely commercial character and frequent effective ownership by companies suggests they are something other than natural or human rights.

Furthermore a significant element of IPRs is not the freedom for an individual rights holder to do something or not to have something done to them, but is rather the 'right' to halt certain behaviour by others. Although these limitations have been circumscribed by the assertion of public benefits in most IPR legislation, these commercial rights may still have a significant effect on the rights exercised by others. Intellectual property rights confer rights within markets, and where IPRs are in tension with other (human) rights, this becomes a wealth issue. Thus, one of the key functions of the narratives that serve to justify intellectual property is their role in reinforcing the necessity of regarding economically valued knowledge as intellectual property in the first place. The importance of the term 'trade-related' is that it makes explicit the central concern of various political interests from the developed states: the need to legislate for a commodity form in knowledge and information. This means the line between pubic and private in the realm of knowledge, most specifically in the Trade Related Aspects of Intellectual Property Rights (TRIPs) agreement, overseen by the World Trade Organisation (WTO), is the distinction between trade related intellectual property (rights) and a residual category, presumably *non-trade related* intellectual property, or knowledge that is not intellectual property.

There is implicitly a moment when something that has previously (potentially) been in the public domain (as non-ownable knowledge) is re-coded as trade-related and thus amenable to the 'protection' afforded other *trade-related* (intellectual) property: the moment when its trade-relatedness is asserted. This movement, as a succession of such moments, is one which is broadly parallel to the enclosures of common land in Great Britain and elsewhere during the sixteenth to nineteenth centuries (May and Sell, 2005). What might once have been public or commonly 'owned' is rendered trade related and thus private.

This notion of trade relatedness is often presented as both common-sense and unproblematic. Nevertheless the line between public and private knowledge objects is subject to constant reconstruction through the reformulation of concepts of tradable knowledge working in conjunction with changes of the material technology that can utilise it. Therefore, although the form of the distinction remains fixed, as a binary division between non-tradable knowledge and tradable knowledge, the content of each sector varies with the development of technologies which enable the capture and profiting from different types of informational item or knowledge object. Equally, shifts in what might be intellectual property drive new attempts to capture attendant rights through trade-relatedness, what we can perhaps more simply refer to as commodification.

The TRIPs agreement's text is careful not to prescribe definite limits to intellectual property; it is permissive of new forms of knowledge becoming property. Exclusions are allowed but not mandated, and there are few, if any, classes of knowledge that are expressly forbidden to IPR related legislation. Through membership of the WTO, such legislative requirements have become an important driver of national legislative (re)construction. The notion of rights as conceptualised within the TRIPs agreement would indicate that the whole purpose of these rights is that they can be traded, that they are alienable. This then suggests that the most important role that IPRs play generally, and specifically of importance in any scientific realm (specifically here, biotechnology), is the formal construction of scarcity (related to knowledge and information use) where none necessarily exists.

Unlike material things, knowledge and information are not necessarily rivalous; co-incident usage does not detract from utility; with certain exceptions (such as the use of trademarks to identify makers of goods) the deployment of knowledge and information resources by multiple users does not reduce its social usefulness, nor diminish the quality or quantity of such resources. In this sense, usually knowledge (before it is made property) does not exhibit the characteristics of material things before they are made (legal) property: knowledge is not naturally scarce in the same way as materially existing things are. Where there are information asymmetries then advantage may be gained by keeping information 'scarce' (that is, reducing its circulation), but this seldom serves the wider social good. Thus, as it is difficult to extract a price for the use of non-rival (knowledge) goods, a legal form of scarcity (IPRs) is introduced to ensure that a price can be obtained for use.

As noted above, although predicated on the notion of individual creators' and innovators' rights, most IPRs are owned and exploited not by innovating individuals but rather by commercial enterprises. Individual innovators either find they are unable to exploit their innovations due to the 'work-for-hire' provisions of their employment contract which ensures their work is owned by their employer; or, if independent, they find that in a modern and complex economic system, their only hope to exploit their invention or creation for monetary gain is to transfer the rights over its reproduction (or manufacture) to a corporation with extensive financial and organisational resources. In both cases corporate actors gain control of these IPRs.

It is also important to remember that even material property in a legal sense can only be what the law says it is; it does not exist waiting to be recognised as such, but rather is the codification of particular social relations, those between owner and non-owner, reproduced as (property) rights. Many years ago Walter Hamilton remarked that it has always been 'incorrect to say that the judiciary protected property; rather they called that property to which they accorded protection' (quoted in Cribbet, 1986: 4). But, whereas material property rights merely codify the existing materiality of things (and their relations of possession), IPRs transform the existence of that which they encompass. There is an important difference between property in knowledge and information, and material property. As Arnold Plant put it, unlike 'real' property rights, patents (and other IPRs)

> [a]re not a *consequence* of scarcity. They are the deliberate creation of statute law; and, whereas in general the institution of private property makes for the preservation of scarce goods, tending (as we might somewhat loosely say) to lead us 'to make the most of them', property rights in patents and copyright make possible the *creation* of scarcity of the products appropriated which could not otherwise be maintained. Whereas we might expect the public action concerning private property would normally be directed at the prevention of the raising of prices, in these cases the object of the legislation is to confer the power of raising prices by enabling the creation of scarcity [Plant, 1934: 31].

This protection of rights for the express purpose of raising prices is, of course, central to the 'problem' of commodification in biotechnology.

Therefore, as Martin Kenney has pointed out, one of the key processes that prompted the commodification of bio-medical science was the recognition by a number of groups (venture capitalists and multinational corporations most importantly) that the fruits of such research could be the basis of a new industry. This transformation of bio-medical science into the biotechnology industry is the transformation of science into technology (Kenney, 1998: 141). Patenting can only be formally achieved for industrial technologies, hence the narrative of trade-related-ness and the narrative of technology meet in establishment of the ability to patent the fruits of bio-medical science; the result can only be achieved by the rendering of science as technology, the shift from scientific endeavour to commercial industrialisation of the sector. Indeed, as James Boyle has pointed out, the establishment of the industry (and its reliance on patents for its continuing operation in its current form and character) is often taken as an important argument (whatever other considerations may be brought to the debate) for the justification of biotechnological patents – without them an important industrial sector would fail (Boyle, 2003: 13). Whether this is the case is a question of judgement, not fact, however, not least as the commodification that industrialisation has entailed can hardly be said to be unproblematic in itself.

INNOVATION AND THE ANTI-COMMONS

The commodification of biotechnological resources and research tools has become a significant issue in the debates around the expansion of the biotechnology sector. Since the mid-1990s at least, questions about the ownership of human genetic material have ranged from disputes over the concerns about the ownership of innovations developed from tissue belonging to patients (Boyle, 1996: 98–101) to the well rehearsed discussions about an anti-commons which may undermine scientific endeavour by reducing access to key research resources (Heller and Eisenberg, 1998). These debates have ensured that biotechnology has increasingly become the focus for much demonology about the patent system, as well as being an area where supporters of intellectual property have endeavoured to demonstrate its social worth, through its support for innovation and scientific advance. As is so often the case, neither side of the debate has been able conclusively to demolish its opponent's position, and thus political disputes linked to patenting bio-medical advances and research tools continue to rage.

Certainly there are a number of institutional causes to the emergence of the leading edge of the biotechnology primarily in the United States, at least in part linked to the ability of entrepreneurs to guard and profit from their investments through a biotech-friendly intellectual property system (Loeppky, 2005). More generally, as Graham Dutfield concludes:

> Ever since the late nineteenth century (and earlier in a few countries) patent-owning firms have been politically active in patent regulation in every economically advanced country, as have other groups such as lawyers and organisations representing pro-patent business interests. By and large, they have been quite successful in capturing – or at least closely collaborating with – state and regional regulatory agencies dealing with international commerce and industrial policy, to the point that the national interest is often treated as being identical to the demands of corporations [Dutfield, 2003: 237].

And because of the even more accentuated commercialisation of the sector in the US, many of the issues round commodification and the impact of establishing property relations in the results of scientific endeavour have had the most fervent exposure in debates about American biotechnology. When universities were broadly research only institutions where much of the early bio-medical work which underpinned the early industrialisation of bio-tech took place, they mostly were ignored by patent owners. This was partly because in many jurisdictions something similar to the US experimental use exception obtains (in one form or another), and partly because the infringements were unlikely to impact on the commercial markets patent holders operated in (indeed experimentation often provided commercial benefits when new uses or applications were discovered for already existing technologies or processes).

However, as universities have moved into commercial markets themselves, and perhaps most importantly have started to both assert their own patent rights and seek to exploit them for gain, this previous cosy lack of concern has started swiftly to evaporate. The difficulty is that many institutions find themselves torn between a norm of scientific community (and therefore the sharing of advances in biotechnology or another field for that matter) and the requirement to utilise their resources as sources of income. As Rebecca Eisenberg puts it, 'Institutions tend to be high-minded about the importance of unfettered access to the research tools they want to acquire from others, but no institution is willing to share freely the materials and discoveries from which they derive significant competitive advantage' (Eisenberg, 2001: 228). Certainly this may lead to hybrid agreements which allow some circulation of tools for non-commercial experimentation among communities of scientists (running in parallel with the experimental exception for patents), but it has also led to some frustration among university scientists finding budgets being eaten up by licence payments. Indeed although two fundamental biotechnological advances (the Cohen–Boyer patent on recombinant DNA and the patent on polymerase chain reactions) were both widely licensed on favourable terms in the US (by Stanford University and the University of California, in the former case, and Hoffman–La Roche in the latter), such practices have not been widely followed, and indeed the logic of licensing has moved in the opposite direction (Eisenberg, 2001: 230).

This confirms that for American publicly funded research the Bayh-Dole Act of 1980 was of profound importance. By giving universities the right to retain title to and license inventions which stemmed from federally funded research, research based institutions could (and did) develop and maintain strong patenting strategies (Maskus, 2000: 202). As a result in the US in the 1990s around 40 per cent of human DNA-related patent applications were filed by public sector researchers (Thomas, 1999: 137). Although not as developed, a similar tendency soon became apparent in Europe. Unsurprisingly, as the trade applicability of 'pure' bio-science has become more obvious so more science has become 'trade related' and thus open to the pressures to be made property.

There is also a question regarding the justice of patenting aspects of a resource that might be (arguably) regarded as the property of all humans. If we all share the vast majority of genetic code (as the Human Genome project has underlined) then should companies be allowed to 'own' particular parts of sequenced genes? These issues have been at the forefront of much of the popular disquiet regarding patenting of human (and other biological and agricultural) genetic information. However, as one might expect, the biotechnology companies involved have repeatedly invoked the labour desert justification for patenting: without such a reward how would they earn a return on the investment, and without the promise of a return how would they secure further funding to continue research? Any moral problems have been regarded as secondary to the interests of discovering new uses for genetic profiling and information, and the reward for such work. But many of the promised benefits of this genetic revolution have actually been slow to arrive (indeed many have proved, at least currently, more chimerical than actual), and this has led to both questions regarding the commercial worth of various holdings of biotechnological patents, and questions about the claims of utility that have been made during the patent application process itself.

The early history of biotechnology is not a story of private sector enterprise, but rather the continuing and long term support of scientific endeavour by various states' governments.[5] But, Rebecca Eisenberg notes, 'When public policy promotes private appropriation of research results as intellectual property even when they emerge from public sector research, it is easy to lose sight of the public and private benefits of disseminating information in the public domain' (Eisenberg, 1996: 572). The patenting of expressed sequence tags (ESTs) and the effective limitation on their use constricts their immediate deployment by other (competing) users and biotechnology innovators. This may have social costs, not least as swiftness to patent may not necessarily indicate that the most socially beneficial user or developer owns any particular sequence.

This expansion of patenting has had a number of effects, two of which have a significant potential impact on the development of biotechnology. As I noted above, we might suppose that there is a case to be made for the patent system's effects being broadly beneficial as regards 'real' innovations, and that this can be distinguished from the system's misuse by powerful market actors. Specifically in biotechnology David Adelman distinguished 'common-method' from 'problem specific' research tools (Adelman, 2005: 1020ff). For the former, there is likely to be some detrimental impact of limitations on access to research tools due to patenting; in the latter case this is less likely to be a major problem in Adelman's assessment, because specific 'blocks' on research will actually foster inventiveness and innovation, producing alternative solutions to specific problems.

Furthermore, as Philip Grubb points out, the question of the inventive step, as regards the use of these tools to produce candidates for patenting, raises a question of the speed of technological change in the field:

> [T]he rate of progress in this field is so extraordinarily rapid that what was once revolutionary very quickly becomes standard practice. The fact that the state of the art changes so dramatically within the time a patent application is pending makes it very difficult to judge the invention in light of what was the state of the art at the filing date, and different tribunals can easily reach different conclusions [Grubb, 1999: 235].

In this sense, one of the central difficulties in a fast moving field (and this is a problem that has also been encountered in software patents) is that what seems innovative may become standard practice very swiftly, and as such the notion of an inventive step can very quickly disappear and thus render any award of patent as a limitation on standard practice rather than as an encouragement for invention. Thus, in many cases the issue of non-obviousness has been rendered null and void as regards patent disputes (Kane, 2004: 716).

This means that for some observers rather than an anti-commons emerging, what is prompted is a wider range of innovative activity (similar to 'inventing around' a patent, or more generally avoiding congestion due to the costs such congestion might involve). Thus, Adelman points out that a recent report on the industry concluded that 'the redundancy and intricacy of biological processes allow for multiple lines of research that enable scientists to circumvent existing problem-specific patents' (Adelman, 2005: 1022). While common-method tools certainly may be subject to concerns about strategic patenting and the emergence of an anti-commons this is not necessarily the case (or potentially the case) across all aspects of biotechnological patenting.

However, part of the problem here is that only through a constant and (politically) costly practice of oversight can the distortions which might lead to the drawing of Adelman's line between the two sorts of research tools in the 'wrong' place be avoided, or at least limited. It may be extremely difficult to decide at the margins when common-method tools become merely problem-specific tools and vice versa; indeed, as demands made upon biotechnology by social actors change, so may this distinction. This can also be compounded by how specific courts interpret the 'doctrine of equivalents' to ascertain whether an infringement of a prior patent has taken place (Grubb, 1999: 236). Patent breadth (or scope) can become a central issue as regards the impact of biotechnological patents on subsequent research and scientific activity.

Where 'reach through' licence agreements are demanded by patent owners, biotechnology processes that use more than one patented research tool can see their potential profitability adversely effected. While reach-through-royalty rates on any specific tool may be negligible, given the often complex methods of biotech advances, these soon mount up if a number of tools need to be deployed (Jaffe and Lerner, 2004: 64). Control of patented technologies is being clearly enhanced through the patent system but this is also producing two clear problems:

a) there is an increased risk of costly litigation, and thus many technology users will pay royalties (rather than dispute the legitimacy of what might be a marginally applicable patent). This may act either to make specific paths of investigation more costly, or indeed may halt their development in the first place – a 'chilling' effect on research;
b) connected to these costs, rather than making research and technological development less risky, the effect of patenting in biotechnology has been to produce a lack of certainty and heightened risks of litigation.

Neither of these effects can be regarded as wholly positive. Indeed, given that many biotechnology patents are effectively 'submarine' applications, the risks to future development may not be accidental. A submarine patent is written as widely as possible and is intended to lurk in the patent system to snare later unsuspecting innovators with a claim that the advance they have made is actually covered by the prior patent.

Given that in many cases biotechnological patents are being awarded for aspects of biotechnological discovery that have as yet had no use attached to them, the real difficulty for an innovator spending time developing a use for a specific biotechnological 'discovery' are clear; once a profitable use has been discovered there is a real risk that this advance may find itself claimed by a prior patent's applicant. Subsequent technological development may have the effect of shifting patents from a relatively narrow scope to a wider scope (and therefore effectively allowing them to act like submarine patents even if that was not how they were originally intended). This is to say that biotechnological patents are often of indeterminate scope as post-grant work may generate results which widen the effective scope of a specified 'molecular portfolio' (Kane, 2004: 724). Thus, even without intent, the advance of the field itself may have a subsequent effect on already issued patents, enhancing their scope and possibly inhibiting further developments outside the owners' network of licensees.

More generally these sorts of effects lead to what Michael Heller and Rebecca Eisenberg have famously depicted as an 'anti-commons' in scientific endeavour. Here:

> too many patent rights on 'upstream' discoveries can stifle 'downstream' research and product development by increasing transaction costs and magnifying the risk of bargaining failures . . . The greater the number of people who need to be brought into agreement in order to permit a research project to proceed, the greater the risk that bargaining will break down or that transactions costs will consume the gains from exchange [Eisenberg, 2001: 224].

This is especially the case in biotechnology, where on one side the rapid commercialisation and commodificaton of the sector have led to widespread patenting and, on the other, the rapidly increasing complexity of the technological deployments necessary to further advance in the field mean that complex networks of previous advances need to be deployed.

As the foregoing has briefly demonstrated there is little agreement about how the patenting of various elements of biotechnology may impact on the practice of biological science more widely. At least, to be more accurate, there is little agreement on how the practices of the biotechnological industry may shape and limit the scientific activities on which it depends. However, if we conceive of these issues a little more widely then this debate looks a little different.

CONCLUSION: THE COMMODIFICATION OF SCIENCE?

Despite the desire by many commentators to limit the discussion of biotechnological innovation and patents to the technical issues I have rehearsed above, lurking just below the surface, a more fundamental issue can easily be discerned: how do we wish to organise the future of scientific endeavour? This is a question about commodification and thus relates directly to the central issue of political economy: what are the effects of, and issues raised by, the continuing development of the modern capitalist system? Thus, we must ask what costs and what benefits accrue to society from treating aspects of biotechnology (if not the entire field as yet) as property within a capitalistically ordered contemporary (global) society? Here certainly there are emerging bodies of evidence, but the real question is one of normative commitments: do we think that the products of science would be better supported through market or public structures? Is the rendering of science as property merely the application of successful organisational methods to a new realm of contemporary technology, or does this raise the danger of enclosure of the 'knowledge commons' and the corporate control of the stuff of life?

Legal mechanisms can be subtle, flexible and well suited to differential governance; however to secure these advantages there needs to be overt and extensive political engagement with the problems that commodification raises. And this is where specific problems begin to emerge. There is a well developed group of commercial interests that wishes to argue that exactly this sort of engagement represents an unwarranted and unnecessary politicisation of intellectual property and of biotechnology. This position suggests that IPRs are merely a neutral and technical solution to the social problem of encouraging and rewarding innovation. Elsewhere Susan Sell and I have gone to some lengths to dispute this ahistorical depiction of intellectual property (May and Sell, 2005). What our work suggests is that actually the whole half millennium of intellectual property's formal history has been (necessarily) political, and indeed continues to be. Thus, the debates about patent scope and the applicability of the characterisation of 'technology' to various advances in biotechnological (and related bio-medical) science are not technical debates, but are rather a realm in which the extensive political resources of various powerful commercial interests are deployed to further their legitimate but *partial* interests. What is now required is a further articulation and advancement of the public-regarding interest, as regards advances in biotechnology, to balance this fervent and well articulated declaration of the overarching importance of private rights.

As regards technical solutions to the general difficulties in the patent system, Jaffe and Lerner (2004: 171–172) propose three areas of action which, despite being focussed on the US experience, make considerable sense as part of an agenda of reform for the (global) realm of intellectual property:

a) *improve patent quality* – here the key issue is to ensure that those patents which are awarded are clearly defensible. The notion that it is sufficient to allow patents to be appealed neglects the costs and difficulties (as regards court proceedings) that this control mechanism introduces into any specific industrial sector. Hence, it is far from clear that dubious patents will always be appealed, or that there are sufficiently wealthy groups in whose interests appeal lies to offer a significant safeguard. However, it is also clear that gaining a patent should not be so onerous that individual inventors or innovators are priced out of the system;

b) *reduce uncertainty* – as Jaffe and Lerner point out, the only area of patenting which has become more certain in recent years has been the application process itself, because patent offices are largely disposed towards granting applications. However, for other actors in any specific sector, the expansion of litigation regarding innovations that broadly written (and granted) patents might encompass has led to all manner of perceived risks in the process of innovative development. Because of lapses in patent quality it may not be clear that a specific case will go one way or another, and thus may involve significant litigation costs, and this uncertainty may have adverse effects on the direction of research and development;

c) *control costs* – thus, overall, especially if patent systems are likely to remain similar to those operated today, there is a clear need to control the costs of litigation and examination, and to ensure that patent office income for operations is no longer directly linked to the award of patents themselves (which introduces a clear moral hazard as regards marginal or dubious applications).

Thus, an important element of reform would be to introduce a pre-grant process of objection or opposition. While this would need to have some mechanism for dissuading frivolous or vexatious objections, nonetheless this would then ensure that the presumption of patent validity which many systems (most obviously the US) work to would be better grounded than is currently the case. This might also be linked to the use of re-examination, again so organised to avoid vexatious appeals, but that would allow patent offices to weigh evidence and submissions that the grant of patent itself had encouraged (Jaffe and Lerner, 2004: 206). It might also be possible to develop a form of 'fair use' or 'fair dealing', perhaps by making experimental exception more explicit, or moving to other forms of relaxation on complete prohibitions on use as have traditionally been deployed in copyright. Of course such measures assume that overall the patent system remains of value as a structure for rewarding innovation in any specific industrial or scientific sector.

If we regard the problems with the development of the science of biotechnology as being linked to (or flowing from) the issue of commodification, then this reveals the deeper problem as regards the scope of patents in this area. Here the political difficulties are compounded by the impact of the global governance of intellectual property and the current activities of various governments at the World Intellectual Property Organisation. Globally active groups such as Greenpeace have argued that 'the patenting of life undermines the patent system, because . . . when it comes to genes, plants and animals, they are not invented but discovered' (Meister, 2000:

244). Here, any subsequent scientific intervention, experimental isolation or development does nothing to *denaturalise* the product, and hence, for Greenpeace, the very use of patents in biotechnology must by definition be contradictory. Therefore, as Eileen Kane notes, the 'intuitive assumption that the natural world is off-limits to the patent system most readily correlates with the legal requirement for novelty in securing a patent' (Kane, 2004: 732). Only when the social value of commercialisation can outweigh this assumption can patents in biotechnology become legitimate. However, very clearly this is an issue of judgement based on an assessment of the advantages and disadvantages of commodification, not merely a technical question of applying a form of regulation to a new sphere.

Rather than responding to the assessed needs of an emerging industrial sector, legislators have often responded to the (essentially commercial) *demands* of corporations in the sector. Certainly in some cases these demands do broadly coincide with a wider depiction of economic needs (to foster innovation, to allow commercial risk to be properly assessed), but these demands also stretch into the realm of market advantage and the common anti-competitive desire of leading corporations to stifle new entrants to markets, and by thus doing maintain or raise prices (Dutfield, 2003: 165–170). Therefore discussion and debates about various technical aspects of the award of patent rights (and the linked issues of examination and scope, for instance) miss the wider political issue of whether much of the realm of biotechnology should be included within the increasing globalised institutionalised regulatory structure of IPRs at all.

Conversely, although the attempt to remove biotechnology from the (global) patent system may be intellectually elegant and indeed relatively easy to justify in terms of the legitimated and formal aspects of intellectual property (as envisaged in various agreements), this is not currently a practical political option. To question the inclusion of biotechnological products/processes in the contemporary patent regime is actually to open a much wider debate about the appropriateness of patenting, which will threaten too many entrenched and, to be fair, reasonably legitimate interests (commercial and otherwise). However, the patent system, as I have already suggested, offers considerable scope for useful and acceptable reform, even if this does not actually end the use of patents in biotechnology.

As the recent debates at the World Trade Organisation and at the World Intellectual Property Organisation have started to demonstrate, there is an emerging realisation among the developing countries' representatives in both these organisations of the threat to the public domain (to the common heritage of mankind, as it is sometimes expressed) by the biotechnology industry. Certainly the industry presents itself as offering solutions to the 'problems' of developing countries, but this is (as many are now realising) a way of gathering support for the patent model both organisations are committed to globalising. If the political economy of biotechnology in the rich and developed countries is already subject to considerable criticism, as noted above, then there is a political opportunity for critics across the world to join together. The depiction of these issues as technical problems is now beginning to be rejected by the developing countries' representatives in Geneva (in part encouraged by a number of NGOs, perhaps most importantly the Quaker United Nations Office), but likewise needs to be (re)politicised in the developed world. Although many patent scholars may reject this politicisation, it is a necessary move to re-establish the political and social control over the institution of intellectual property which in recent decades has been slipping away. This may not halt patenting in biotechnology, but can constrain it within more acceptable limits by once again recognising that patents (like all intellectual property) are mechanisms for balancing private rights with the maintenance of the public domain.

NOTES

1. In this chapter I use the term 'commodification' to indicate the rendering of those things not previously regarded as being property as property. Once so rendered, as noted later in the chapter, these things are made scarce and thus can be exchanged and licensed in markets. Although the term also raises issues about the labour relation in biotechnology (the relationship between innovating individuals and their employers, most obviously), in this chapter I leave these issues to one side.
2. In this and subsequent paragraphs, comments regarding the history of intellectual property rights are drawn from May and Sell (2005).
3. Issues of global governance of intellectual property are discussed in the last section of this book.
4. See my discussion in May (2000: Chapter 1).
5. An excellent brief but informative history of biotechnology can be found in Lievrouw (2004: 147–158), and a longer treatment is provided by Dutfield (2003: Chapter 6).

REFERENCES

Adelman, David E. (2005). 'A Fallacy of the Commons in Biotech Patent Policy', *Berkeley Technology Law Journal* 20: 985–1030.

Bakan, Joel (2004) *The Corporation: The Pathological Pursuit of Profit and Power*. New York: The Free Press.

Boyle, James (1996). *Shamans Software and Spleens: Law and the Construction of the Information Society*, Cambridge, Mass.: Harvard University Press.

Boyle, James (2003). 'Enclosing the Genome: What the Squabbles over Genetic Patents Could Teach Us', available at: http://law.duke.edu/boylesite (22 September 2005).

Cribbet, J.E. (1986). 'Concepts in Transition: The Search for a New Definition of Property', *University of Illinois Law Review* 1: 1–42.

Drahos, Peter and John Braithwaite (2002). *Information Feudalism: Who Owns the Knowledge Economy?* London: Earthscan Publications.

Dutfield, Graham (2003). *Intellectual Property Rights and the Life Sciences Industries: A Twentieth Century History*. Aldershot: Ashgate.

Eisenberg, Rebecca S. (1996). 'Intellectual Property at the Public–Private Divide: The Case of Large-scale cDNA Sequencing', *University of Chicago Law School Roundtable* 3: 557–573.

Eisenberg, Rebecca S. (2001). 'Bargaining Over the Transfer of Proprietary Research Tools: Is this Market Failing or Emerging?', in R.C. Dreyfuss, D.L. Zimmerman and H. First (eds), *Expanding the Boundaries of Intellectual Property: Innovation Policy for the Knowledge Society*. Oxford: Oxford University Press, pp. 223–249.

Gleick, James (2000). 'Patently Absurd' *New York Times Magazine*, 12 March, 44–49.

Grubb, Philip W. (1999). *Patents for Chemicals, Pharmaceuticals and Biotechnology: Fundamentals of Global Law, Practice and Strategy*. Oxford: Clarendon Press.

Heller, Michael A. and Rebecca S. Eisenberg (1998). 'Can Patents Deter Innovation? The Anticommons in Biomedical Research', *Science* 280 (May): 698–701.

Jaffe, Adam B. and Josh Lerner (2004). *Innovation and its Discontents: How Our Broken Patent System is Endangering Innovation and Progress, and What to do About it*. Princeton, NJ, Princeton University Press.

Kane, Eileen M. (2004). 'Splitting the Gene: DNA Patents and the Genetic Code', *Tennessee Law Review*. 71, 4: 707–767.

Kenney, Martin (1998). 'Biotechnology and the Creation of a New Economic Space', in A. Thackray (ed.), *Private Science: Biotechnology and the Rise of the Molecular Sciences*. Philadelphia: University of Pennsylvannia Press.

Landes, William M. and Richard A. Posner (2004). *The Political Economy of Intellectual Property Law*. Washington, DC: American Enterprise Institute/Brookings Joint Centre for Regulatory Studies.

Lievrouw, Leah A. (2004). 'Biotechnology, Intellectual Property and the Prospects for Scientific Communication', in S. Braman (ed.), *Biotechnology and Communication: The Meta-Technologies of Information*. Mahwah: Lawrence Erlbaum Associates.

Loeppky, Rodney (2005). 'History, Technology and the Capitalist State: the Comparative Political Economy of Biotechnology and Genomics', *Review of International Political Economy*. 12, 2 (May): 264–286.

Maskus, Keith (2000). *Intellectual Property Rights in the Global Economy*. Washington, DC: Institute for International Economics.

May, Christopher (2000). *A Global Political Economy of Intellectual Property Rights: the New Enclosures?* London: Routledge.

May, Christopher and Susan K. Sell (2005). *Intellectual Property Rights: A Critical History*. Boulder, Colo: Lynne Rienner Publishers.

Meister, Isabelle (2000). [presentation of Greenpeace's position] in S. Sterckx (ed.), *Biotechnology, Patents and Morality*. Second Edition, Aldershot: Ashgate, pp. 243–246.

Plant, Arnold (1934). 'The Economic Theory Concerning Patents for Inventions', *Economica* 1 (February): 30–51.

Polanyi, Karl (1944 [2001]). *The Great Transformation: The Political and Economic*

Origins of Our Time [reprinted with new introduction] Boston, Mass: Beacon Press.

Ramesh, Randeep (2004). 'Monsanto's Chapati Patent Raises Indian Ire', *The Guardian*, 31 January: 19.

Samuelson, Pamela (1989). 'Innovation and Competition: Conflicts over Intellectual Property Rights in New Technologies', in V. Weill and J.W. Snapper (eds), *Owning Scientific and Technical Information: Value and Ethical Issues*. New Brunswick, NJ: Rutgers University Press.

Shiva, Vandana (2001). *Protect or Plunder? Understanding Intellectual Property Rights*. London: Zed Books.

Thomas, S.M. (1999). 'Genomics and Intellectual Property Rights', *Drug Discovery Today* 4, 3 (March): 134–138.

12. On the comparative institutional economics of intellectual property in biotechnology

F. Scott Kieff[1]

INTRODUCTION

Despite numerous reforms over the past century,[2] important problems continue to plague the IP systems of today, generating numerous proposals for further reform tomorrow. For example, recent high profile cases like the patent litigation threatening to shut down the BlackBerry® service[3] have drawn sharp criticism in the business community[4] as being prime examples of the pernicious impact of protecting intellectual property (IP) rights with strong property rules, backed up by injunctions, rather than weaker liability rules, which would give rise only to a right to payment.[5] Controversial examples more closely linked to biotechnology include the litigation over a potential experimental use exemption for infringement.[6] Various forms of liability treatment have been offered. For example, Ayres and Klemperer advocate a patent litigation system characterized by uncertainty and delay, which they show could serve as a form of compulsory license, or liability rule.[7] Others simply advocate various exemptions to infringement, such as for what they call fair use.[8] Some suggest that the open-source model be applied more generally, including in all bioscience.[9]

Underlying these critiques of IP is a view that property rights either restrict access or cause anticompetitive effects. The arguments raised today are quite similar to those raised throughout most of the past century; and, as usual, the reform efforts target all three branches of the federal government – legislature, executive agencies, and courts.[10] This chapter endeavors to show how addressing these concerns with conventional approaches[11] is likely to exacerbate the problems of access and anticompetitive effect, and how these problems can be mitigated by adopting an unconventional strong-property-rule approach informed by property's coordination effects.

Put simply here for introductory purposes, the chapter suggests a goal that IP can achieve effectively and efficiently. The chapter endeavors to shift the dialog over property rights in general to include in its focus the problem of coordination in addition to the long-standing focus on the problems of externalities and rent dissipation.

In the context of IP in particular, the chapter explores reasons why, although IP regimes should not be expected to be effective in achieving a narrowly tailored reward function by providing direct incentives for specific inventive or creative efforts, they should be expected to be effective in facilitating the coordination among complementary users of the subject matter IP rights protect that is needed to facilitate commercialization of that subject matter. This type of coordination, which is important for increasing competition and access, hinges on whether the IP rights themselves are enforced by strong property rules backed up by a right to exclude. The basic intuition behind this view is motivated by the recognition that while the private ordering needed to achieve commercialization can lead to textured contracts having many terms including price but also including a host of seemingly esoteric and unique provisions – such as technical support, field-of-use or territory limitations, grant-backs, cross-licenses, payment schedules, most-favored-nation provisions, etc. – a court imposed damage award, which is emblematic of liability rule treatment, is in all but the rarest of cases reduced to a simple monetary amount. In this regard, property rule treatment is seen as a criterion of efficacy.

Recognizing that enforcing IP rights with property rules also would require them to be designed in ways that would mitigate the various social costs generally associated with property rule treatment, the chapter explores ways this can be done. The rights must have their contours staked out at their time of creation by claimants instead of being set immutably by statute so as to mitigate problems of rent-dissipation and information cost. They must not frustrate reasonable investment backed expectations of others when staked out so as to mitigate the problems of the asset specificity and opportunism. They must give clear and predictable notice about what they cover after being staked out so as to mitigate transaction costs. In addition, to mitigate monopoly effects and anticommons effects, the ownership of these rights must be in the hands of an openly identifiable residual claimant, which is an individual market actor who can negotiate over them and extract value – the residual claim – from electing to give up either permission via a license or title via an assignment. This owner must be given broad flexibility to divide them up and aggregate them. Together, these several parameters can be seen as criteria of efficiency.

Setting forth these various criteria of efficacy and efficiency in this summary format is designed only to highlight some reasons why some different institutional features of the different IP regimes of patent, trademark, and copyright may be working well and why others may not. The ultimate net economic performance of each regime is highly multi-factorial, and indeed most industries do not interact with just one IP regime. But while judging overall net economic performance is beyond the scope of this chapter, what this chapter does endeavor to add to the analysis is an elucidation of the types of impacts different institutional choices are likely to make for a given system, as well as how some types of positive impacts might be achieved and some types of negative impacts might be mitigated.

This chapter proceeds as follows. Employing a comparative analysis of alternative goals such as providing direct incentives, mitigating rent dissipation, or internalizing externalities, and alternative institutions and organizations such as norm communities like an open source project, the next section develops a theory of the institution of property rights as playing a particular role in coordinating among the many complementary users of an asset. The third section shows how this coordination theory of property in general is particularly well suited as a theory for IP, especially when compared to other dominant IP theories in the literature such as the reward and prospect, or rent dissipation, theories. The coordination theory also is shown to explain why some particular aspects of IP regimes may be working well and why others may be candidates for change. the fourth section concludes.

NIE, COORDINATION, AND PROPERTY RIGHTS

The field of New Institutional Economics (NIE) pays particular attention to the economic significance of institutions and organizations, as distinct from other factors, such as technology, capital, or labor.[12] As described by North:

> Institutions are the humanly devised constraints that structure human interaction. They are made up of formal constraints (rules, laws, constitutions), informal constraints (norms of behavior, conventions, and self imposed codes of conduct), and their enforcement characteristics . . . .
>
> It is the interaction between institutions and organizations that shapes the institutional evolution of an economy. If institutions are the rules of the game, organizations and their entrepreneurs are the players.
>
> Organizations are made up of groups of individuals bound together by some common purpose to achieve certain objectives. Organizations include political bodies (political parties, the Senate, a city council, regulatory bodies), economic bodies (firms, trade unions, family farms, cooperatives), social bodies

(churches, clubs, athletic associations), educational bodies (schools, universities, vocational training centers).[13]

This chapter focuses on two of the most salient alternative tools for facilitating coordination in basic science: norm communities, and property rights. Although applying the above definitions strictly would make some of these tools look more like organizations than institutions, for purposes of this chapter it is sufficient to note that they all have important institutional aspects that benefit from NIE's comparative institutional analyses.

NIE emphasizes the use of comparative institutional analyses to look at the different characteristics of institutions and what impact they have on individuals and organizations over time.[14] Following such an approach suggests we should ask not only what we want, but also which mix of formal and informal institutions will work better in achieving our set of goals. The approach recognizes that because no institution is perfect – they all pose problems – our choices must be informed by our understanding of the problems we most want to solve and the problems we can best mitigate or bear.

Coordination and Property

Although property rights have long existed, evolution continues in the views about why property rights emerge within communities and what role property rights can and should play. This chapter elucidates one view that has only recently emerged in the literature: a view that sees the role of property rights as facilitating coordination. Property rights in general and IP in particular are not offered as perfect solutions to every problem. The case is not being made for property or IP, *über alles*. Rather, the point is that property rights can provide an important additional and middle-ground tool for optional use by individuals engaged in private ordering beyond those offered by the extreme poles of either the free, open market without them on the one hand or the hierarchies of a norm community, firm, or government on the other hand. But, to play this role, property rights must be designed to facilitate private ordering in a way that increases output of, and as a result access to, the subject matter they protect. To achieve this role effectively, they must operate as rights of exclusion around which coordination can take place. This provides incentives to complementary users of the asset protected by the property right to engage in trades with each other. To do so efficiently, property rights must be structured to mitigate the costs of rent dissipation, information, transactions, and public choice. The discussion below shows how each of these goals can be met.

The conventional view of property rights in the literatures of both law and economics follows the 1967 work by Demsetz that sees property rights as tools for internalizing externalities.[15] Demsetz built on the 1960 work on externalities by Ronald Coase,[16] which itself was a response to work on externalities from the beginning of the 1900s by A.C. Pigou.[17] Interestingly, the majority view of IP rights is premised on the same externalities focus as this literature, but seems to follow only its beginnings relating to Pigouvian taxes and subsidies, while overlooking its refinements relating to property rights.

Demsetz argued that property rights emerge when the benefits of internalization outweigh its costs – when the good of concentrating benefits and costs on owners so they deploy resources more efficiently outweighs the bad of the transaction costs associated with recognizing those rights.[18] According to Demsetz, property rights emerged among the historical native North American population he was studying because, with the lack of property rights, 'the underuse of animal husbanding and land management resources (skills and labor) led to near exhaustion [or overuse] of animal resources (food and clothing) . . . [while the presence of property rights] provided incentives for individuals to make more use of the one set of resources so as to not waste, and indeed to replenish, the other'.[19]

But this left open two important questions. The first concerns the exact mechanism by which property rights operate to achieve this internalization benefit. The second concerns the ways the costs and benefits of using property rights compare to the costs and benefits of alternative institutions.

This chapter contributes understanding of these open questions by focusing on the issue of coordination. The chapter elucidates how property rights can operate to achieve coordination in a way that involves a unique mix of the benefits and costs associated with relying solely on other institutional and organizational tools.

The type of coordination emphasized here refers to the process by which many diverse individuals interact with each other for a particular activity to be achieved effectively. This helps them not only achieve that common goal, it also helps them to be more diverse from each other and specialized in what skills and other resources they each bring to the collective enterprise were they otherwise unable to coordinate.[20] For IP, coordination can be particularly important because the subject matter protected by IP ideally is not yet the subject of successful commercialization, or perhaps not even known yet.[21] Although it has long been recognized that such uncertain and risky endeavors have a particularly strong need for coordination,[22] it also has been long recognized that this is a call for 'collective action, but not necessarily state action'.[23]

Coordination of this type is more about a beacon effect than control, and more about stability and certainty of complex deals (a bargaining effect) than about simple rate of return investment. The key is the incentive for diverse complementary users of the asset to come together (the beacon effect) and transact with each other (the bargaining effect). Both effects are facilitated by the credibility of the threat of an injunction, which is the signature attribute of a legal entitlement that is backed up by a property rule, and frustrated by the alternative of a liability rule.

The basic Calabresi–Melamed framework for deciding between property and liability focuses attention on which locus of decision-making about the true value of the underlying asset is the lowest cost provider of a correct decision.[24] Under this rubric, a liability rule can be more efficient if a collective, public, or governmental determination of the true value of the asset would be cheaper than a private evaluation reached by agreement of the parties; and conversely a property rule should be used if the private evaluation would be cheaper.[25] Rob Merges points out that one implication of this approach is that property rules are better for IP because private parties have a comparative advantage over courts in valuing IP.[26] But the coordination approach explored here pushes beyond this rubric's focus on the problem of relative advantages information-processing and extends the analysis to these two other concerns – the beacon effect and the bargain effect.

The beacon effect is achieved because the more credible the threat of the injunction behind every patent, for example, the more it creates incentives for diffuse individuals having an interest in the subject matter protected by the patent to decide individually to act in a way that ends up being coordinated. While the infringe-and-pay-damages approach of liability rule treatment may provide sufficient incentive for the individual patentee to come to court to receive payment, it does little compared to the right to exclude in drawing towards that individual all of the other potential complementary users of the patented subject matter so they may work together to engage successfully in the complex commercialization process.

But the bargaining effect associated with property rule treatment is even more important. Where there are large numbers of potential infringers, liability rules make bargaining between the patentee and the potential infringers, licensees, or assignees more difficult.[27] The problem afflicts both sides of the potential transactions.

The patentee's incentives are disrupted in at least two ways. First, the patentee may face decreased incentives because of the ordinary under-compensation error that is typically associated with liability rules.[28] Second, the use of a liability rule may create a prisoner's dilemma or collective action problem among potential infringers in which each individual's dominant strategy is to infringe in order to garner more of the potential gains from exchange for itself. That is, the use of a property rule is particularly important for the impact it has in limiting each potential infringer's incentive to infringe *ex ante*. Otherwise, under a liability rule, the patentee will not have adequate incentive to bargain with these infringers because such bargaining will not yield effective protection from others. In effect, the patentee faces what property skeptics call an anticommons problem because the patentee must get a binding commitment from each of the many possible takers; but because none of them can sell such a binding commitment under a liability rule regime, they leave the owner facing a thicket.[29]

This is not just a question of setting damages too low, but rather is a question of relying on damages rather than injunctions. Several negative impacts follow from the lack of injunction to block the behavior that would otherwise be practiced by the set of infringers.

First, uncoordinated acts of infringement may cause collective profits – those reaped by the patentee directly and through damages awards from infringers – to fall below the total costs of creating and commercializing the invention, resulting in a destruction of wealth.[30] Second, as recognized by Haddock, McChesney and Speigel, the threat of this potential onslaught of infringements induced by a liability rule will discourage investments in the subject matter covered by the IP right *ex ante*.[31] But this drop of individualized investment by the IP owner alone, is not the only result and could be mitigated by higher damages amounts.[32]

While the patentee's incentives can be maintained under a liability rule regime by awarding enhanced damages or letting the patentee engaging in self help,[33] the many complementary users of the patented subject matter also will face a drop in their incentives to even be drawn to the beacon effect otherwise associated with a property rule regime's right to exclude; and this incentive cannot easily be maintained. The heart of this problem is tied to the limited nature of a court imposed liability rule. Often price is not the only important term in these deals and courts are woefully inadequate compared to the marketplace for determining and enforcing these other terms. While private ordering among parties can lead to textured contracts having many terms including price but also including a host of seemingly esoteric and unique provisions – such as technical support, field-of-use or territory limitations, grant-backs, cross-licenses, payment schedules, most-favored-nation provisions, etc. – a court imposed damages award is in all but the rarest of cases reduced to a simple monetary amount.

This effect is increased the more that those investments cannot otherwise be deployed to other uses, the more they cannot be insured, the more they cannot be hedged, and the more they cannot be diversified. What this means is that the more the players approach each investment decision simply as one item in a large portfolio, the more they will not care whether the investment is in an activity associated with a liability rule or a property rule.[34] Put differently, the more the activity is associated with a liability rule, the more only large portfolio players will elect rationally to invest in it and the less it will be rational to invest unique skills and unique assets. In this way, liability rule treatment, which often is urged in the name of increasing competition, has the paradoxical effect of favoring large players, not market entrants.

Although the role of property rights as tools for facilitating coordination recently has been mentioned in the NIE literature, it has not been elaborated until now. As discussed previously, the classic work by Demsetz on the emergence of the institution of property rights focuses on their role in internalizing benefits and costs. Within the field of IP, prior work by the present author suggested the role of property rights as focal points in facilitating coordination among complementary users of an invention.[35] Independently, newer work by Demsetz also highlighted this coordination function of property rights.[36] Under the Demsetz new view, the key is 'coordination in the sense of bringing forth control decisions that are consistent with each other but that emanate from different persons'.[37] This is consistent with the approach that is more fully elaborated above, which shows how coordination is achieved by property through two effects. It brings parties together (beacon effect) and it helps them interact with each other once brought together (bargain effect).

One implication of the coordination view explored here[38] is that property rights backed up by property rules increase access, rather than decrease it. This is explored in greater depth below.

A second implication of the focus on this type of coordination is that when considering the alternative tools available for facilitating it, as discussed immediately below, it is important to recognize that the alternatives to property are more closely associated with large, established market actors, while property rights will be at least also associated with smaller market entrants. More particularly, the alternative of norm communities is closed to outsiders, by definition, while the alternative of government organizations is more responsive to public choice pressures from existing larger firms. In this way, the choice to rely on these other alternatives for coordination instead of property rights has the impact of increasing anticompetitive effect.

A third implication is that coordination, like all things, can be both good and bad. While facilitating coordination among complementary users of under-deployed assets increases access and competition by easing market entry and commercialization, coordination among large, existing firms will facilitate anticompetitive effects. Regrettably, as discussed below, there are several aspects of the modern IP regimes that are having this bad coordination effect.

Contrasting Property with Open Source as an Alternative Tool for Facilitating Coordination

Because the other tools that often get used in place of property – such as norm communities like open source projects – also can facilitate coordination of the type needed to increase access, it is important to consider how property compares to these other tools. Most germane to basic science is the particular alternative of a norm community like an open source project.

Coordination may occur among individuals who are linked to each other through some social group such as family, friendship, ethnic or religious identity, or some other norm community such as an open source software project. The NIE literature looks at the different approaches to informal rules, often called norms, because, as Williamson argues, what otherwise would be a focus on positive law regimes would reflect a type of 'legal centrism' that fails to account adequately for dispute resolution and enforcement activities that occur without the formal legal system.[39] Indeed, the first significant connection between the literatures of NIE and IP centered on the role of norms.[40] Norms can be thought of in at least two ways: as 'prescriptive norms', also called 'normative norms', which refer to beliefs about what people should do, and as 'descriptive norms', or 'regularities', which refer to how people tend to behave.[41]

Much of the NIE literature on informal or non-legal ordering has focused on enforcement and dispute resolution. One recent example is the work by Lisa Bernstein on relational contracting within homogeneous communities, which has focused on what it calls 'private ordering' as a mechanism by which individuals in the market can interact with lower administrative costs than with formal legal institutions through the use of more informal institutions for enforcement and dispute resolution such as norms, reputation, etc.[42] Similarly, recent work by Barak Richman comes closer to the theory of the firm literature and focuses on the importance of the private enforcement and dispute resolution techniques as means for ensuring not just lower administrative costs, but also better contractual enforcement, and enhanced transaction certainty.[43]

The view of property rights offered in this chapter differs from both of these perspectives by seeing private ordering in the more general sense than simply private enforcement.[44] Instead, private ordering is seen as the set of interactions among individuals that are more reliable because they are enforced in some way, whether by private informal institutions, such as norms, or by formal legal institutions, such as the coercive power of the state. This view is consistent with traditional liberal views of the rule of law and role of government as the monopoly over the coercive powers – such as force – to back property rights and contractual arrangements because such backing enhances the overall market economy by enhancing individual liberty to elect to deploy one's resources in whatever way best suits that individual.[45]

While recent work by Richman has shown that private enforcement mechanisms may, under appropriate conditions such as small and homogenous communities, provide even more transactional security at a lower administrative cost than public enforcement,[46] the point here is that having the option of public enforcement is a benefit to those under other more generalized or diverse conditions than such homogenous communities. Put differently, one disadvantage of such closed markets is that they are closed to outsiders. As Troy Paredes explains within the context of corporate and securities laws, 'when laws are in place, parties can rely less on personal and family relationships when transacting, allowing them to engage in transactions with strangers'.[47]

Keeping transactions entirely within a particular organization like a firm or norm community also raises the disadvantages that Stephen Haber calls the problems of 'crony capitalism'.[48] The enforcement benefits within closed organizations are due to the specificity of investments the community's members must make in it, which in turn bring along the inevitable concerns about opportunism. What is more, the attributes that underlie the unique connection to the community typically are non-fungible – such as family, religious, or ethic affiliation, or a close relationship with the community leadership.

The development of software like Linux within a community that adheres to an open source philosophy can be seen as one example of a coordinated activity that occurred within a norm community around a coordinating device akin to fame rather than around more formal property like patents. Under this view, the fame of Linus Torvalds allows him to control development of the Linux kernel to ensure that it occurs in a coordinated fashion. The ability of fame or other focal points to achieve coordination is consistent with the beacon view of property discussed earlier.[49]

While open participation would seem to be a touchstone promise of open source, several empirical studies of several different open source software projects have shown that this openness is not experienced in reality, in that changes to the actual projects in these cases actually are limited to a very small number of individuals in a different cohesive control group for each case studied.[50] While it makes sense as a practical matter of information processing needs for there to be a small control group,[51] this stark difference from the rhetoric of the legend matters a great deal. Unlike the formal property rights in something like patents, the fame that is the key to open source type of centralized coordination is less easily transferred, divided, or bundled. It also is specific to that community. In addition, fame can be more difficult to obtain in general than property. And its exclusivity makes it more difficult for diverse individuals to obtain. At bottom, the element that allows control by the leader within a norm community, whether it be fame or some other special community attribute, is only available to those who are insiders – in the case of Linux, that includes only Torvalds and his chief lieutenants – not those wanting to enter. Simply put, the above discussion elucidates some reasons why relying on norm communities like open source projects to the exclusion of property would have the effect of generally biasing against new entrants and in favor of those who are members of the establishment.

Mitigating the Problems of Property Rights

The above discussions explored the benefits of property. Coordination was explored as a goal that property rights can achieve. So, too, were the ways in which property rights can achieve this goal differently from alternatives. Like all things though, property has both benefits and costs. The discussions that follow unpack some of the leading criticisms of property rights that have been offered in the NIE literature, along with some of the tools that have emerged for mitigating these problems. The discussions also show how these tools actually are being put to use in present IP regimes.

Rent dissipation

One of the first problems that arises when property rights are made available to individuals within a community – or indeed when any government benefit is made available – is the problem of rent dissipation. Rent can be thought of as the benefit that is gained by engaging in a certain activity. Private rents are those accruing to the individual. Public rents are those accruing to society as a whole. Private and public rents may be different. The potential differences in both magnitude and sign between public and private rents may cause private incentives to engage in a given rent-generating activity to be either too little or too big than would be socially optimal. Where the availability of private rents provides incentives for an individual to engage in efforts designed to gain those private rents that are too strong, the resulting efforts may turn out ultimately to dissipate the social rents.[52] This is the problem of rent dissipation.

Rent dissipation itself can take at least two forms. One type of rent dissipation involves over investment in the race to obtain the rent. Another type of rent dissipation involves investment in alternative but socially undesirable techniques to win that race. One way to conceptualize the over investment type of rent dissipation is in the context of a race towards a common prize. If the community is characterized by a prize having a known value and an uncoordinated group of individuals who are each seeking the prize, then each individual rationally *might* elect to spend up to just less than the value of the prize to get it, which would mean that as a group they are spending more in aggregate than the value of the prize.[53] In the context of innovation, the effect has been demonstrated by economic models of multiple firms seeking the same invention in a race to patent, which show that investment overall may be too great.[54]

One way to conceptualize the improper alternative investment type of rent dissipation is also in the context of a race towards a common prize, but this time where some types of racing are viewed by society as good and others are viewed as bad. In the context of sports, for example, the use of practice sessions is often viewed as good while the use of performance-enhancing drugs is often viewed as bad. In the context of regulated markets, the use of innovation is generally considered to be a good form of competition (making better products or services) while the use of agency capture is generally considered to be a bad form of competition (getting the government to differentially regulate a competitor). This type of rent dissipation can be seen as a component of the problems studied under the rubric called public choice.

Importantly, the NIE literature does suggest ways that rent dissipation can be mitigated. Anderson and Hill have shown that rent dissipation problems associated with the creation of property rights can be mitigated if the potential owners of the rights are able to tailor them at the time of creation.[55] The intuition underlying this result is that this approach allows the owners to shape the right based on the best information available at the time about its value, including the different parameters that will impact the value in different ways – for example, its precise contours. The greater the wedge between the definition of the right and its actual creation, the greater the chance there will be a mismatch against actual needs. Anderson and Hill point out that the two central public choice problems will both contribute to the size of this wedge. A simple 'land-grab' approach will lead to overinvestment in racing to grab and over-grabbing of actual parcels, simply because the opportunity to claim later will be forgone.[56] In this regard, nobody is able to claim the residual that would be left behind by waiting until an actual need developed – there is no 'residual claimant'.[57] In addition, once government actors see the private interest in obtaining the rights, the bureaucracy will have an incentive to withhold the rights unless they determine a particular claimant is 'worthy', which will in turn provide a convenient excuse for the bureaucracy to amass the resources it claims are needed to judge 'worthiness'.[58]

At bottom, the more the regime allows those who ultimately hold the rights to craft the rights at the time of creations, other things being equal, the more likely rent dissipation effects will be mitigated. Even a quick comparison of different IP regimes reveals a stark difference in this regard. For example, patent applicants shape their own property rights through the drafting of the claim.[59] Similarly, the contours of the rights staked out by trademarks are largely set by the rights-holders themselves through actual use.[60] In contrast, the contours of a copyright are set as immutable rules (not even default rules) through the central regime rather than by the individual claimants.

Two general conclusions are suggested. First, the rent dissipation effect would be expected to be greater in this regard for copyright than for patent and trademark because of the difference in the way the rights are staked out. Second, this effect is something that cuts against changing any of the regimes – patent, trademark, or copyright – in a way that would have the effect of fixing the contours of the rights where they otherwise could be staked out by claimants at the time of creation.

Transaction costs

Property rights, just like any other type of entitlement, raise the problem of transaction costs, because to work well they must be able to be sold and licensed to those who value them most at any given time. The term 'transaction costs' plays a central role in the literature on property rights in general and IP in particular; but it is a term that often is misunderstood. Transaction costs are particularly important to the field of NIE because 'transaction-costs economics is the original centerpiece of what Williamson . . . called the New Institutional Economics'.[61] There has since been substantial empirical support for the validity of the transaction costs implications of NIE, as studied by Paul Joskow and others.[62]

The term 'transaction cost' generally refers to all the costs associated with contracting among individuals, including the hassle those parties experience in finding and dealing with each other, the costs of lawyers and other professionals to arrange the deals, and the bargaining process itself.[63] Transaction costs also can be thought of as including information costs because information must be gathered and processed before those individuals decide to interact with each other.[64] The term encompasses the costs of successful transactions – such as time and money – as well as the costs of failed transactions – such as lost opportunities – to the extent those failed transactions are good things that would have occurred but for the costs of transacting.

The comparative analysis of NIE reminds us that there is both a good and bad side to transactions and that as a result the costs of transactions may be worth bearing. For example, while on the bad side transactions impose costs, on the good side they are associated with the very specialization and division of labor that generally are thought to be good things.[65] That is, the availability of transactions to obtain from others the goods and services beyond those an individual is most interested in or most adept at providing itself facilitates each individual's ability both to have and to hone those specialized skills and tastes, as well as to bear individualized distributions. The link between specialization and transactions allows even large numbers of individuals to achieve complex tasks by coordinating with each other directly or indirectly.

In addition, given such individualism in the form of diverse skills and preferences, transactions, and thus their concomitant costs, have other important beneficial side effects that often are overlooked. First, transactions are associated with the privately beneficial exchanges among individuals that are essential for achieving private gains from trade.[66] Second, transactions are associated with the publicly beneficial socialization that occurs as individuals come to interact with each other.[67] This socialization effect occurs because for transactions to achieve gains from trade it must be the case that individuals having diverse resources and preferences learn enough about each other's resources and preferences to exploit them. This process of learning about each other's values is part of socialization. Third, the bargaining process – for both consummated transactions and for failed ones – inherently elicits important information about not only the particular transaction being negotiated, including intensity of preferences and budget constraints, but also relative values compared to other available transactions. That is, transactions can mitigate information costs.

It would be great if transactions could achieve their benefits without their costs. It also would make sense to strive both to increase their benefits and decrease their costs. But to the extent efforts to eliminate the transaction costs associated with direct exchanges between individuals in the market are by replacing them with court or agency mandated exchanges – such as by replacing property rules with liability rules – they will decrease some of the benefits of having those transactions occur directly between individuals. For example, the availability of court or agency mandated exchange may decrease the incentives, opportunities, and abilities for individuals to directly interact with each other.

What is more, reliance on court or agency mandated exchanges triggers its own set of costs, including the transaction costs and public choice costs associated with these organizations, as well as their comparatively increased costs of obtaining and processing certain types of information, their general tendency to err on the side of setting price too low, and the potential negative impact they can have on *ex ante* incentives and private ordering by injecting general uncertainty and overall instability with respect to non-price terms.

Indeed, just as transactions themselves can have positive and negative effects, so too can those employed to facilitate transactions, such as lawyers and other professionals. On the one hand, these professionals often are portrayed as a large component of the negative side of transaction costs.[68] On the other hand, because they help the transactions occur,[69] they also can be seen as part of the positive side of transactions to the extent that the transactions are a good thing. Therefore, a decision to replace lawyer mediated transactions with court or agency mediated transactions would require a comparison of the net costs associated with each. But because both are likely to require professional expenses (some mix of lawyers and lobbyists) at roughly comparable levels, it is difficult to imagine a saving of this cost associated with a shift towards government-mediated transactions.

What is more, the likelihood and extent of the pernicious impact of most transaction costs is recognized to be, all-in, generally worse in political markets than in economic markets.[70] The intuition behind this view is that for political markets the assets being traded – such as promises to vote a certain way for example – are both harder to evaluate and harder to enforce in that they are less certain at the time of negotiation, less predictable, less fungible, less dividable, and less bundle-able.[71]

An additional impact of NIE's comparative analysis is that is reveals why, all-in, the likelihood and extent of the pernicious impact of many types of transaction costs generally is worse in what are known as thinner markets as compared with thicker markets, where thinner and thicker refer to the amount and diversity of resources and participants, including their diverse evaluative techniques and preferences.[72] There are two basic intuitions behind this lesson. The first is that thickness increases the chance some individual in the market will find it profitable to arbitrage what otherwise would be a gap in information flow by finding and acting on that information, to offer an attractive option for what otherwise might be a hold-up problem, etc. The second is that the increase in bargaining associated with a thicker market mitigates information costs.

There is reason to think that some types of transaction costs may be worse in markets that are thicker, in at least some sense. For example, the behavioralism logic behind the problem of groupthink suggests that as the group gets bigger the problem gets worse.[73] But, to the extent that thicker is taken to mean not only bigger but more diverse, then even the problem of groupthink may also decrease with market thickness.

The transaction costs effects of patents in the field of basic biotechnology research are instructive. While there of course is some pernicious impact of the transaction costs associated with a state of affairs that includes patents, the degree of that impact must be compared against the similar problems that arise without patents. Prior work by the present author explores in some depth why the addition of patents to what otherwise was a market characterized only by academic kudos should be expected to make the market thicker, not thinner, and thereby decrease overall transaction costs.[74]

While it is easy to imagine the difficulty facing a scientist who just wants to gain access to a patented technology but does not want to spend the time and money to hire a team of expensive lawyers, the patentees figure this out for themselves. Remarkably low transaction cost business models are devised and implemented. For example, in the 'freezer program' business model that has long been in common use, the patent is assigned to a business that arranges for the patented biological material to be regularly brought fresh and frozen direct to the scientist's university department or even laboratory and then charges the scientist's research account for those quantities actually used.[75] The transaction costs that the scientist experiences for such a model are even less than typically associated with buying a can of soda from a soda machine. In contrast with this type of direct billing, the typical soda machine requires the buyer to use coins or low denomination bills – a higher transaction cost that is nonetheless well tolerated by society. Indeed, the freezer programs may provide a host of additional benefits. They save the scientist from having to spend the time and other resources needed to obtain the material herself. They also help the scientific community at large by providing a uniform source of inputs that decreases variability across scientific experiments.

A related point about transaction costs is that they are borne, at least in part, by both the party wanting to buy or license and the party wanting to sell or license – both the infringer and the owner. This is important because it helps explains why many property owners elect not to aggressively enforce the property rights against certain users by granting broad licenses rather than suing to exclude.[76] Indeed, recent empirical data shows that far from being subject to endless holdups and blockades, in both industry and universities, researchers have beaten whatever problems patents in this area might have imposed by adopting strategies of 'licensing, inventing around patents, going offshore, the development and use of public databases and research tools, court challenges and simply using the technology without a license (*i.e.*, infringement)' to achieve their particular goals.[77]

And the law correctly makes sure that property owners cannot avoid their share of these transaction costs. When property owners are not willing to incur the transaction costs associated with policing their own rights, the law exposes them to the risk of varying degrees of forfeiture. For example, if a patentee sits back for too long while letting others infringe, then later actions for infringement may be barred by laches.[78] And, if the patentee, instead of sitting back, actually leads the infringer to infringe, then an action for infringement may be barred by equitable estoppel.[79] Importantly, however, neither laches nor estoppel fundamentally threatens the IP system because each leaves it within the power of the IP owner to avoid the loss.

What is more, certain features inherent in the commercial law system impose much higher costs on property owners than might at first be apparent. Put differently, in the real world perfectly strong property rule protection for IP is not possible in the context of the existing system of commercial law for several reasons. First, as Ayres and Klemperer point out, uncertainty in how the rights will be enforced in court, functions the same as enforcing those rights with liability rules,[80] and largely because of the reward theories themselves there is substantial uncertainty in the rules governing the rules for obtaining IP rights,[81] transacting over IP rights,[82] and enforcing IP rights.[83] Second, 'the ability for an infringer to be kept effectively judgment proof through corporate and bankruptcy laws may also operate as a form of liability rule gloss on the present property rule regime'.[84] Third, '[o]therwise infringing uses that are by or for the federal government enjoy sovereign immunity protection that effectively results in a compulsory licensing regime'.[85] Therefore, total restriction on access under a property rule always can be avoided to some extent because at least some liability rule treatment always is available for IP.

Behavioralism

Related to the problem of transaction costs is the problem of 'behavioralism', which refers to all of the ways in which human beings are not perfectly rational in making decisions and instead are said to be only 'boundedly' rational in that they suffer cognitive biases, framing effects, employ heuristics, etc.[86] While some scholars, such as Posner, have suggested that decision-making under conditions of behavioralism can be thought of as the same thing as perfectly rational decision making in a world of positive information costs,[87] other scholars, such as Williamson, suggest behavioralism really refers to something more complex.[88] As explained by Williamson, the problems of behavioralism include situations that simply are impossible to think through,[89] the problems of misconception, like short-sightedness and incorrectly assessing probabilities, the problems of being rushed to make decisions,[90] and the limitations of language.[91] According to Williamson, an especially productive way to conceptualize the set of problems associated with behavioralism is taught by Simon as the 'idea of the mind as a scarce resource'.[92]

Regardless of precise etiology, the problems of behavioralism have a number of manifestations. Decision-making processes reveal strategies that, using the terminology of Simon, seek to 'satisfice' rather than 'optimize'; or in the more modern parlance, employ 'heuristics', as explored more recently in the work by Amos Tversky, Daniel Kahneman and Paul Slovic.[93] Other manifestations include risk and loss aversions,[94] and various cognitive biases such as primacy and recency,[95] framing,[96] anchoring,[97] and overoptimism, overconfidence, and egocentricism.[98]

Another component of the behavioralism problem is the problem known as 'groupthink'.[99] There are several components to the groupthink problem. One involves the heuristic individuals use to avoid having to re-think problems that they think already have been thought through sufficiently by trusted others, thereby creating what Cass Sunstein describes as an 'information cascade'.[100] Presumably, the opposite effect is also seen, whereby the heuristic is one of mistrust, not trust, and so the information content takes on the opposite sign.[101] A related component, also explored by Sunstein, is that individuals may appear to change or even actually change their views and behaviors in response to perceived peer pressure.[102] What is more, once group think has set in, there may be a lock-in effect, as pointed out by Arrow:

> [Social and political] agreements are typically harder to change than individual decisions. When you have committed not only yourself but many others to an enterprise, the difficulty of changing becomes considerable . . . [103]

An additional component of the groupthink effect is tied to the phenomenon of fashion. Sometimes a particular behavior, view, slogan, manner, or appearance is desired in its own right, as an affirmative expression of a discrete fashion preference – a fashion statement.[104] And, as evidenced by the cyclical nature of changes over time in the width of men's neck ties, fashion is fickle and so the fashion effect may be either to conform to the groupthink or to deviate from it. That is, an individual might either adopt or eschew groupthink as an affirmative fashion statement. Sometimes the culture is in fashion and sometimes the counter-culture is in fashion.

The behavioralism literature does add a great deal to our understanding. But some of the policy prescriptions that might at first blush seem to follow from it may not be so prudent. Consider, for example, switching to liability rule treatment as a strategy for avoiding irrational hold ups.

Several countervailing concerns must be addressed. First, if the ability to avoid the property rule treatment hinged upon the failure of a deal getting done, then there would be a markedly increased incentive for those wanting to obtain use through court-ordered terms to resist striking licensing deals. A legal test that rewards a failure to cooperate would lead to a decrease in cooperation, not an increase. Second, the legislators, administrators or judges who would be asked to determine when this should take place are themselves individuals who also face their own behavioralism limitations. Third, because they are government actors they trigger the public choice concerns discussed earlier.

Monopoly effects

Whenever property rights are used they trigger at least to some extent the problem of monopoly effects. But, as with the problem of transaction costs, the problem of monopoly effects often is misunderstood. One consensus lesson of economics, NIE and otherwise, is that markets are not perfect and they do fail. Each of the problems explored in this chapter can be, and often is, viewed as a type of 'market failure'.[105] Nevertheless, applying a theory of second-best, the mere identification of market failure does not in, and of, itself justify a call for resolution because it is the all-in comparative analysis among truly available options that should drive policy. Put differently:

> Traditional economics ascribes departures of actual market organizations from the ideal type of *perfect markets* to monopolistic practices. The approach of [NIE], on the other hand, holds that because of transaction costs, and thus informational problems, such departures *may* serve economizing purposes.[106]

For example, rules limiting competition, such as those limiting access to the stock exchange, can have many positive, or efficiency-promoting, effects: '[t]he exchange organizes not only the conclusion of contracts but also all associated transaction activities (from search to enforcement) . . . .'[107]

This does not mean that all market failures should be embraced. Rather, the general point is that when thinking about market failures it is essential to keep track of the real costs and benefits of all available options. In addition, there are two more specific points that need to be kept in mind when thinking about the ways markets work or do not work within the context of property rights in general and IP in particular. The first is the distinction between *ex ante* and *ex post*, or the distinction between dynamic and static efficiency. The second is the precise nature of the *inefficiency* (in contrast with what some see as *unfairness*) associated with monopolies.

While there is debate about exactly how rational or irrational individuals are when they make decisions about whether and how to act there is consensus that individuals do make such decisions and do plan. The term '*ex ante*' refers to the time period before a decision is made about a given action. The term '*ex post*' refers to any of the times afterwards. That is, the information and other resources an individual has *ex ante* will impact the decision-making process.

This includes not only what is known, but what is expected. As a result, there can be feedback between the *ex ante* and *ex post* worlds because individuals interpret events in the world around them as having some predictive value for the way events in the future will unfold. As studied in the work on rational expectations by Robert Lucas,[108] individuals constantly update and reinterpret information presently available to make best estimates about the future.[109] In game theory terminology, the point is that life is a multi-cycle game, not a single-cycle game, and individuals may use information from past cycles of the game when making decisions about how to play future cycles.[110] Individuals may change their expectations about what may happen to a given state of affairs in the future based on what they perceive to be happening to similar states of affairs in the present and past. If individuals perceive that property rights and contracts are not being enforced, they may have less faith in property rights and contracts being enforced in the future, all other things being equal. As investment in such property rights and contracts becomes less attractive, ordinary incentive analysis suggests that individuals will shift investments towards other activities. Indeed, the literature on private ordering places great emphasis on the role of *ex ante* predictability and certainty in property and contract enforcement for facilitating efficient investment and other decision-making over time, or in the dynamic sense.[111]

This dynamic approach can be in tension with other more static approaches to efficiency, which may see resource distributions at any point in time as sub-optimal. For example, a promise to make my car available to you at a particular time if you elect to use it then may put us in a position when that time arrives in which the car is not in use by anyone.[112] In the static sense, at that moment in time, it may indeed look as though the car is being allowed to go to waste, which would be inefficient.[113] Yet, if I am allowed to deploy the car to other uses to avoid the risk that it might go unused, then your expectation that it will be available for your use if you so choose will be dashed. What is more, if you know this *ex ante*, then you may not even be willing to enter into the contract to reserve the car in exchange for some other compensation, such as money, or you will be willing to pay only a lesser amount. Thus, in the dynamic sense, the expected future abrogation of the contract to provide the car that presumably would make both you and me better off because we each would elect to enter into it in the first instance, may make the contract one that is less likely for us to consummate *ex ante*. As a result, over time we cannot engage in as many productive exchanges as otherwise. Put differently, there would be dynamic inefficiency.[114]

It is recognized that recent work by Ian Ayers and Eric Talley, and by Jason Scott Johnston, shows how, due in large part to many of the behavioralism problems explored earlier, uncertainty in enforcement may in some cases improve the ability to negotiate over property rights and contracts by decreasing hold-out problems through a feed-back mechanism in which uncertainty makes more credible the threat of infringement or breach *ex post*, which may cycle back to decrease the incentive *ex ante* for the rights-holder to hold out in the first instance.[115] Nevertheless, other recent empirical work by Rachel Croson and Johnston shows that in other cases uncertainty degrades the ability to reach dynamic efficiency.[116] Indeed, other work by Ayres and Robert Gertner highlights the importance of at least some certainty through the use of what they term 'penalty default rules' because they will have the impact of bringing to light information about potential negotiations and help avoid opportunism by one party attempting 'to get a larger piece of the smaller contractual pie'.[117] At bottom, at least in many cases private bargaining over property rights can be more efficient if the right is clearly defined *ex ante* according to a predictable rule, rather than made *ex post* by a judge applying a standard.[118]

The difference between *ex ante* and *ex post*, or dynamic and static efficiency, also matters beyond the narrow setting of individual transactions discussed above – although that is not irrelevant – because in many ways change is desirable in and of itself. For example, as resources such as fossil fuels become depleted, we must change to make use of alternative energy sources. Innovation that occurs over time can improve the size of the pie for everyone by making available more options.[119] Put simply, the distinction between dynamic and static efficiency is particularly important for IP because IP is focused on innovation over time.

The problem of monopolies is another specific point that must be kept in mind when thinking about the ways markets work or do not work within the context of IP. Because monopolies can create important inefficiencies, they have been the subject of substantial attention by both lawyers and economists. Indeed, the core purpose of antitrust law is 'to root out unreasonable restraints of trade and transactions that substantially lessen competition or tend to create monopoly'.[120] The central inefficiency associated with monopolies is the creation of dead weight loss by the monopoly's ability to set price above marginal cost, or to have power over price.[121]

But, NIE suggests several reasons why the extent of this inefficiency may not be the same in practice as it is in theory.

First, monopoly is a term that relates to a market, not to any particular goods or services sold in that market.[122] There often is a difference between a product or service market and an IP asset. Consumers often buy computers that essentially involve the licensing of hundreds of licensed IP rights – for hard drive, processors, DRAM, other chips, etc. – without acting as direct customers with respect to any of the IP owners.

While in a certain sense every property right can be thought of as a monopoly, only those that convey effective control over an entire market can have the troubling economic inefficiencies associated with monopolies. For example, the owner of a hypothetical piece of real estate, Blackacre, can exclude use of that particular parcel, but must compete with other parcels of land in the market for land generally. Indeed, while the amount of real estate in the world actually is limited by the surface area of the planet, unless it turns out that the scope of human intellectual content is presently so close to the limit of its full potential that there is no reason to think that for IP the long run monopoly impact of a given property right is likely to be any worse than for real property; and instead it is likely to be much less. Nevertheless, in the short run for at least some goods or services the broad scope of some IP rights may convey what at least some would see as market power with respect to consumers having a particularly dire need – such as medical patients in need of a particular patented drug.

Second, the economic inefficiency that is associated with a monopolist's power over price is not inevitable. More specifically, the inefficiency is tied to the potential for a decrease in quantity (not an increase in price) as compared with the perfectly competitive model. If the monopolist is able to engage in perfect price discrimination, then the quantity produced will be the same as if there were competition, and while the price charged to at least some consumers will be higher, there will be no dead weight loss inefficiency.[123] While perfect price discrimination, like perfect anything, is not possible in the real world, the extent to which the monopolist can engage in price discrimination may mitigate the practical extent of the theoretical static inefficiency associated with monopoly dead weight loss.[124]

Anticommons, patent thickets, and patent trolls

A final problem that some think arises when property rights are used, especially when property rights in IP are used, is the problem Michael Heller termed the 'anticommons',[125] and others term a 'patent thicket'.[126] This chapter suggests that the anticommons problem really is not a problem of property rights and instead is a problem associated with using other types of barriers. From this perspective, the new anticommons literature can be viewed as in a sense at best providing merely another term for what previously have been known the problems of 'permit thickets' and 'license Raj' and at worst both facilitating monopolization and frustrating those aspects of property rights that work well for the private ordering and coordination that help increase access.

Heller's contribution to the property literature regarding anticommons was originally based on his study of real property in the post-socialist economies of Eastern Europe, but he has also applied it to IP. As described by Heller:

> Consider new areas for property law, such as the problem of spurring private investment in biomedical research or creating well-functioning markets in post-socialist economies . . . . By drawing the wrong property boundaries around resources, by fragmenting ownership too much, it turns out that privatization can destroy resource productivity in enduring ways. To capture these unexpected results from excessive privatization, I have proposed the idea of anticommons property, an image that goes beyond the old trilogy [private, commons, and state] and crystallizes emerging real-world property relations that had previously remained invisible . . . .[A] resource is prone to underuse in a tragedy of the anticommons when multiple owners each have a right to exclude others from a scarce resource and no one has an effective privilege of use. In theory, in a world of costless transactions, people could always avoid common or anticommons tragedy by trading their rights. In practice, however, avoiding tragedy requires overcoming transaction costs, strategic behaviors, and cognitive biases of participants, with success more likely within close-knit communities than among hostile strangers. Once an anticommons emerges, collecting rights into usable private property is often brutal and slow. . .. I developed the idea initially from closely observing privatization in post-socialist economies. One promise of transition to markets was that new entrepreneurs would fill stores that socialist rule had left bare. Yet after several years of reform, many privatized storefronts remained empty, while flimsy metal kiosks, stocked full of goods, mushroomed up on the streets. Why did the new merchants not come in from the cold? One reason was that transition governments often failed to endow any individual with a bundle of rights that represents full ownership. Instead, fragmented rights were distributed to various socialist-era stakeholders, including private or quasi-private enterprises, workers' collectives, privatization agencies, and local, regional, and federal governments. No one could set up shop without first collecting rights from each of the other owners.[127]

Heller seems to suggest that what he terms 'fragmentation', or excessive numbers of rights holders, is key to the anticommons effect because the transaction costs of dealing with so many claimants will dominate.[128]

But fragmentation itself is not the key to the anticommons effect that is observed in post socialist economies. What really drives the problem is the lack of what Alchian and Demstez call a 'residual claimant'.[129] To provide a brief summary definition at the outset, in the context of the anticommons problem caused by many holders of a right to respond 'no' to requests for permission, a residual claimant is essentially an individual who is able to extract private value from such a request by electing to respond with a 'yes'. But to more fully understand the nature of the issue further elaboration is required.

As Buchanan and Yoon explain, there actually exists 'a formal symmetry between the over-usage of a resource because of common (multiple) access and the under-usage because of multiple exclusion rights'[130] In highlighting this symmetry, they then point out that in both cases (commons and anticommons) the heart of the problem can be tied to the nature of the holders of the right (to use or exclude, depending on whether the tragedy is one of commons or anticommons). More particularly, according to Buchanan and Yoon, the problem lies in whether the holders have 'noneconomic motivations' in that they are those 'who cannot or may not desire to, capture directly pecuniary gains', meaning that their goals may not be 'primarily distributional but instead may reflect different objectives'.[131] Indeed, Buchanan and Yoon warn of the potentially pernicious impact in either case (commons or anticommons) of the 'genuine zealot . . . [who] may be insensitive to proffered compensations'.[132] Therefore, the concern Buchanan and Yoon highlight is that the crux of the problem for both commons and anticommons relates to the ability of those engaged in the group activity to coordinate with each other, but when the individuals have noneconomic motivations they are unlikely to so coordinate unless they happen to share some other coordinating attribute, such as being close-knit.[133]

In contrast, as discussed previously, coordination is a central problem studied by NIE and one general response to coordination problems can be property rights. While at first blush, given the way Heller presents the anticommons problem, it would seem that property rights are more a part of the problem than a part of the solution, it turns out this just is not so. Property rights provide individuals with the economic motivation to engage in trades with each other. Indeed, the easier it is for the holder of a property right to engage in such a trade and the greater the value that the individual can extract from the trade (the greater the residual claim), the greater the motivation and ability the individual has to engage in it.[134]

What actually drives the anticommons problem in the post-socialist economies is both the lack of residual claim and the lack of clarity and certainty that are associated with the pertinent rights of exclusion.[135] Richard Epstein and Bruce Kuhlik begin the discussion by pointing out in response to the perceived anticommons problem relating to IP that one distinguishing feature of the anticommons in the post-socialist economy is that efforts by the bureaucrats to engage in open trading of their permission for personal gain are likely to trigger various forms of criminal liability for graft, bribery, public corruption, etc.[136] But the differences are even greater than they indicate. In such a sequential bribe situation there is a greater degree of uncertainty that each bribe will be either needed or effective. This is in part because those being bribed cannot openly coordinate. It is also because some of those whose permission would be needed might not even be open to being bribed. They may justifiably be steadfastly acting to prevent an activity they see as bad.[137] Alternatively, they may derive more benefit – perhaps sense of control or power or even just some other kind of perhaps perverse pleasure – from simply being able to say 'no' than from what otherwise might be obtained in exchange for saying 'yes'.[138]

The anticommons problem in the post-socialist environment – indeed the anticommons problem, period – is tied to the inability of those who hold rights of exclusion to negotiate openly for a way to extract value from a decision to give reliable permission rather than withhold permission or give faulty permission. There may be no residual claimant who can openly sell a 'yes'. There may be no clarity about who to even approach to buy a 'yes' or what to give in order to get a 'yes'. There may be no certainty about whether a 'yes' will even be effective. Therefore, there is a huge difference between the openly tradable nature of property on the one hand and the anticommons on the other hand.

Simply put, the anticommons problem can be seen as just another label for what Epstein earlier referred to as a 'permit thicket'.[139] Earlier still the problem was labeled in India, after the removal of British rule, which was also called 'Raj', where it was said that Raj had been replaced by 'License Raj' in the form of excessive and unpredictable requirements for permits and licenses from the many branches of the central government in order to conduct many important business activities.[140] In essence, the anticommons problem can be seen as a coin having two poisonous sides: the pernicious 'permit thicket' or 'License Raj' implications for taxing and retarding development on the one hand; and the 'tollbooth' implications of extortion by agencies on the other hand.

Another version of the anticommons problem for IP appears to be what some call the problem of 'patent trolls'.[141] The argument seems to be that 'patent trolls' hold their patents neither for development nor for prospective licensing, but solely to hold up others who accidentally stumble in their path. To the extent the concern about trolls reflects anxiety about the uncertainty of the scope and validity of patents, as well as the high cost of patent litigation – both of which would provide potential opportunities for 'trolls' to exploit even weak or low-value patents – then the problem can be best addressed using various tools for policing bad patents.[142]

But the pernicious impact of the troll is limited to a large extent by very practical economic factors. First, all patents are wasting assets, in that they have a life capped at less than 20 years, and are subject to defenses based on laches and estoppel. Second, a decision to lie in wait causes the troll to lose income that would have to be recouped in the future. But just like in the context of predatory pricing, the promise of that future gain is risky.[143] Indeed, just like a fallow plot of land may attract offers for development, so a patent posted on the PTO web page and searchable for free, as all are, provides sufficient information to attract anyone seriously interested in practicing the covered technology. A patentee who is not looking to sell or license is not beyond reach of those who wish to buy or license. Those sets of economic forces on both parties help explain why, once the court made clear an injunction was imminent, even the infamously bitter litigation over the BlackBerry® service settled before any disruption of service took place. What is more, the settlement price in that case is both significantly below independent estimates that reflect the holdout risk, and even more significantly below the licensee's reserves of cash and cash equivalents.[144]

Indeed, the raw numbers suggest that one underappreciated element of the delay in settlement in that case may have been restrictions on the market for corporate control, not the problems of anticommons, patent thickets or patent trolls. The actual settlement price suggests that the infringer either was acting rationally in holding out because of the uncertainty that there was going to be an injunction (in keeping with the view that property rules can encourage deals and liability rules can frustrate them), or it was acting irrationally in not closing a deal at such an attractive price – a price in line with market estimates and lower than its own private estimates as evidenced by the size of its reserves of cash and cash equivalents. If the market for corporate control were working better, there might have been enough gains to be had by settling the case sooner than a raider would have done a takeover, fired the leadership, and struck a deal with the patentee.[145] Earlier settlement also would have saved more goodwill for the infringer, RIM, the maker of BlackBerry®, which now has more competition.

NIE AND THEORIES OF IP

Because the conventional normative theories of IP – reward and prospect or rent dissipation – fail to address the helpful coordination effect explored above, the policy prescriptions they generate fail to facilitate the downstream access that can be achieved through coordination. What is more, by generally weakening IP rights, the prescriptions that flow from conventional theories only serve to facilitate the type of coordination that increases anticompetitive effect. The focus on coordination offered here helps explain why particular features of the positive law IP regimes of patent, trademark and copyright are working and why others are not. This coordination view thereby can inform policy debates about which aspects of these regimes are best candidates for change.

Conventional IP theories are focused either on providing direct incentives as a tool for increasing access or on controlling rent dissipation, and these various theories are critiqued at some length in prior work by the present author.[146] The commercialization theory and its component registration theory also are explored at some length in earlier works by the present author,[147] and so only an overview is provided here, along with a discussion of newer implications and applications. The commercialization theory of IP views IP rights backed by property rules as important tools for facilitating the downstream commercialization of the subject matter that is protected by IP rights, after that subject matter has been made.[148] This downstream commercialization requires coordination among the many complementary users of the IP subject matter including developers, manufacturers, laborers, managers, investors, advertisers, marketers, etc.[149] Providing a focal point, or beacon, the publicly recorded IP right helps each of these individuals to find each other,[150] and then by cracking the Arrow Information Paradox otherwise facing them, helps them negotiate with each other.[151] At the same time, therefore, IP rights facilitate the creation and maintenance of both diversity and socialization among individuals within the market by providing these diverse individuals with incentive and means for coordinating with each other. In addition, as studied by the registration component of the commercialization theory, the positive law rules for determining when a valid IP right may be obtained protect reasonable investment-backed expectations (and thereby decrease the risk of asset specific investments and opportunism) by making sure that the right to exclude does not block activities individuals otherwise are doing, and they do so with relatively low administrative and public choice costs.[152] In this regard, the commercialization and registration theories are essentially two components of the coordination view explored here.

What is perhaps most striking about the commercialization theory, given that it is not either the majority or the minority views within the conventional literature on the law and economics of IP, is that, as a matter of historical fact, it was the central motivation behind the framing of at least the present patent system, the 1952 Patent Act, as well as part of the motivation behind the present trademark system, the 1946 Lanham Act.[153] Moreover, while the commercialization theory is discussed by the conventional literature, it is often misperceived in at least two ways. First, the theory is often misperceived on its own terms. Second, the solutions it offers for many of the problems generally identified with IP rights often are overlooked. Both types of misperceptions are discussed below.

Correcting Conventional Takes on Commercialization Theory

The focus of the commercialization theory is on the incentives for diffuse individuals to decide individually to act in a way that ends up being coordinated.[154] While rewards may provide an incentive to act to the individual reward recipient, rewards do little compared to property rights to bring that individual together with all other complementary users to engage successfully in the complex commercialization process.[155] Regrettably, this simple mechanism of the commercialization theory's coordination function is often misunderstood in the literature in several respects.

First, the link between the commercialization theory and the prospect or rent dissipation theories often is confused.[156] Put simply, the *commercialization* theory focuses on the ability for IP to coordinate efforts among *complementary* users of the asset to *increase* (or avoid insufficient) use of resources, whereas *prospect* theory focuses on the ability for IP to coordinate efforts among *competing* users of an asset to *decrease* (or avoid excessive) use of resources.[157] Therefore, efforts to respond to the prospect and rent dissipation theories' concerns about overuse (rent dissipation) are inapposite to commercialization theory.

Second, the link between the commercialization theory for IP and the theory of property rights generally is often overlooked. That is, much of the conventional literature overlooks the coordination function in its entirety and simply lumps the property rights aspects of the prospect theory by Kitch with the property rights aspects of the work by Demsetz on internalizing externalities.[158] But, as discussed above, property acts as a tool for facilitating coordination among complementary users of the assets protected by IP in a way that is not explored in the early Demsetz work or in the work by Kitch but is explored in the work by the present author on the commercialization theory.[159]

Third, the commercialization theory also has been confused erroneously with the work of Schumpeter in being focused on the IP holder's assertion of control.[160] While the commercialization theory is focused on who will have an incentive and ability to negotiate with whom, it is agnostic on the question of who will end up controlling those negotiations. In fact, control will be a function of a great many factors other than who owns the patent. For example, the parties' relative wealth effects, bargaining positions, negotiating skills, other resources, holdout prices, alternative options, etc., will each impact the bottom line issue of control. In a world in which each market player may bring its own skill sets, patent sets, technology sets, and other assets and opportunities to bear on development of a particular patented subject matter, the end result of who will control subsequent development and use of that subject matter is unclear, and indeed is left to the market and private bargains. For this reason, for example, the concern raised by Robert Merges and Richard Nelson about control by the owner of an IP right that they consider to be too broad is also overstated.[161] The mere fact that a particular IP right is broad does not mean that its owner will control negotiations with others in that same technology. In this regard, the coordination function of IP is distinct from the two extremes of open competition on the one hand and control on the other hand. The IP right facilitates coordination among both competing and complementary users of the asset without determining who will control in any given case. The commercialization view of IP focuses on the importance of IP backed by a property right as a tool for facilitating such a division of labor and other forms of specialization.

Fourth, the importance the commercialization theory places on the distinction between *ex ante* and *ex post* may be confused by the different use of those terms recently by Mark Lemley.[162] Under the commercialization theory, for IP to serve the commercialization function, the rules about how IP can be obtained and enforced must be knowable to all market actors *ex ante*, in advance of their decisions about whether to act. This means that regulation and liability rule treatment may be suspect to the extent they have the effect of re-writing agreements or changing rules *ex post*.[163] When used in this context, the terms '*ex ante*' and '*ex post*' are used in their general sense, which is different from how they are used in the recent work by Mark Lemley.[164] Lemley uses the term '*ex ante*' in a special narrow sense to refer to the time period before any specific creative work is made.[165] Similarly, he uses the term '*ex post*' in a special narrow sense to refer to a time period after any specific creative work is made.[166] The commercialization theory relies on the term '*ex ante*' in the more general sense to refer to a time period before any given act occurs, with a focus on the importance of predictability. For example, this view of *ex ante* would focus on the period before the textured contracting needed to facilitate commercialization takes place. Similarly, it relies on the term '*ex post*' in the more general sense to refer to a time period after any given act occurs, again with a focus on predictability. For example, this view of *ex post* would focus on the period after the contracting has taken place. That is, as these terms are used for purposes of the commercialization theory, the focus is on the ability for private actors to predict a legal result before deciding whether, or how, to act on any specific issue. Under the commercialization view of IP, predictability *ex ante* is essential in facilitating private ordering.

Fifth, some have suggested that 'if patent law's concern is to ensure commercialization of inventions, then it is both overinclusive and underinclusive'.[167] The point is well taken, as far as it goes; but it may not account for the full reach of the commercialization theory. On the question of overinclusiveness, Abramowicz points out that 'sometimes first-mover advantages will outweigh second-mover advantages'.[168] This is correct. But only where a sufficient number of the complementary users of the asset believes *ex ante* that this is the case with sufficient conviction to take on the coordinating role will coordination so easily take place without the property right. This can and likely does happen. But the point of the commercialization theory is that IP rights can make it easier for this to happen in many more settings. On the question of under-inclusiveness, Abramowicz points out the need for commercialization of subject matter that does not meet the positive law rules for IP protection.[169] But the point of the registration component of the commercialization theory of IP is that the positive law rules for obtaining IP are normatively important for protecting the reasonable investment backed expectations of potential commercialization efforts by third parties.[170] Put simply, these positive law rules about IP validity are essential for making the IP system work well. The extent to which they leave behind some subject matter is a reason to explore the use of other tools to help coordination in those areas, such as perhaps the firm, or maybe the government. IP does not solve all problems and it is only offered as an additional tool for helping to solve some.

Commercialization Theory's Overlooked Solutions

The commercialization theory also provides several overlooked solutions for the underlying problems often associated with IP. These include the problems of transaction costs, anticompetitive effect and access.

The commercialization theory sees the IP right backed by the credible threat of an injunction as playing an essential coordinating role for all the players in the commercialization process.[171] Those wishing to buy title to or permission under the IP right must negotiate with the IP holder. As long as the existence of the IP right and the identity of the IP holder are readily discernible,[172] each of the putative participants in the commercialization process will have an individual incentive to seek out and negotiate with that person, and through that person with each of the others.

While the reward literature in particular has emphasized the concerns about output restrictions, or problems of access, the below discussion points out why such concerns are significantly less severe than perceived, and indeed why in some cases property rights may be essential for mitigating them. It also shows both why the concerns about government and public choice must not be overlooked as well as the ways in which these problems either can be magnified or mitigated by particular aspects of positive law IP regimes. As a result, it shows several aspects of the present positive law regimes that are candidates for change because they only exacerbate the problems of anticompetitive effect and access.

As discussed above in the context of reward theories, much of the literature on IP rights is consumed with concerns about limiting the potential monopoly power associated with property rights in IP. But actual empirical data are inconclusive on whether, for example, patents have been used to facilitate cartel behavior.[173] Although a dominant concern of the reward literature on IP is that IP rights can confer power over price of the type generally associated with monopolies, the connection this literature draws between IP and monopolies in essence is backwards in several respects. That is, as discussed below, IP rights often just do not confer monopoly power; and yet they can be essential anti-monopoly weapons – their availability can serve as an effective anti-monopoly vaccine for a market.

IP rights often do not confer monopoly power in large part because there is rarely a one-to-one correlation between any particular IP asset and a market.[174] In addition, IP rights face competition from alternative technologies, extant and potential.[175] At bottom, for example, even a patent on the better mousetrap faces competition from existing spring and glue traps, the threat of future traps, and, of course, cats.[176]

What is more, IP rights can facilitate market entry, at least so long as they are backed by property rules. As a result, they can be powerful anti-monopoly weapons.[177]

For example, the commercialization theory suggests that if meaningful IP rights had been available in the computer software industry in the 1970s and 1980s,[178] by the time of the Microsoft antitrust suit the industry likely would have been characterized by a medium number of medium-sized players rather than a single large player.[179] 'According to Judge Frank, in this context the David Co. v. Goliath, Inc., competition is dependant upon investment in David Co., which will not occur unless it is armed with the patent slingshot.'[180]

As another example, consider the impact on competition of the 1980 shift in positive patent law that opened patents to the field of modern biotechnology. Only in the US and only since 1980 have patents been available in modern biotechnology.[181] While the US, Europe, and Japan each had large biotechnology companies often collectively called 'Big Pharma'[182] before 1980 and still have them after 1980, only in the US and only since 1980 has the biotechnology industry also included a steady pool of roughly 1,400 small and medium-sized companies that is also consistently turning over.[183]

In addition, the gains IP rights offer for competition and market entry across markets at any one time as well as across time offset the potential for individual dead weight loss in cases where an IP right truly conveys a monopoly at some point in time for some market. In part, this point is tied to the distinction between dynamic and static efficiency, which is to say that the static inefficiency associated with monopoly dead weight loss may be outweighed by the dynamic efficiency gains associated with innovation and entry.[184]

What is more, IP rights can and often do operate to facilitate price discrimination, which can mitigate the dead weight loss efficiency considerations of monopolies.[185] That is, the use of property rights in IP

> is also consistent with another basic work by Demsetz in which he demonstrated that (1) private producers can produce public goods efficiently given the ability to exclude nonpurchasers and (2) price discrimination is consistent with competitive equilibrium for such public goods.[186]

Indeed, because of the doctrines of indirect infringement, IP rights facilitate price discrimination through tying in a great many more cases than otherwise, including for example where tying is not facilitated by technological constraints.[187]

At bottom, while IP rights do give some power over price and therefore are associated with some dead-weight loss in theory, the actual monopoly effects of IP are overstated and the anti-monopoly benefits of IP are overlooked. In the real world, the benefits of this type of market power for capital formation and dynamic competition must be weighed against its theoretical cost in the form of static dead-weight loss. Indeed, the lessons of the literature on second-best and the basic comparative institutional analysis of NIE are that there are many reasons why it may be prudent to avoid letting anti-monopoly concerns drive us to respond too aggressively to every occasion of power over price. In this sense, for example, the reward literature's concern over mitigating monopoly effect of IP can be seen as unduly exalting static efficiency over dynamic efficiency.[188]

While the commercialization theory sees the nature of IP as a right backed by the credible threat of an injunction to be the core benefit of IP in providing coordination, it recognizes that this coordination requires transactions. But transactions have both good and bad components to them, as do their realistic alternatives, and one lesson of NIE is to engage in comparative institutional analyses.

One of the central focuses of the reward theories is on the transaction costs associated with IP compared to a commons. Thus, it is appropriate to compare the transaction costs of exchanges over property rights in IP against the transaction costs of exchanges over what otherwise would be the subject matter of IP but instead were within a realistic commons, such as the putative commons of basic academic knowledge.[189] Prior work by the present author has shown that even this so-called 'commons' is riddled with its own form of less commercial but nonetheless important property rights known informally as 'kudos', which include more personal and less fungible assets generally associated with academic and public sectors such as reputational benefits, fame, promotions, awards, titles, etc.[190] A comparative institutional analysis reveals why, for exchanges in that setting of a putative commons compared with the same setting having added IP rights, the transaction costs of exchanges are likely to be worse without IP than with IP because IP brings increased wealth and diversity to that market.[191] One of the lessons of NIE explored earlier is that transaction costs are likely to be more pernicious in thinner markets than in thicker markets,[192] and the use of IP thickens the market.[193] As also discussed earlier, recent work by Buchanan and Yoon adds to this analysis by pointing out that exchanges in such a commons also are more likely to fail because of what they call the 'non-economic motivations' associated with such assets.[194] Put simply, there are reasons to think that transaction costs are likely to be higher for a commons compared to for IP, and recent empirical work by John Walsh, Charlene Cho, and Wesley Cohen did not find transaction costs problems associated with patents in basic science, essentially because potential infringers engaging in low value uses were simply being allowed to infringe with approval, albeit tacit, from patentees.[195]

Somewhat related to the concerns over transaction costs in the reward literature are similar concerns about behavioralism. More specifically, in response to concerns about behavioralism leading to failures in transactions over IP rights, commentators have called for regulation of IP rights through the imposition of liability rule treatment and greater antitrust enforcement.[196] To be sure, like all actors in the real world, IP owners are not perfectly rational. That is, people are only 'boundedly' rational, in that they suffer cognitive biases, framing effects, employ heuristics, etc.[197] On the one hand, identification of behavioralism concerns does suggest reasons to be skeptical about the ability for individuals actually to achieve for themselves what is in their own best interest, and behavioralism has justified resort to liability rules, regulation, immutable contract terms, etc. On the other hand, the individuals the government will use to affect these responses – legislators, regulators, and judges – are, of course, human beings, too, and so will also suffer the limits of behavioralism.[198] What is more, to the extent these government decisions will occur *ex post*, they will interfere with *ex ante* incentives. Finally, regulation brings with it the inevitable costs of government, including the tollbooth and rent-dissipation problems of agency capture, as well as the real concomitant problems of permit thickets, License Raj or anticommons; and, as discussed below, these indeed can be real problems within the context of IP.

Ironically, much of the recent literature advocating enhanced regulation of IP rights is tied to anticommons concerns.[199] In contrast to the real anticommons problem of the post-socialist economy discussed earlier, an IP owner extracts private value from the IP right to exclude use by openly trading permission for use in exchange for money or other consideration.[200] As discussed earlier, the economic motivations associated with such 'residual claims' are precisely what mitigate anticommons concerns.[201] While an IP owner may have some incentive to suppress the subject matter protected by IP, this incentive is countered by the uncertainty that higher untapped value may lie in wait.[202] Put simply, the resulting social value of IP rights is that they encourage their owners and others with whom the owners can coordinate to discover and market methods for pushing use towards the full competitive level so the IP rights will not create anticommons problems, in biotechnology, software technology, or even for more mundane technologies like nails and screws.[203] Indeed, recent empirical work by Ronald Mann has found that even in the controversial area of business method patents, there turns out not to be any serious 'patent thicket' problem.[204]

CONCLUSION

Although many different useful perspectives have been offered in the literature about the goals society should have in mind before deciding to create property rights in general and IP rights in particular, too often an overlooked goal is coordination. This chapter suggests that coordination of the type needed to facilitate commercialization is a goal that can be achieved by property rights in general and IP rights in particular. Coordination of this type is useful in helping diverse members of society remain diverse from each other in terms of skills, assets and preferences, etc., while at the same time interacting with each other as complementary users of assets in a way that helps bring those assets to market. Focus on coordination is offered as an alternative to focus on other goals that have been suggested in the literature including internalizing externalities, avoiding rent dissipation, and providing direct incentives. And property rights are offered as a tool for achieving this goal that is an alternative to other institutions and organizations including norm communities like open source projects. Recognizing that each institution and organization will have benefits and costs the chapter also highlights strategies for helping to ensure that the benefits of property rights are enhanced while the costs of property rights are mitigated or otherwise structured to be easily borne. The chapter also shows ways in which the liability rule prescriptions that dominate the literature can have the counter-productive effect of facilitating the type of coordination that frustrates competition.

NOTES

1 The author gratefully acknowledges financial support for this work from the 2003–2004 and 2004–2005 W. Glenn Campbell and Rita Ricardo-Campbell National Fellowships and Robert Eckles Swain National Fellowships at Stanford University's Hoover Institution, the 2003–2004 Israel Treiman Faculty Research Fellowship at Washington University School of Law, as well as intellectual contributions from numerous colleagues.
2. *See, e.g.*, Donald S. Chisum, Craig A. Nard, Herbert F. Schwartz, Pauline Newman, and F. Scott Kieff, Principles of Patent Law 6-42 (2nd ed. 2001) (reviewing history of changes to patent law); Willaim F. Patry, Copyright Law and Practice, Washington, DC: BNA Press, 1-120 (1994) (reviewing history of changes to copyright law); Beverly W. Pattishall, et al., Trademarks and Unfair Competition, Albany: Matthew Bender & Co., 1-5 (4th ed. 1998) (reviewing history of changes to trademark law); Frank I. Schecter, The Historical Foundations of Trademark Law (1925) New York: Columbia University Press. *See also* F. Scott Kieff, *Property Rights and Property Rules for Commercializing Inventions*, Minnesota Law Review 85, 697–754 (2001); F. Scott Kieff, *The Case for Registering Patents and the Law and Economics of Present Patent-obtaining Rules*, Boston College Law Review 45, 55–123 (2003).
3. Indeed, there are two similar cases that have been sharing the high profile. One involves the BlackBerry® service, and the Supreme Court ultimately did not grant review of this one, and one involving the eBay service, which presently is before the Court. See, *NTP, Inc. v. Research In Motion, Ltd.*, 418 F.3d 1282 (Fed. Cir. 2005), *cert denied*, 126 S.Ct. 1174 (2006) (BlackBerry®); *MercExchange, LLC v. eBay, Inc.*, 401 F.3d 1323 (Fed. Cir. 2005), *cert granted* 126 S.Ct. 733. For more on the BlackBerry® case see *infra* notes 144-145 and accompanying text.
4. *See, e.g.*, Patently Absurd, *Wall St. J.* (Mar. 1, 2006), at A14 (criticizing a set of cases including NTP); Bruce Sewell, Troll Call, *Wall. St. J.* (Mar. 6, 2006), at A14 (criticizing both the NTP and eBay cases).
5. The label 'property rule' is used here as it is used in the classic Calabresi-Melamed framework under which an entitlement is said to enjoy the protection of a property rule if the law condones its surrender only through voluntary exchange. The holder of such an entitlement is allowed to enjoin infringement. An entitlement is said to have the lesser protection of a liability rule if it can be lost lawfully to anyone willing to pay some court-determined compensation. The holder of such an entitlement is only entitled to damages caused by infringement. See Guido Calabresi & A. Douglas Melamed, 'Property Rules, Liability Rules, and Inalienability: One View of the Cathedral', 85 Harv. L. Rev. 1089 (1972). *But see*, Jules L. Coleman and Jody Kraus, 'Rethinking the Theory of Legal Rights', 95 *Yale. L.J.* 1335 (1986) (offering a 'reinterpretation of the Calabresi-Melamed framework' under which property rules and liability rules merely represent two pieces of a broader 'transaction structure' in that they are two different approaches for setting forth 'conditions of legitimate transfer').
6. Compare *Madey v. Duke University*, 307 F.3d 1351 (Fed. Cir. 2002) (very narrow reading of generally-available experimental use exemption as a matter of common law) with *Merck KGaA v. Integra*, 125 S.Ct. 2373 (2005) (very broad reading of special,

statutory, experimental use exemption for uses that are reasonably foreseeably related to obtaining Food and Drug approval).

7. *See*, Ian Ayres & Paul Klemperer, 'Limiting Patentees' Market Power Without Reducing Innovation Incentives: The Perverse Benefits of Uncertainty and Non-Injunctive Remedies', 97 *Mich. L. Rev.* 985 (1999) (arguing that sufficient incentive to invent can be provided without the monopoly power associated with a property right).
8. *See, e.g.*, Maureen A. O'Rourke, *Toward A Doctrine of Fair Use in Patent Law*, 100 Colum. L. Rev. 1177 (2000) (offering fair use exception because of excessive transaction costs causing too many market failures in the transactions over IP rights as property rights).
9. *See, e.g.*, http://sciencecommons.org (advocating open-source approaches to science).
10. Representative examples from different times throughout the past century include the effort by Congress to create the Temporary National Economic Committee (also known at the TNEC), S.J. Res. 300, 75th Cong. § 2 (1938), the President's Commission on the Patent System, 'To Promote the Progress of . . . Useful Arts' In an Age of Exploding Technology (1966), and the year-long set of hearings jointly held in 2001 by the Federal Trade Commission and the Justice Department's Antitrust Division. See Notice of Public Hearings Competition and Intellectual Property Law and Policy in the Knowledge-Based Economy, 66 Fed. Reg. 58,146, 58,147 (Nov. 20, 2001).
11. For a discussion of conventional approaches see *infra*, the third section.
12. For a good introduction to NIE, see John Drobak and John Nye, Frontiers of the New Institutional Economics (Burlington, MA: Academic Press 1997) (volume of chapters honoring Douglass North and his contribution to the field of NIE). *See also* Thrainn Eggertsson, Economic Behavior and Institutions (Cambridge: Cambridge University Press 1990) (survey of NIE, or as Eggertsson refers to it: 'neo-institutional economics'); Philip Keefer and Mary M. Shirley, *Formal versus Informal Institutions in Economic Development*, *in* Claude Menard, Institutions, Contracts, and Organizations: Perspectives from New Institutional Economics (Cheltenham, UK and Northampton, MA, USA, Edward Elgar 2000, pp. 88–107).
13. Douglass C. North, *Prize Lecture,* available on-line at http://www.nobel.se/economics/laureates/1993/north-lecture.html. The logical relationship between organizations and institutions can be conceived topologically in at least two ways. The first sees organizations as operating within institutions, such as a firm following society's laws and rules. The second sees institutions as operating within organizations, such as the internal set of rules, norms, and enforcement characteristics that govern those within the organization and, in effect, define who is within the organization and who is outside of it.
14. For detailed explorations of NIE, see Eirik G. Furubotn and Rudolf Richter, Institutions and Economic Theory: The Contribution of the New Institutional Economics (Ann Arbor: University of Michigan Press 2003) (reviewing field and collecting sources) and Masahiko Aoki, Towards a Comparative Institutional Analysis (Cambridge, MA: MIT Press 2001) (applying game theory to comparative institutional analysis and NIE).
15. *See* Harold Demsetz, *Toward a Theory of Property Rights*, 57 Am. Econ. Rev. 347, 356 (1967) (explaining the emergence of property rights when benefits of internalization outweigh transaction costs of recognition of property rights).
16. *See* Ronald Coase, *The Problem of Social Cost*, 3 J. L. & Econ. 1 (1960) (pointing out how a fully defined set of property rights can allow for externalities to be internalized).
17. Pigou saw factory chimney soot as a problem of externalities imposed on others in the environment around the factory and argued that taxes or subsidies could be used properly by government to encourage such factories to account properly for the benefits and harms they project on those around them. According to Pigou, 'resources devoted to the prevention of smoke from factory chimneys' provide an 'uncompensated service', or what some would call a positive externality, while smoke 'inflicts a heavy uncharged loss on the community', or provides what some would call a negative

externality. *See generally* Arthur C. Pigou, The Economics of Welfare 160–61, 166–68 (1920). *See also* Arthur C. Pigou, Wealth and Welfare (1912).

18. *Id.* at 353 (noting that property rights did not emerge among those living on the southwest plains because the benefits would have been less since there were no animals of commercial importance comparable to the furry animals of the north whose pelts were tradable and because the costs would have been more since the animals that were there tended to wander more).
19. Kieff, *Commercializing Inventions*, *supra* note 2, at 718, n.95.
20. For more on the link among specialization, division of labor, and coordination, see generally Gary S. Becker & Kevin M. Murphy, *The Division of Labor, Coordination Costs, and Knowledge*, *in* The Essence of Becker, 609 (Ramon Febero & Pedro Schwartz, eds., Stanford: Hoover Institution Press 1995). In the context of IP, for example, the process of bringing a new invention to market after that invention has been made – a process called commercialization – often requires the coordination of inventors, financiers, labor, management, advertisers, and marketers. *See generally* Kieff, *Commercializing Inventions*, *supra* note 2 at 707–12 (discussing role of patents in commercialization of inventions). That is, without the ability to coordinate, in the case of an invention for example, the inventor hoping to achieve commercialization would need to serve simultaneously as financier, laborer, manager, advertiser, and marketer. The recognition of this problem was indeed one of the motivating factors behind the present U.S patent system, which focuses on the importance of coordination to achieve invention commercialization. *See* Giles S. Rich, *The Relation Between Patent Practices and the Anti-Monopoly Laws*, 24 J. Pat. Off. Soc'y 159, 177 (Mar., 1942) (discussing incentive aspects of patent system and noting that one of its most important components 'applies to the inventor but not solely to him, unless he is his own capitalist'). *See also* Kieff, *Commercializing Inventions*, *supra* note 2 (discussing the commercialization theory of the patent system).
21. The IP regimes have rules governing what may be protected by IP that operate to make sure this is true and the basic reason these rules must operate in this way is to protect the reasonable investment backed expectations of third parties. *See* Kieff, *Registering Patents*, *supra* note 2 (exploring normative case for positive law rules for validity within the context of patents).
22. *See, e.g.*, Frank Knight, Risk, Uncertainty, and Profit, 268 (1965).
23. *See, e.g.*, Furubotn & Richter, *supra* note 14, at 64. (citing Kenneth Arrow, *The Organization of Economic Activity: Issues Pertinent to the Choice of Market Versus Nonmarket Allocation*, in Joint Econ. Comm., 91st Cong., The Analysis and Evaluation of Public Expenditures: The PPB System 47-64, at 62 (Comm. Print 1969).
24. To be sure, important additional considerations not directly applicable here have also been offered. *See, e.g.*, Richard R.W. Brooks, *The Relative Burden Of Determining Property Rules And Liability Rules: Broken Elevators In The Cathedral*, 97 Nw. U. L. Rev. 267, 268, n. 8 (2002) (elucidating analytical framework for assessing 'the relative burden (or costs, or difficulty) faced by judges when attempting to determine property rules and liability rules').
25. Calabresi & Melamed, *supra* note 5 at 1106. *Also compare* Richard A. Posner, Economic Analysis of Law 29 (1st ed. 1972) ('where transaction costs are high, the allocation of resources to their highest valued uses is facilitated by denying property right holders an injunctive remedy against invasions of their rights and instead limiting them to a remedy in damages') *and* James E. Krier & Stewart J. Schwab, *Property Rules and Liability Rules: The Cathedral in Another Light*, 70 N.Y.U. L. Rev. 440, 459–64 (1995) (arguing that property rules are better when administrative costs are high) *with* A. Mitchell Polinsky, *Resolving Nuisance Disputes: The Simple Economics of Injunctive and Damage Remedies*, 32 Stan. L. Rev. 1075, 1111 (1980) (pointing out that where decisions by a court are more costly the case for property rules is stronger).
26. *See* Robert P. Merges, *Of Property Rules, Coase, and Intellectual Property*, 94 Colum. L. Rev. 2655, 2664 (1994).

27. Kieff, *Commercializing Inventions*, *supra* note 2, at 733 (citing David Haddock et al., *An Ordinary Economic Rationale for Extraordinary Legal Sanctions*, 78 Cal. L. Rev. 1, 17 (1990)).
28. *See* Richard A. Epstein, *A Clear View of The Cathedral: The Dominance of Property Rules*, 106 Yale L.J. 2091 (1997).
29. *See infra* at p. 282 (discussing anticommons thicket).
30. Kieff, *Commercializing Inventions*, *supra* note 2, at 733 ('As Ayres and Klemperer recognize, if there are fixed costs of entry or exit, or if infringers have higher marginal cost than the patentee, then market entry by infringers will generate extra costs for society.' (citing Ayres & Klemperer, *supra* note 7, at 1015)). *See also* Henry Smith, *Property and Property Rules*, 76 N.Y.U. L. Rev. 1719 (2004) (noting that in addition to information costs, property rules also make sense because they deter opportunism by potential takers and discourage owners from engaging in wasteful self help)
31. Kieff, *Commercializing Inventions*, *supra* note 2, at 733 (citing Haddock et al., *supra* note 27, at 16–17).
32. Although, these higher damages amounts would still raise the ire of those who see property rights as tools that enable unfair holdout strategies.
33. *See* Smith, *supra* note 30 (discussing the problems of requiring self-help as an alternative to property rule protection).
34. The portfolio effect explains why property rules can dominate liability rules (in situations where assets and information are highly individualized and non-portfolio) even in the face of the very elegant and insightful projects that have shown how across a portfolio of decisions there are ways in which liability rules dominate. *See, e.g.*, Ian Ayres, Optional Law (2005).
35. *See* Kieff, *Commercializing Inventions*, *supra* note 2, at 717–18, 727–41 (emphasizing the coordination function and citing Demsetz, *supra* note 15). *See also* Kieff, *Registering Patents*, *supra* note 2, at 67–68.
36. Harold Demsetz, *Toward a Theory of Property Rights II: The Competition between Private and Collective Ownership*, 31 J. Legal Stud. S653, S664–5, S656–7 (2002).
37. *Id.* at S664.
38. *See also* Kieff, *Commercializing Inventions*, *supra* note 2.
39. *See*, Oliver E. Williamson, The Economic Institutions of Capitalism, 20–21 (1985) (discussing 'legal centrism').
40. *See, e.g.*, Robert Merges, *Contracting into Liability Rules: Intellectual Property Rights and Collective Rights Organizations*, 84 Cal. L. Rev. 1293 (1996) (exploring the role of norms in establishing private institutions to coordinate IP transactions); Arti Kaur Rai, *Regulating Scientific Research: Intellectual Property Rights and the Norms of Science*, 94 Nw. U. L. Rev. 77 (1999) (applying law and norms theory to intellectual property).
41. For a recent discussion of these two types of norms within the context of IP *see, e.g.* F. Scott Kieff, *Facilitating Scientific Research: Intellectual Property Rights and the Norms of Science – A Response to Rai & Eisenberg*, 95 NW. U. L. Rev. 691, 693, 696 (2001). To be sure, norms of each type may influence the other. What is more, when it comes to prescriptive norms about how individuals should behave, they may be driven by either external, or internal pressures. *See, e.g.*, Troy A. Paredes, *A Systems Approach to Corporate Governance Reform: Why Importing U.S. Corporate Law Isn't the Answer*, 45 Wm. & Mary L. Rev. 1055, 1087–88 (2004) ('By "norms" I do not mean those steps that managers take to please the market or to avoid shame or a lawsuit, although sometimes 'norms' is used broadly this way. Rather, I am referring to a sense of right and wrong – a sense of duty and responsibility – that directors and officers internalize and enforce on themselves simply because it is the right thing to do') (citing, *inter alia*, Lynn A. Stout, *On the Export of U.S.-Style Corporate Fiduciary Duties to Other Cultures: Can a Transplant Take?* 10 (UCLA School of Law, Working Chapter No. 02–11, 2002), available at http://ssrn.com/abstract=313679 ('In lay terms, corporate insiders act like fiduciaries not only because they fear external sanctions, but also

because they have internalized a sense of obligation or responsibility toward others . . .'); Edward B. Rock, *Saints and Sinners: How Does Delaware Corporate Law Work?*, 44 UCLA L. Rev. 1009, 1104, 1013 (1997) ('All of us internalize rules and standards of conduct with which we generally try to comply. We do this not only because we may fear some sanction, formal or informal, but also because doing so is important to our sense of self-worth, because we believe that doing a good job is the right thing to do.')).

42. *See, e.g.*, Lisa Bernstein, *Opting out of the Legal System: Extralegal Contractual Relations in the Diamond Industry*, 21 J. Legal Stud. 115 (1992) (showing how some communities opt for informal private enforcement mechanisms for contractual relationships instead of formal legal approaches because the administrative costs can be lower). *See also* Lisa Bernstein, *Private Commercial Law In The Cotton Industry: Creating Cooperation Through Rules, Norms, And Institutions*, 99 Mich. L. Rev. 1724 (2001). Bernstein's use of the term 'private ordering' to refer to private enforcement is consistent with the use by Williamson, which is narrower than the use in this chapter, which encompasses all private interactions voluntarily entered. *See infra* note 44 (contrasting Williamson's use of the term 'private ordering'). *See also* Steven L. Schwarcz, *Private Ordering*, 97 Nw. U. L. Rev. 319 (2002) (also using the term 'private ordering' to refer to private enforcement or regulation).
43. Barak D. Richman, *Firms, Courts, and Reputation Mechanisms: Towards a Positive Theory of Private Ordering*, at 4 (104 Colum. L. Rev. 2329–51 (2004) (available on-line at http://ssrn.com/abstract=565464) ('This chapter argues that concerns over transactional assurance and contractual enforcement, not administrative costs, drive merchant communities to private ordering (and to vertical integration as well).')
44. The term 'private ordering' is used more broadly in this chapter than it is in some of the NIE literature. Williamson, for example, often uses the term 'private ordering' to refer to the various informal mechanisms to privately enforce contractual relationships as compared with formal legal process. *See, e.g.*, Williamson, The Economic Institutions of Capitalism, *supra* note 39, at 163–68 (suggesting that repeat play and reputation can serve as 'private ordering' tools for enforcement). Here, the term is used to refer to all private interactions entered into voluntarily by individuals as compared to those coerced by a hierarch, such as cooperation directed by management among different divisions within a firm or tax transfers directed by law among members of a state. For uses of the term private ordering as it is used here *see, e.g.*, Henry E. Smith, *Exclusion and Property Rules in the Law of Nuisance*, 90 Va. L. Rev. 965, 983 (2004) (using the term 'private ordering' to refer to private voluntary exchanges, not to private enforcement); Thomas W. Merrill & Henry E. Smith, *Optimal Standardization in the Law of Property: The Numerus Clausus Principle*, 110 Yale L.J. 1, 8 (2000) (using the term 'private ordering' in the context of individual choice and freedom of contract); Richard A. Epstein, *All Quiet on the Eastern Front*, 58 U. Chi. L. Rev. 555, 569 (1991) ('Within the context of Eastern Europe, property and economic protections are critical to the ability to turn nations and economies around from central planning to private ordering').
45. *See, e.g.*, Douglass North, Institutions, Institutional Change, and Economic Performance (1990) (elucidating the importance to economic growth of the reliable enforcement of property rights and contracts by formal public legal institutions); Douglass C. North & Robert P. Thomas, The Rise of the Western World (1973) (putting property rights at the center of the explanation of economic performance); Avner Greif & Eugene Kandel, *Contract Enforcement Institutions: Historical Perspective and Current Status in Russia, in* Economic Transition in Eastern Europe and Russia: Realities of Reform (Edward P. Lazear, ed., Stanford: Hoover Institution Press 1995) (same). *See also* Friedrich A. von Hayek, *The Principles of a Liberal Social Order, in* The Essence of Hayek, (Chiaki Nishiyama & Kurt R. Leube, eds., 1984) (providing general discussion of the theory of liberal government including its use of coercive powers to enforce law).

46. *Compare* Barak D. Richman, *Community Enforcement of Informal Contracts: Jewish Diamond Merchants in New York*, (John M. Olin Center for Law, Economics & Business, Discussion Chapter No. 384, 2002), at 24 (contrasting benefits and costs of, *inter alia*, private and public enforcement mechanisms under different conditions).
47. Paredes, *supra* note 41, at 1064 (also noting that 'Strong legal protections for shareholders expand the available pool of capital for businesses and entrepreneurs and facilitate contracting by shoring up shareholder rights.').
48. Stephen Haber, Crony Capitalism and Economic Growth in Latin America: Theory and Evidence (2002) [hereinafter 'Haber, Crony Capitalism'].
49. *See also* Randy Calvert, *The Rational Choice Theory of Social Institutions: Cooperation, Coordination, and Communication, in* Modern Political Economy: Old Topics, New Directions 216, 244 (J.S. Banks and E.A. Hanushek eds., Cambridge: Cambridge University Press 1995) ('[r]ecognizing or creating focal points is one important way in which the players can successfully coordinate.').
50. *See* Jai Asundi, et al. Examining Change Contributions in an OSS Project: The Case of the Apache Web Server Project, (Working Chapter, 2005) (providing data for the Apache project and discussing numerous examples of empirical studies of other projects).
51. The smaller the control group the more intense can be the information content of the communications among them. As Henry Smith has pointed out there is a fundamental informational tradeoff:

    'As audience size increases, the marginal benefits of intensive communication are likely to decrease and the marginal costs are likely to increase. Thus, to minimize the sum of communication costs, any communication system faces a tradeoff between information intensiveness on the one hand and information extensiveness on the other.'

    Henry E. Smith, *The Language of Property: Form, Context, and Audience*, 55 Stan. L. Rev. 1105, 1111 (2003).
52. It also may be possible for the private rents to be too small compared to the social rents. For example, what an inventor gets for herself often is less than what her invention generates for society. *See* Steven Shavell and Tanguy van Ypersele, Rewards Versus Intellectual Property Rights (National Bureau of Econ. Research Working Chapter No. 6956, 1999) (suggesting a system of government-sponsored cash rewards instead of or in addition to a system of patents as a tool for improving the match between the private and public rents associated with an invention). The rent the invention generates for society takes into account a wide range of benefits. The rent something like a patent generates for society properly accounts for a narrower range of benefits tied to the contribution that patent made towards bringing the underlying invention to society more broadly or earlier than what would have occurred absent the patent. As another example, an inventor may develop something only slightly better than available options in a way that turns out to cause waste overall. A.K. Dixit & Joseph E. Stiglitz, *Monopolistic Competition and Optimum Product Diversity*, 67 Am. Econ. Rev. 297 (1977) (showing how it may be profitable for the one firm to come to market to get the customers but yet total industry profits can decline by more than consumer welfare increases). Recent work by Brett Frischmann and Mark Lemley has explored the way both of these examples also can be thought of as the problem of externalities, or what they call 'spillovers'. Brett M. Frischmann & Mark A. Lemley, Spillovers (working chapter, 2005).
53. If the value of the prize is X and the group of individuals is Y in number then each individual might rationally elect to spend up to just less than X to obtain the prize, say some amount equal to X minus a small discount, say ε or $(X-\varepsilon)$. Yet, if all individuals spend that amount, then the community has spent the amount equal to $[(X-\varepsilon) \times Y]$ to obtain something worth only X. The rub is that the expression $[(X-\varepsilon) \times Y]$ will be greater than X itself as long as X and Y are positive numbers greater than one and ε is a positive number less than one. Put simply, the amount spent in that community

as a whole to obtain the prize is greater than the amount the community as a whole got by obtaining the prize, which would be a waste of resources. Of course, there are a number of reasons to think that even an uncoordinated set of individuals might also chose to compete for the common prize in a way that avoids or mitigates the extent of rent dissipation or even fails to achieve the common prize. If each individual knows of the others, then each individual may discount the expected value of the prize to reflect the chance that one of the others will win. Facing such a decreased prize after adjusting for risk each individual may spend less. This may lead to a decrease in aggregate spending that in turn leads either to mitigation of the amount wasted or perhaps even failure to achieve the common prize. Alternatively, the uncertainty each individual has in this low, risk-adjusted payoff may not cause individuals to sufficiently decrease spending to mitigate overall rent dissipation effect. The large profits state-run lotteries earn from individuals paying far more than risk-adjusted payouts would advise is evidence of this type of rent dissipation behavior in practice despite obviously low risk-adjusted payouts. *See* Kieff, *Commercializing Inventions supra* note 2 at 711, n.68.

54. *See, e.g.*, Yoram Barzel, *Optimal Timing of Innovations*, 50 Rev. Econ. & Stat. 348 (1968) (showing how overinvestment can lead to invention occurring too early); Glenn C. Loury, *Market Structure and Innovation*, 93 Q. J. Econ. 395 (1979) (model showing overinvestment under appropriate conditions); P. Dasgupta & Joseph E. Stiglitz, *Industrial Structure and the Nature of Innovative Activity*, 90 Econ. J. 266 (1980) (same).
55. *See* Terry Anderson & Peter J. Hill, *Privatizing the Commons: An Improvement?*, 50 S. Econ. J. 438, 441, 447 (1983).
56. *Id.* at 442.
57. Anderson & Hill attribute the term 'residual claimant' to the work by Armen Alchian and Harold Demsetz on the theory of the firm. *Id.*, at 439 (citing Armen Alchian & Harold Demsetz, Production, Information Costs, and Economic Organization, 62 Am. Econ. Rev. 777 (1972)).
58. Anderson & Hill, *supra* note 55, at 443.
59. Indeed, because patentees are the ones who are lowest cost processors of the information needed to assess validity information costs are mitigated when property the owners themselves are given such strong incentives to make these determinations, and recent empirical models suggest these incentives do work. *See* Amalia Yiannaka & Murray Fulton, Privately Optimal Patent Breadth under the Threat of Patent Validity Challenges, presented to the 8th International Consortium on Agricultural Biotechnology Research (ICABR): International Trade and Domestic Production, held in Ravello (Italy), July 8–11, 2004 (available on-line at http://www.economia.uniroma2.it/conferenze/icabr2004/papers/Yiannaka.A.pdf) (showing how patentees integrate concerns about validity challenges into their own decision-making *ex ante*). At the same time, the rules for patentability over the prior art protect third parties' reasonable investment backed expectations by preventing valid patents from issuing where there have been any verifiably prior investments. *See* Kieff *Registering Patents supra* note 2, at 76-99. Importantly, these patentability rules are all enforced with rules biased in such a way that they involve remarkably low administrative, public choice, and both Type I and Type II error costs. *See*, F. Scott Kieff, *How Ordinary Judges and Juries Decide the Seemingly Complex Technological Questions of Patentability over the Prior Art*, *in* F. Scott Kieff, Perspectives on Properties of the Human Genome Project 125 (New York: Academy Press 2003).

   In addition, the disclosure rules of patent validity may also help protect third parties' reasonable investment backed expectations by helping these third parties to avoid inadvertent trespass. *Id.* at 99–105. But it is not clear that these rules are working as well as they could be for at least two important reasons. First, the uncertainty governing the process of patent claim construction may be frustrating the patent system's important *ex ante* incentives for private ordering by both patentees and infringers. For an excellent collection of recent empirical work on claim construction by R. Polk

Wagner, see www.claimconstruction.com. Importantly, the uncertainty here is not the individualized uncertainty associated with what some see as high reversal rates on appeal but rather the lack of coherence, or predictability, that the entire body of claim construction law seems to be generating. Ironically, the empirical work by Wagner suggests that although the body of legal rubrics that are available for claim construction may not yield predictability, simply knowing the identities of the members of the appellate panel at the Federal Circuit may yield at least case specific predictability at the time of oral argument. Second, the increasing reliance in infringement cases on the doctrine of equivalents is similarly frustrating the patent system's important *ex ante* incentives for private ordering by both patentees and infringers. The point here is that patentees may be able to gain more flexibility in claim scope while at the same time providing more certainty to infringers by relying on the established disclosure rules of Section 112 than on the doctrine of equivalents. *See* Kieff *supra* note 2, at 1726–27 (discussing Warner-Jenkinson Co., Inc. v. Hilton Davis Chemical Co., 520 U.S. 1153, 117 S.Ct. 1352 (1997) (holding that patent claims that are not infringed literally may still be infringed under the judge-made rule called the 'doctrine of equivalents' to allow patentees flexibility) and Festo Corp. v. Shoketsu Kinzoku Kogyo Kabushiki Co., Ltd., 535 U.S. 722, 122 S.Ct. 1831 (2002) (holding that the doctrine of equivalents is cabined by the judge-made rule called 'prosecution history estoppel' to allow third parties more certainty in knowing what will infringe a patent)).

60. And as with patents the rules on validity for trademarks help protect third party investments to at least some extent through the limited scope of trademark rights in the first instance by, for example, the doctrine that prevents trademark rights from covering functional elements. *See, e.g.*, In re Morton-Norwich Products, Inc., 671 F.2d 1332, 1339 (C.C.P.A. 1982) (Rich, J.) (reviewing functionality doctrine and collecting sources). In addition, *ex ante* investments by third parties are protected through a rule giving a cause of action to a prior user of a mark that is made famous by a subsequent user. *See, e.g.*, Big O Tire Dealers, Inc. v. Goodyear Tire & Rubber Co., 561 F.2d 1365 (10th Cir. 1977) (protecting small prior user's mark using theory sometimes called 'reverse confusion' because the public is led to confuse the first-user's mark with the more famous second-user's mark and think that the first is the second rather than the more typical confusion case in which a second user's mark is confused with that of a first user). In some cases, both users may be allowed to operate in different markets. *See, e.g.*, Burger King of Florida, Inc. v. Hoots, 403 F.2d 904 (7th Cir.1968) (holding that the national chain Burger King is allowed exclusive use of the mark throughout the nation except in the town of Matoon, IL, where a prior user in that particular location is allowed to continue exclusive use). The rules for enforcing and determining validity of trademarks facilitate *ex ante* private ordering because, as with patents, they turn on facts equally knowable to all market actors in advance. Key evidence on trademark validity typically takes the form of survey data from ordinary customers.
61. Furubotn & Richter, *supra* note 14, at 176. For more on transaction costs generally, see Douglas W. Allen, *What are transaction costs?*, 14 Res. L. & Econ. 1 (1991); Douglas W. Allen, Transaction costs *in* 1 Encyclopedia of Law and Economics, (Boudewijn Bouckaert & Gerrit De Geest, eds., Cheltenham, UK and Northampton, MA, USA: Edward Elgar 2000).
62. *See, e.g.*, Paul L. Joskow, *Vertical Integration and Long-Term Contracts: The Case of Coal-Burning Electric Generating Plants*, 1 J.L. Econ. & Org., 33 (1983); Paul L. Joskow, *Asset Specificity and the Structure of Vertical Relationships: Empirical Evidence*, 7 J.L. Econ. & Org., 95 (1988). *See also* Howard A. Shelanski and Peter G. Klein, *Empirical Research in Transaction Cost Economics: A Review and Assessment*, 11 J.L. Econ. & Org., 335 (1995) (survey of empirical evidence on transaction-costs economics assessing roughly 100 references on empirical research in transaction-cost economics published before 1993).
63. Furubotn & Richter, *supra* note 14, at 291.

64. *See* Armen A. Alchian, *Information Costs, Pricing, and Resource Unemployment*, 7 W. Econ. J. 109 (1969). *See also* George J. Stigler, *The Economics of Information*, 69 J. Pol. Econ. 213 (1961) (noting that acquiring and processing information about potential exchange opportunities are costly).
65. John J. Wallis & Douglass C. North, *Measuring the Transaction Sector in the American Economy, 1870–1970*, *in* Long-Term Factors in American Economic Growth, 95 (Stanley L Engerman & Robert E. Gallmann, eds.) (Studies in Income and Wealth, No. 51, 1986). The connection between division of labor and transaction costs, including the inevitable limit that transaction costs place on the extent of the division of labor, was articulated earlier by Adam Smith. *See* Harold Demsetz, *The Cost of Transacting*, 82 Q. J. Econ. 33, 35 (1968) (empirical evidence of transaction costs in the market of the New York Stock Exchange and quoting Adam Smith: '[a]s it is the power of exchanging that gives occasion to the division of labor, so the extent of this division must always be limited by the extent of that power, or, in other words, by the extent of the market.').
66. *See* Robert Ellickson, Order Without Law: How Neighbors Settle Disputes 184 (1991) (pointing out that societies tend to develop institutions – such as norms in the case he is studying – that 'minimize the members' objective sum of (1) transaction costs and (2) deadweight losses arising from failures to exploit potential gains from trade.'). *See also* Ronald Coase, *The Problem of Social Cost*, 3 J. L. & Econ. 10 (1960) (noting that the principal condition that must be satisfied for individuals to maximize wealth by engaging in an exchange is that the transaction costs of the exchange must not exceed the gains from trade.); Terry L. Anderson & Donald R. Leal, *Free Market Environmentalism: Hindsight and Foresight*, 8 Cornell J.L. & Pub. Pol'y 111 113 (1998) ('[h]umans interact to capture potential gains from trade – the knowledge for this interaction is bounded by transaction costs. The gains from trade (a positive-sum game) result because people place different values on goods and services and because people have different abilities to produce those goods and services. Because of these differences, trade has the potential to make the parties exchanging goods and services – of lower value to each respectively – better off.').
67. *See, e.g.*, Milton Friedman, *Value Judgments in Economics*, in The Essence of Friedman 3, 3–8 (Kurt R. Leube, ed., 1987) (discussing the 'role of the market as a device for the voluntary cooperation of many individuals in the establishing of common values' and concluding that '[i]n many ways, this is the basic role of the free market in both goods and ideas – to enable mankind to cooperate in this process of searching for and developing values.').
68. *See, e.g.*, Ronald J. Gilson, *Seeking Competitive Bids Versus Pure Passivity in Tender Offer Defense*, 35 Stan. L. Rev. 51, 62–63 (1982) ('Let me start with two important elements of transaction costs in the acquisition setting: information costs necessary to identify the opportunity; and mechanical costs – for example, lawyers', accountants', and investment bankers' fees – necessary to effect the transaction and cope with regulatory or other barriers (including defensive tactics by the target).').
69. *See, e.g.*, Troy A. Paredes, *A Systems Approach to Corporate Governance Reform: Why Importing U.S. Corporate Law Isn't the Answer*, 45 Wm. & Mary L. Rev. 1055, 1110–12 (exploring Gilson's analytical framework of the lawyer as transaction cost engineer and citing Ronald J. Gilson, *Value Creation by Business Lawyers: Legal Skills and Asset Pricing*, 94 Yale L.J. 239, 255 (1984) (describing lawyers as 'transaction cost engineers'); Bernard Black & Reinier Kraakman, *A Self-Enforcing Model of Corporate Law*, 109 Harv. L. Rev. 1911, 1923 (1996) (pointing out that in addition to lawyers 'savvy investors and issuers' also help facilitate transactions); Curtis J. Milhaupt & Mark D. West, *The Dark Side of Private Ordering: An Institutional and Empirical Analysis of Organized Crime*, 67 U. Chi. L. Rev. 41, 58 (2000) (also using term 'transaction cost engineers' for lawyers)). *See also* Lisa Bernstein, *The Silicon Valley Lawyer as Transaction Costs Engineer?*, 74 Or. L. Rev. 239, 241 (1995) (further exploring Gilson's analytical framework of the lawyer as transaction cost engineer

and, in addition to Gilson, also citing Lawrence M. Friedman et al., *Law, Lawyers, and Legal Practice in Silicon Valley: A Preliminary Report*, 64 Ind. L.J. 555, 562 (1989) ('[t]he Silicon Valley lawyer not only works with engineers, he thinks of himself as a kind of engineer – a legal engineer . . . his job is to solve problems, to take a principle, a task and engineer it legally')).

70. For an in-depth treatment of the topic see Douglass C. North, *A Transaction Cost Theory of Politics*, 2 J. Theoretical Pol. 355 (1991).
71. *Id. See also* Douglass C. North, *Institutions and Credible Commitment*, 149 J. Institutional & Theoretical Econ., 11, 18 (1993) ('Political markets are far more prone to inefficiency'). While North explains why the net effect of a shift from economic market to political market is expected to be an increase in transaction costs there are of course some ways in which some aspects of transaction costs may be lower. For example, to the extent political markets are constrained by norms, political markets can to some extent function like small norm communities and, as explored later, in the discussion of norm communities, enforcing deals in small norm communities may be in some ways less expensive and more effective than across an open market.
72. The so-called efficient market hypothesis (also known as 'EMH') is based on the view that in a perfectly thick market, assets will be perfectly priced. The basic theoretical foundation for the EMH was laid by Paul Samuelson and Benoit Mandelbrot. *See* Paul A. Samuelson, *Proof That Properly Anticipated Prices Fluctuate Randomly*, 6 Indus. Mgmt. Rev. 41, 48 (1965); Benoit Mandelbrot, *Forecasts of Future Prices, Unbiased Markets, and Martingale Models*, 39 J. Bus. 242, 248 (1966). Empirical support was added by Eugene Fama. *See* Eugene Fama, *Efficient Capital Markets: A Review of Theory and Empirical Work*, 25 J. Fin. 383, 392 (1970).
73. For more on groupthink see *infra* note 99, and accompanying text.
74. *See* Kieff, *supra* note 41.
75. One on-line shopping guide for basic scientists provides this description:

    'Vendor Freezer and Cabinet programs offer a freezer or cabinet with a customized inventory of the products you use. Companies may provide a complimentary cabinet, freezer, or refrigerator, stock it, and often apply discounts to the host lab.'

    www.biocompare.com/freezer.asp (web site advertised as 'The Buyer's Guide for Life Scientists', which lists the details of several companies' programs and provides links). *See also* www.bio.umass.edu/biology/genomics/freezer.phtml. (advertising interdepartmental freezer program at the University of Massachusetts); www.narf.vcu.edu/abi.html (Virginia Commonwealth University).
76. *See id.*, at 705 (explaining why patentees rationally elect not to enforce aggressively).
77. J.P Walsh et al., *Working through the Patent Problem*, 299 Science 1021 (2003). *See also* J.P. Walsh et al., View from the Bench: Patents and Material Transfers, 309 Science 2002 (2005) (reporting empirical results demonstrating that 'access to patents on knowledge inputs rarely imposes a significant burden on academic biomedical research.').
78. *A.C. Aukerman Co. v. R.L. Chaides Constr. Co.*, 960 F.2d 1020 (Fed.Cir.1992) (en banc) (discussing laches). This does not mean that the patentee must go after every infringer right away. The laches effect may be put on hold with respect to some infringers where the patentee is kept busy tracking down others and bringing lawsuits against them. *Accuscan v. Xerox Corp.*, 1998 WL 273074 (S.D.N.Y. May 26, 1998) (presumption of laches rebutted where patentee delayed filing infringement suit in order to avoid the burden of conducting two simultaneous infringement suits and to attempt to negotiate a license agreement with the defendant).
79. *Wang Labs. v. Mitsubishi Elecs.*, 103 F.3d 1571, 1582 (Fed.Cir. 1997) (discussing equitable estoppel).
80. *See* Ayres & Klemperer, *supra* note 7.
81. *See, e.g.*, Kieff, *Registering Patents*, *supra* note 2 (criticizing impact of reward theories on rules for obtaining patents).
82. *See, e.g.*, F. Scott Kieff & Troy A. Paredes, *The Basics Matter: At the Periphery of*

Intellectual Property, 73 G.W. Law Rev. 174 (2004) (criticizing impact of reward theories on rules for transacting over IP and on antitrust enforcement of IP-based transactions).

83. *See, e.g.*, Kieff, *Commercializing Inventions*, *supra* note 2 (criticizing impact of reward theories on rules for enforcing patents and elucidating the importance of property rights protected by property rules for enforcing patents).
84. F. Scott Kieff, *Patents for Environmentalists*, 9 Wash. U.J.L. & Pol'y 307, 313 (2002) (Invited symposium piece for National Association of Environmental Law Societies annual meeting entitled 'Sustainable Agriculture: Food for the Future', held March 15–17, 2002, at Washington University School of Law) (citing Kieff, *Commercializing Inventions*, *supra* note 2, at 733–34, n.154). For more on the interface between IP and bankruptcy see F. Scott Kieff & Troy A. Paredes, *Toward an Understanding of Intellectual Property, Bankruptcy, and Corporate Control*, 82 Wash U. Law. Q. 1313 (2004).
85. *Id.* (citing 28 U.S.C. § 1498 (1994), under which the government provides a limited waiver of its sovereign immunity for acts of infringement by or for the federal government and instead allows suits against the government in the US Court of Federal Claims for a reasonable royalty). State governments similarly enjoy immunity under the 11th Amendment. *See* Florida Prepaid Postsecondary Educ. *Expense Bd. v. College Sav. Bank*, 527 U.S. 627 (1999) (state immunity from patent infringement suits); *College Sav. Bank v. Florida Prepaid Postsecondary Educ. Expense Bd.*, 527 U.S. 666 (1999) (state immunity from Lanham Act trademark infringement and unfair competition suits); *Chavez v. Arte Publico Press*, 204 F.3d 601 (5th Cir. 2000) (state immunity from copyright infringement suits). The point here is that anyone interested in achieving liability rule treatment for an IP right can achieve that result by prevailing on a government agency to arrange for the infringement.
86. For an excellent recent review of the behavioralism literature *see, e.g.*, Russell Korobkin, *Bounded Rationality, Standard Form Contracts, and Unconscionability*, 70 U. Chi. L. Rev. 1203 (2003) (collecting sources). As noted by Troy Paredes, 'Explaining and understanding these deviations from perfect rationality make up the core of [the field known as] behavioral law and economics.' Troy A. Paredes, *Blinded by the Light: Information Overload and Its Consequences for Securities Regulation*, 81 Wash. U. L.Q. 417, 434–444 (2003) (citing Behavioral Law & Economics (Cass R. Sustein ed., 2000); Christine Jolls et al., *A Behavioral Approach to Law and Economics*, 50 Stan. L. Rev. 1471 (1995); Russell Korobkin, *A Multi-Disciplinary Approach to Legal Scholarship: Economics, Behavioral Economics, and Evolutional Psychology*, 41 Jurimetrics J. 319 (2001); Russell B. Korobkin & Thomas S. Ulen, *Law and Behavioral Science: Removing the Rationality Assumption from Law and Economics*, 88 Cal. L. Rev. 1051 (2000); Jennifer Arlen, *Comment: The Future of Behavioral Economic Analysis of Law*, 51 Vand. L. Rev. 1765 (1998)). *Cf*, Richard A. Posner, *Rational Choice, Behavioral Economics, and the Law*, 50 Stan. L. Rev. 1551 (1998) (commenting on the behavioralism literature in general and in particular Jolls *et al. supra*); Gary S. Becker, *Nobel Lecture: The Economic Way of Looking at Behavior*, *in* The Essence of Becker (Ramon Febero & Pedro Schwartz, eds. 1995).
87. Richard A. Posner, *The New Institutional Economics Meets Law and Economics*, 149 J. Institutional & Theoretical Econ., 73, 80 (1993). This view of behavioralism is consistent with a view that sees information costs associated with obtaining and processing information, which traces its routes back to the work of Herbert Simon. *See, e.g.*, Herbert A. Simon, *A Behavioral Model of Rational Choice*, Q. J. Econ., at 241 (1955) ('the task is to replace the global rationality of economic man with a kind of rational behavior that is compatible with the access to information and computational capacities that are actually possessed by . . . man.'). *See also* Press Release: The Sveriges Riksbank (Bank of Sweden) Prize in Economic Sciences in Memory of Alfred Nobel for 1978 (available on-line at http://www.nobel.se/economics/laureates/1978/press.html).

88. Oliver E. Williamson, Markets and Hierarchies: Analysis and Antitrust Implications (1975) at 109–10.
89. *Id.* (citing Herbert Simon, *Theories of Bounded Rationality*, in Decision and Organization, 161 (C.B. McGuire & R. Radner, eds., 1972)).
90. *Id.* (citing Oliver E. Williamson, *Calculativeness, Trust, and Economic Organization*, 36 J. Law & Econ. 453 (1993) (problems of being rushed to make decisions)).
91. *Id.* (citing Michael Polanyi, Personal Knowledge: Towards a Post-Critical Philosophy (New york: Harper Torch Books 1962)).
92. *Id.* (citing Herbert Simon, *Rationality as Process and Product of Thought*, 68 Am. Econ. Rev., 1, 12 (1978)).
93. Paredes *supra* note 86, at 435–36 (citing Herbert A. Simon, A Behavioral Model of Rational Choice, 69 Q.J. Econ. 99, 262–64 (1955); Judgment Under Uncertainty: Heuristics and Biases (Daniel Kahneman et al. eds., 1982); John W. Payne et al., The Adaptive Decision Maker 1–2 (1993); Herbert A. Simon, Models of Bounded Rationality: Economic Analysis and Public Policy (1982)). *See also* Press Release: The Bank of Sweden Prize in Economic Sciences in Memory of Alfred Nobel 2002, available on line at http://www.nobel.se/economics/laureates/2002/press.html.
94. For the basic exploration of methods for measuring risk aversion see Kenneth J. Arrow Aspects of the Theory of Risk-Bearing (1965); John W. Pratt, *Risk Aversion in the Small and in the Large*, 32 Econometrica, 122 (1964).
95. Jeffrey J. Rachlinski, *The Uncertain Psychological Case for Paternalism*, 97 Nw. U. L. Rev. 1165, 1169–70 (2003) ('psychologists have found that when individuals are asked to memorize a long sequence of words, they are more likely to remember the first few words (the "primacy" effect) and the last few words (the "recency" effect) much better than the words in the middle of the list') (citing Eugene B. Zechmeister & Stanley E. Nyberg, Human Memory: An Introduction to Research and Theory 60–71 (1982) (reviewing research on primacy and recency effects in memory)).
96. For empirical evidence of framing effects see, e.g., Daniel Kahneman & Amos Tversky, *Choices, Values, and Frames*, 39 Am. Psychol. 341 (1983) (framing effects observed in decisions involving lotteries and other risky monetary payoffs); Amos Tversky & Daniel Kahneman, *The Framing Effect of Decisions and the Psychology of Choice*, 211 Science 453 (1981) (same).
97. Rachlinski *supra* note 95 at 1171 ('When making numeric estimates, individuals will tend to rely heavily on reference points and then adjust from these reference points') (citing Tversky & Kahneman *supra* note 96 1128–30 (explaining anchoring and the related process of adjustment)).
98. Rachlinski *supra* note 95 at 1172 (defining 'overoptimism, which consists of overestimating one's capabilities; overconfidence, which consists of overestimating one's ability to predict outcomes; and egocentricism, which consists of overstating the role that one has played in events in which one has participated'). *See also* Paredes *supra* note 86, at 481 ('Some of the most well-known sources of these deviations from rationality include loss aversion, framing, the representativeness heuristic, the availability heuristic, overoptimism, and overconfidence.')
99. *See* Troy A. Paredes, *Too Much Pay, Too Much Deference: Is CEO Overconfidence the Product of Corporate Governance?*, at 60, n. 227 (2004) (discussing groupthink in the context of corporate governance and as a contributing factor to CEO overconfidence) (working chapter, copy on file with author) (citing Irving L. Janis, Groupthink (2d ed. 1982) and Marleen O'Connor, *The Enron Board: The Perils of Groupthink*, 71 U. Cin. L. Rev. 1233 (2003)).
100. *See id.*, at 12 (citing Timur Kuran & Cass R. Sunstein, *Availability Cascades and Risk Regulation*, 51 Stan. L. Rev. 683, 683–91, 720–23 (1999) (describing the problem as one of 'informational cascades' through which a view cascades through a pool of individuals as each individual adopts the view of those believed to be better informed); Cass R. Sunstein, Conformity and Dissent, U. Chicago Law & Economics, Olin Working Chapter No. 164 (available on-line at http://ssrn.com/abstract=314880)).

101. While this opposite component of the effect can be seen to be encompassed by the elucidation from Kuran and Sunstien, it is stated here separately to make sure it is not overlooked.
102. Kuran & Sunstein, *supra* note 100 at 723–25. *See also*, Paredes *supra* note 99, at 13.
103. Kenneth Arrow, The Limits of Organization, (New York: Wahl Norton and Company 1974) at 128.
104. The desirability of a slogan as a fashion statement in and of itself is tied to a controversial issue in trademark law relating to marks that are desired in and of themselves, unconnected to a goods and services.
105. *See, e.g.*, Stephen Breyer, *Analyzing Regulatory Failure: Mismatches, Less Restrictive Alternatives, and Reform*, 92 Harv. L. Rev. 549, 553–58 (1979) (exploring various market failures including externalities, monopoly, and information costs); Ian Ayres & Eric Talley, *Solomnic Bargaining: Dividing a Legal Entitlement to Facilitate Coasean Trade*, 104 Yale L.J. 1027, 1029 (1995) (listing information costs, transaction costs, and externalities – what they refer to as 'free riding' – as examples of market failure).
106. Furubotn & Richter, *supra* note 14, at 291 (citing Williamson, The Economic Institutions of Capitalism, *supra* note 39) and citing Coase, *Industrial Organization: A Proposal for Research*, in Victor R. Fuchs, Policy Issues and Research Opportunities in Industrial Organization 59–73 National Bureau of Economic Research ((NBER), 1972) ('One important result of this preoccupation with monopoly is that if an economist finds something – a business practice of one sort or other – that he does not understand, he looks for a monopoly explanation.').
107. Furubotn & Richter, *supra* note 14, at 302 (citing other examples such as 'the evolution of "privately ordered" medieval trade organizations as explored in [the following works:]' Avner Grief, *Reputation and Coalitions on Medieval Trade: Evidence on the Maghribi Traders*, 49 J. Econ. Hist., 857 (1989) (among long-distance Jewish traders in the Mediterranean during 11th century called the Maghribi); Roger Milgrom, et al., *The Role of Institutions in the Revival of Trade: The Law Merchant, Private Judges, and the Campagne Fairs*, 2 Econ. & Pol. 1 (1990) (law merchant system of the Champaign Fairs of the 12th and 13th centuries)). *See also* Jonathan H. Adler, *Conservation through Collusion: Antitrust as an Obstacle to Marine Resource Conservation*, 61 Wash. & Lee L. Rev. 3 (2004) (elucidating how antitrust enforcement may interfere with environmental conservation and other goals).
108. *See* Press Release: The Sveriges Riksbank (Bank of Sweden) Prize in Economic Sciences in Memory of Alfred Nobel for 1995, available on line at http://www.nobel.se/economics/laureates/1995/press.html.
109. *See, e.g.*, Robert E. Lucas, *Expectations and the Neutrality of Money*, 4 J. Econ. Theory 103 (1972). *See also* Sanford J. Grossman, *An Introduction to the Theory of Rational Expectations Under Asymmetric Information*, 48 Rev. Econ. Stud. 541, 543 (1981) (describing rational expectations equilibrium).
110. Games that are not static are sometimes said to have multiple cycles, rounds or iterations, or are said to repeat. For a general overview of game theory see, *e.g.*, John von Neumann & Oskar Morgenstern, Theory of Games and Economic Behavior (1944) (first formal treatment of game theory as a part of economics); Douglas G. Baird, et al., Game Theory and the Law (1994) (more modern treatment of game theory with focus on legal implications).
111. *See generally* Paredes, *supra* note 69, at 1133–34 ('Legal certainty, which is part and parcel of well-defined property rights, is a valuable asset that facilitates business and investing, aside from how the law actually allocates rights and responsibilities')
112. To be sure, this is a highly stylized example and in the real world every contract can have detailed insurance, futures and options components. Indeed, the availability of these provisions provides justification for treating contracts among sophisticated parties as though they do indeed speak to these issues, even when silent on their face.
113. This gives rise to the approach in some contract cases to allow for what is termed 'efficient breach'. See Richard A. Posner, Economic Analysis of Law 117–19 (4th ed.

1992) (discussing efficient breach approach). *See also* Oliver Wendell Holmes, Jr., The Common Law 301 (Legal Classics Library 1982) (1881) (originating the approach). *But see* Daniel Friedmann, *The Efficient Breach Fallacy*, 18 J. Legal Stud. 1 (1989) (criticizing the approach).

114. *See generally*, David D. Haddock *et al.*, *An Ordinary Economic Rationale for Extraordinary Legal Sanctions*, 78 Cal. L. Rev. 1, 16–17 (1990) (showing how uncertainty in enforcement discourages investment *ex ante*).
115. Ian Ayres & Eric Talley, *Solomonic Bargaining: Dividing a Legal Entitlement to Facilitate Coasean Trade*, 104 Yale L.J. 1027–1118 (1995); Jason Scott Johnston, *Bargaining under Rules versus Standards*, 11 J.L. Econ. & Org. 256–281 (1995).
116. Rachel Croson & Jason Scott Johnston, *Experimental Results on Bargaining under Alternative Property Rights Regimes*, 16 J.L. Econ. & Org. 50, 67–70 (2000).
117. Ian Ayres & Robert Gertner, *Filling Gaps in Incomplete Contracts: An Economic Theory of Default Rules*, 99 Yale L.J. 87, 127 (1989).
118. Robert Cooter & Thomas Ulen, Law and Economics 100 (1988). For a discussion of the broader debate between legal systems based on rules and those based on standards, see generally Mark Kelman, A Guide To Critical Legal Studies 15–63 (1987) (describing basic framework of the debate and collecting sources); Louis Kaplow, *Rules Versus Standards: An Economic Analysis*, 42 Duke L.J. 557 (1992) (exploring the costs implicated by the choice between rules and standards and showing: rules typically are more costly than standards to create; standards typically are more costly for individuals to interpret, both by individuals deciding how to act under them and by government decisionmakers deciding how to apply them; and individuals are more likely to act in accordance with the goals of rules as long as the individuals can determine how they will be applied); Russell B. Korobkin, *Behavioral Analysis and Legal Form: Rules vs. Standards Revisited*, 79 Or. L. Rev. 23 (2000) (reviewing more recent literature and collecting sources).
119. Einer Elhauge, *Defining Better Monopolization Standards*, 56 Stan. L. Rev. 253, 275 (2003) (criticizing forms of antitrust enforcement motivated by concerns for static efficiency but that may negatively impact innovation collecting sources). *See also* Christopher S. Yoo, *Rethinking the Commitment to Free, Local Television*, 52 Emory L.J. 1579 (2003) (reviewing tension between static and dynamic efficiency within the context of public goods and monopolistic competition).
120. F. Scott Kieff & Troy A. Paredes, *The Basics Matter: At the Periphery of Intellectual Property*, 73 G.W. Law Rev. 174 (2004) (generally exploring the interfaces IP law shares with other regimes such as antitrust, and collecting sources including Phillip Areeda & Louis Kaplow, Antitrust Analysis: Problems, Texts, Cases 174–250, 447–77, 785–806 (1997)). For a different take on the interface between IP and antirust see Herbert Hovenkamp, et al., IP and Antitrust (2003).
121. This dead weight loss represents a collective loss of societal wealth, in that it is not merely wealth that has been shifted from consumers to producers but rather wealth that is altogether lost from producers and consumers collectively. The dead weight loss inefficiency associated with power over price is depicted graphically, and its etiology is explained in a manner targeted for a lay audience, in Chisum, et al. *supra* note 2, at 60–66. To be sure, there are other inefficiencies associated with monopolies, including, for example, the rent dissipating effects that competition for monopoly profits may generate. *See generally* Gordon Tullock, *The Welfare Costs of Tariffs, Monopolies, and Theft*, 5 W. Econ. J. 224 (1967) (studying rent-seeking costs of monopoly). Yet, the rent dissipating effects of monopolies, like other rent dissipation, depends on several factors.
122. *See Illinois Tool Works Inc. v. Independent Ink, Inc.*, 547 US S.Ct. 1–17, 2006 WL 468729 (2006) (patent does not give rise to presumption that patentee has market power). *See also*, Kenneth W. Dam, *The Economic Underpinnings of Patent Law*, 23 J.L. Stud. 247, 249–50 (1994) ('the right to exclude another from "manufacture use and sale" may give no significant market power, even when the patent covers a product that is sold in the market').

123. For those who are familiar with the graphical representation of the monopolist's dead weight loss triangle, an example of which is depicted in Chisum, et al. *supra* note 2, at 65, price discrimination allows the monopolist to convert what otherwise would be that dead weight loss triangle into being producer surplus instead.
124. For a basic overview of the economics of price discrimination see Jean Tirole, The Theory of Industrial Organization 133–68 (1997). It is also recognized that in certain cases efforts to engage in price discrimination may lead to decrease in efficiency. For example, recent work by Wendy Gordon, Glynn Lunney and Michael Meurer has shown that while price discrimination by IP owners might lead in theory to more use in certain instances, in practice some price discrimination strategies can result in less output than if such price discrimination were prohibited, depending, in part, on the licensing arrangements employed to discriminate among users). Wendy J. Gordon, *Intellectual Property as Price Discrimination: Implications for Contract*, 73 Chi.-Kent L. Rev. 1367 (1998); Glynn S. Lunney, *Copyright and the Supposed Efficiency of First-Degree Price Discrimination* (2002) (working chapter); Michael J. Meurer, *Copyright Law and Price Discrimination*, 23 Cardozo L. Rev. 55 (2001). However, as summarized by Richard Posner:

    'Perfect price discrimination would bring about the same output as under competition, because no customer willing to pay the seller's marginal cost would be turned away. But perfect price discrimination is infeasible, and imperfect price discrimination can result in a lower or higher output than under competition, or the same output. *See* F.M. Scherer & David Ross, Market Structure and Industrial Performance 494-96 (3d ed. 1990); Paul A. Samuelson, Foundations of Economic Analysis 42-45 (1947); Joan Robinson, The Economics of Imperfect Competition 188-95 (1933). Many economists believe that even crude discrimination is more likely to expand than to reduce output, *see, e.g.*, Robinson, *supra*, at 201; Scherer & Ross, *supra*, at 494–96; Peter O. Steiner, Book Review, 44 U. Chi. L. Rev. 873, 882 (1977), but there does not appear to be a firm basis for this belief. *See* Hal R. Varian, *Price Discrimination, in* Handbook of Industrial Organization, at 597, 629–33 (Richard Schmalensee & Robert D. Willig eds., 1989).'

    Richard A. Posner, *Antitrust in the New Economy*, 68 Antitrust L. J. 925, 932 n.10 (2001).
125. Michael A. Heller, *The Tragedy of the Anticommons: Property in the Transition from Marx to Markets*, 111 Harv. L. Rev. 621 (1998).
126. *See, e.g.*, Carl Shapiro, Navigating the Patent Thicket: Cross Licenses, Patent Pools, and Standard Setting, in 1 Innovation Policy and the Economy 119, 119 (Adam B. Jaffe et al. eds., 2001) (treating a 'patent thicket' to be the case of many patents relating to a single product); James Bessen, Patent Thickets: Strategic Patenting of Complex Technologies (Research on Innovation Working Paper 2003), available at http://www.researchoninnovation.org/thicket.pdf. (same).
127. Michael A. Heller, *The Dynamic Analytics of Property Law* , 2 Theoretical Inquiries L. 79, 87–89 (2001) (drawing the definition of anticommons property from Heller, *supra* note 125, and building on the discussion of IP rights from Michael A. Heller & Rebecca S. Eisenberg, *Can Patents Deter Innovation? The Anticommons in Biomedical Research*, 280 Science 698 (1998) [hereinafter Heller & Eisenberg, *Anticommons*].
128. *See* Heller *supra* note 125, at 624 (arguing that when too many owners hold rights of exclusion in a resource, the resource is prone to under use). *See also* Heller & Eisenberg, *Anticommons supra* note 127, at 700; Michael A. Heller, *The Boundaries of Private Property*, 108 Yale L.J. 1163, 1174–75 (1999).
129. *See supra* note 57.
130. *See, e.g.*, James Buchanan & Yong J. Yoon, *Symmetric Tragedies: Commons and Anticommons*, 43 J. Law & Econ. 1, 12 (2000). For an interesting taxonomy applying this symmetry see Lee Anne Fennell, *Common Interest Tragedies*, 98 Nw. U. L. Rev. 907 (2004).
131. *Id.* Buchanan & Yoon (citing the example of an environmental regulator whose permission is needed to put an asset to use but whose permission should not be bought).

To be sure, without being motivated by direct pecuniary gains, a regulator may be economically motivated along the lines of the political favors discussed in the public choice literature.

132. *Id.*
133. Heller suggests the coordination benefits of being 'close knit'. *See* text accompanying note 127, *supra*. *See also supra* the first part (discussing the coordination benefits of norm communities).
134. *See* R. Quentin Grafton & Dale Squires, *Private Property and Economic Efficiency: A Study of Common Pool Resource*, 43 J. L. & Econ. 679 (2000) (providing empirical data showing how various institutional changes towards the treatment of private property rights as fully tradable assets are essential for facilitating efficient use of common pool resources).
135. The importance of certainty for facilitating private ordering is explored *supra* at notes 108–118, and accompanying text.
136. *See, e.g.*, Richard Epstein & Bruce N. Kuhlik, Navigating the Anticommons for Pharmaceutical Patents: Steady the Course on Hatch-Waxman, Chicago John M. Olin Law & Economics Working Paper (2d Series) No. 209, at 4 ('But the state bureaucrat is not the owner of any asset whose value will remain unlocked unless he brings it to market'). Nevertheless, individual regulators have incentives to try to extract such value, leading to a bad, 'tollbooth', view of regulation.
137. This would be consistent with the public interest view of regulation.
138. Consider for example the well known childhood tease, or prank, in which a peer is offered a lick of an ice cream cone and then after inducing anticipation, but before delivery, the cone is withdrawn to prevent the lick, while the offeror utters 'Gotcha!' or some more colorful equivalent.
139. *See* Epstein & Kuhlik, *supra* note 136 (citing Richard. A. Epstein, *The Permit Power Meets the Constitution*, 81 Iowa L. Rev. 407 (1995)).
140. I thank participants in the faculty workshop held at Wolfson College, Cambridge University, June 28, 2004, for pointing out this term to me. For more on the problem of License Raj in India *see, e.g.*, Jagdish N. Bhagwati, India in Transition: Freeing the Economy 49–51 (Oxford: Oxford University Press 1993) (discussing the system of permits and licenses needed in India for both outside investment and for internal economic development). *See also* Emran, M. Shahe, et al., *After the 'License Raj:' Economic Liberalization and Aggregate Private Investment in India* (2003) (available on-line at http://ssrn.com/abstract=411080) (same); Sunita Parikh & Barry R. Weingast, *A Comparative Theory of Federalism: India*, 83 Va. L. Rev. 1593, 1608 (1997) ('This system, known in India as License Raj, means that the center retains control over the distribution of permits and licenses for new areas of economic development through the relevant central ministry').
141. *See* Brenda Sandburg, You may not have a choice. Trolling for Dollars (July 30, 2001) (available on-line at www.phonetel.com/pdfs/LWTrolls.pdf) (attributing the origin of the term to Peter Detkin, who at the time was counsel at Intel).
142. *See, e.g.*, Kieff, *Registering Patents*, *supra* note 2 (suggesting a decrease in the presumption of validity as a tool for achieving symmetry in fee shifting between patentees and infringers).
143. *See Matsushita Elec. Indus. v. Zenith Radio Corp.*, 475 U.S. 574, 588–94 (1986) (discussing perils of predatory pricing).
144. *See* Mark Heinzl & Amol Sharma, *RIM to Pay NTP $612.5 Million to Settle Blackberry Patent Suit*, Wall St. J., March 4, 2006, at A1 (noting that settlement estimates ranged to above $1 billion and that infringer's reserves of cash and cash equivalents were about $1.8 billion).
145. A quick calculation is instructive. The infringer in that case, RIM, is a publicly traded company whose stock price fluctuated over the year from a low of about 52, to a typical price around 63, and a high of about 88 (US$ per share), a high that it almost immediately regained by the next business day after the settlement. The majority of

the outstanding shares (191 million) were in the public float (141 million). If the entire public float were purchased in a takeover by offering a $10 premium over the prevailing price of 63, it would require about $1.4 billion over that price. This new controlling shareholder could then fire management and settle the case. If the settlement were at the estimated high level of $1 billion dollars, then that takeover investor would have invested a total of $2.4 billion over the prevailing price, plus perhaps another $100 million in professional fees and other costs for a total investment of $2.5 billion. If the price then jumped back to its year high after the settlement – which did occur – then this investor would see an increase in book value of about $3.5 billion, leaving a net gain of about $1 billion. If the deal were done as a leverage buyout using the shares themselves as collateral for a loan then the return on investment would hinge on the valuation used to support the loan, which would determine the size of the loan. If the valuation were set at the generally prevailing price then return on investment would be measured as a $1 billion dollar gain over an investment of $2.5 billion dollars, which yields the attractive floor for the rate of return at about 40%. If the valuation were set higher, then the rater of return would be higher as well. Of course, Wall Street's regular raiders likely did the same math. The point here is that the reasons they may have elected not to dive in likely included anti-takeover provisions in the corporate documents themselves, as well as various regulatory restrictions on the market for corporate control that are designed to decrease takeovers.

146. Kieff, *Registering Patents*, *supra* note 2.
147. Kieff, *Commercializing Inventions*, *supra* note 2, and Kieff, *Registering Patents*, *supra* note 2.
148. *See* Kieff, *Commercializing Inventions*, *supra* note 2, at 707–12 (discussing commercialization role of patents).
149. *Id.* (discussing these many players and their incentives to interact).
150. *See* Kieff, *Registering Patents*, *supra* note 2, at 99–100 (pointing out that the publicly recorded patent documents help coordinate commercialization by giving notice of the property right over wish bargaining or avoidance can occur). *Compare* Richard A. Epstein, *Notice and Freedom of Contract in the Law of Servitudes*, 55 S. Cal. L. Rev. 1353 (1982) (proposing 'that under a unified theory of servitudes, the only need for public regulation, either judicial or legislative, is to provide notice by recordation of the interests privately created').
151. *See*, Kieff, *Commercializing Inventions*, *supra* note 2, at 710 (discussing importance of property right for encouraging 'holder of the invention and the other players in this market to come together and incur all costs necessary to facilitate commercialization of the patented invention.').
152. *See* Kieff, *Registering Patents*, *supra* note 2, at 76–98 (pointing out that the prior art rules for patents protect investments by third parties with low administrative and public choice costs).
153. *Id.*, at 736–47 (reviewing the central role of the commercialization theory in the history of the framing of the 1952 Patent Act, which provided what essentially remains as our present patent system, by the same group that had only soon before framed the Lanham Act, which essentially remains as our present trademark system).
154. Kieff, *Commercializing Inventions*, *supra* note 2, at 707–12 (discussing role of patents in coordinating complementary users of an invention so as to facilitate inventions commercialization).
155. *Id.* Compare the focus on providing direct incentives to the holder of the IP rights under the reward theories. *See, e.g.*, Mark Lemley, *Ex Ante versus Ex Post Justifications for Intellectual Property*, 71 U. Chi. L. Rev. 129, 130 (discussing role of IP as an 'incentive the right gives the owner').
156. *See, e.g.*, Lemley, *supra* note 155, at 141, n. 42 (referring to commercialization theory as 'elaboration' of 'prospect' theory). In addition, unlike the prospect and reward theories, the commercialization theory, and its companion registration theory, has explanatory power for the positive law rules of the of the IP legal institutions.

157. *See supra* note 149, and accompanying text (elucidating the basic coordination function of the commercialization theory). *See also* Kieff, *Registering Patents supra* note 2, at 62–66 (discussing prospect and rent dissipation theories in relation to commercialization theory). For game theory examples of the formal link between the role property rights can have in these two different settings – described in that chapter as racing games and mating games – see Dale T. Mortensen, *Property Rights and Efficiency in Mating, Racing, and Related Games*, 72 Am. Econ. Rev. 968 (1982). One additional point about rent dissipation that bears mentioning is that it also teaches something about the coordination theory of property. More specifically, what is often overlooked in the view of property rights as tools for internalizing externalities is that the free rider, tragedy of the commons, and positive externalities problems each can be thought of essentially as an inverse of the problem of rent dissipation. The problems of free riding, commons, and positive externalities refer to cases in which individuals within a group decide not to invest in a given activity for fear that others will benefit but not compensate and as a result too little of the activity is produced. The problem of rent dissipation refers to a case in which individuals within a group decide to invest in a given activity for fear that others will do the same and win the race for the common prize and as a result too much of the activity is produced. In both sets of cases, the failure to coordinate leads to inappropriate amounts of the given activity being conducted.
158. *See, e.g.*, Julie E. Cohen, Lochner *in Cyberspace: The New Economic Orthodoxy of Rights Management*, 97 Mich. L. Rev. 462, 497 n.121 (1998) (citing work by Demsetz and noting: 'Similar reasoning underlies Edmund Kitch's proposed "prospect" approach to patents.'); Rebecca S. Eisenberg, *Patents and the Progress of Science: Exclusive Rights and Experimental Use*, 56 U. Chi. L. Rev. 1017, 1040 (1989) (citing work by Kitch and Demsetz and noting: 'The prospect theory offers a justification for patents that is in keeping with broader theories of property rights elaborated by Harold Demsetz . . .'); Neil Weinstock Netanel, *Copyright and a Democratic Civil Society*, 106 Yale L.J. 283, 309 n.108 (1996) (citing work by Kitch and Demsetz and noting: 'For neoclassicists, therefore, intellectual property is less about creating an artificial scarcity in intellectual creations than about managing the real scarcity in the other resources that may be employed in using, developing, and marketing intellectual creations.'); Rai, *supra* note 40, at 121 n.236 (citing work by Kitch and Demsetz).
159. *See* Kieff, *Commercializing Inventions supra* note 2.
160. *See, e.g.*, Lemley *supra* note 155, at 139, n. 35 (discussing role of patentee as coordinator due to the control exerted through the patent and citing Kieff *supra* note 2 and Joseph A. Schumpeter, Capitalism, Socialism, and Democracy, 100–02 (discussing control of a monopolist). *See also* Lemley at 140 (suggesting that when the government assigns the IP right it is effectively selecting who will have 'control over an area of research and development rather than trusting the market to pick the best researcher').
161. *See* Robert Merges & Richard Nelson, *On the Complex Economics of Patent Scope*, 90 Colum. L. Rev. 839 (1990) (studying problem of a single firm controlling development of a particular technology). .
162. *See* Lemley, *supra* note 155.
163. For a discussion of a number of such *ex post* changes and the problems they present see Kieff & Paredes, *supra* note 82.
164. *See* Lemley, *supra* note 155.
165. *Id.* at 130.
166. *Id.*
167. Michael Abramowicz, *Perfecting Patent Prizes*, 56 Vand. L. Rev. 115, 174 (2003). *See also* Kieff, *Registering Patents*, *supra* note 2, at 68, n. 57 (noting that 'participants in the Spring 2001Workshop Series of the John M. Olin Program in Law and Economics at the University of Chicago Law School [raised] a similar objection').
168. *Id.*

169. *Id.* at 174–75 ('Patent law is underinclusive because commercializers of unpatentable inventions also face the prospect of copying').
170. *See* Kieff, *Registering Patents*, *supra* note 2, at 68–70, and 76–98 (responding to Abramowicz by noting that the registration theory component of the commercialization theory elucidates the importance of the positive law rules for obtaining IP rights for protecting third party investments in a way that mitigates administrative costs and public choice costs).
171. By focusing on the right to exclude, the commercialization theory of IP differs in important ways from the general theory of property in land and goods, which typically considers more than the right to exclude. Adam Mossoff provides an excellent historical account of property theories that emphasizes the failure of approaches that focus only on the right to exclude. *See* Adam Mossoff, *What Is Property? Putting the Pieces Back Together*, 45 Ariz. L. Rev. 371, 376 (2003) ('The concept of property is explained best as an integrated unity of the exclusive rights to acquisition, use and disposal; in other words, property is explained best by the integrated theory of property.'). *But see, e.g.*, Thomas W. Merrill, *Property and the Right to Exclude*, 77 Neb. L. Rev. 730, 747–48 (1998) (suggesting right to exclude is central feature of property).
172. *See Kieff, Registering Patents*, *supra* note 2 (discussing notice function of IP).
173. *See* C.D. Hall, *Patents, Licensing, and Antitrust*, 8 Res. L. & Econ. 59 (1986).
174. *See* Kieff, *Commercializing Inventions*, *supra* note 2 at 729–731 (reviewing reasons why IP rights confer insufficient market power to be monopolies and collecting sources).
175. *Id.*
176. *See* Chisum, et al. *supra* note 2, at 61 (providing an overview).
177. Kieff, *Commercializing Inventions*, *supra* note 2 at 744 (discussing role of IP rights as anti-monopoly weapons).
178. Patents were not available because of judge-made exceptions to patent law that had crept into the law in the late 1960s. The utilitarian nature of the industry made it an unlikely candidate for benefiting in the anti-monopoly sense from copyright and trademark protection.
179. *Id.* (giving example of computer software industry as one in which the putative monopoly power of Microsoft was correlated with a time of no meaningful IP protection in that industry).
180. *Id.* (citing *Picard v. United Aircraft Corp.*, 128 F.2d 632, 643–44 (2d Cir. 1942) (Frank, J. concurring)).
181. *See* F. Scott Kieff, *Perusing Property Rights in DNA*, *in* F. Scott Kieff, Perspectives on Properties of the Human Genome Project 125 (2003) (discussing shifts in positive law).
182. *NIH: Moving Research from the Bench to the Bedside: Hearing Before the Subcomm. on Health of the H. Comm. on Energy and Commerce*, 108th Cong. 49 (2003), *available at* http://energycommerce.house.gov/108/action/108-38.pdf (statement of Phyllis Gardner, Senior Associate Dean for Education and Student Affairs, Stanford University) (detailing the differences between the biotechnology industry and the pharmaceutical industry).
183. *Id.* at 47. At the same time, both Europe and Japan have demonstrated technological capacities in this industry that are comparable to the U.S. In addition, both Europe and Japan have comparably developed capital markets and even if they did not, businesses could operate in Europe and Japan while still having access to the capital markets in the US.
184. *See supra* note 119 and accompany text (pointing out the importance of exercising restraint for certain forms of antitrust enforcement designed to protect static efficiency so as to facilitate innovation and promote dynamic efficiency).
185. *See supra* note 124 and accompanying text (discussing role of price discrimination in mitigating output-restricting effects of monopolies).
186. Kieff, *Commercializing Inventions*, *supra* note 2, at 727 (citing Harold Demsetz, *The*

Private Production of Public Goods, 13 J.L. & Econ. 293 (1970)). *See also supra* notes 123–124, and accompany text (discussing price discrimination as limit on monopoly power).

187. *Id.*, at 727–30 (exploring use of IP rights as tools for facilitating price discrimination). *See also* F. Scott Kieff & Troy A. Paredes, *The Basics Matter: At the Periphery of Intellectual Property*, 73 G.W. Law Rev. 174 (2004) (pointing out that the diverse contracting that is allowed facilitates both price discrimination and coordination among complementary users). There are several aspects of the positive law IP regimes that facilitate complex transactions and contacting of the type that can both facilitate coordination and decrease output distortions of a property right. For example, the work for hire doctrine in copyright law helps concentrate ownership in the result of a complex production process. As another example, the provisions of Section 271 of the Patent Act insulate patentees from fear of liability for misuse which allows patentees to elect to sue or license anyone would otherwise be liable for direct infringement, induced infringement, or contributory infringement. See 35 U.S.C. § 271(a)-(d). *See also* Kieff, *Commercializing Inventions*, *supra* note 2, at 736–38. Before the 1952 Act, courts had used the misuse doctrine to erode the ability for intellectual property owners to price discriminate or engage in restricting licensing. Section 271(d) expressly states that such conduct shall not be misuse. *See also* Dawson Chem. v. Rohm and Haas Co., 448 U.S. 176 (1980) (recognizing impact of Section 271(d) and its reason for inclusion in the 1952 Patent Act). To be certain this was clear, Congress acted again in 1988 by adding subparts 4 and 5 to Section 271(d) of the Patent Act to expressly provide that neither a refusal to license nor a tying arrangement in the absence of market power is patent misuse. *See* 35 U.S.C. § 271(d)(4–5) (added by Pub. L. No. 100-703, 201, 102 Stat. 4676 (1988)). The trademark regime allows similar contracting, but the because the need to make commercial use of the subject matter protected by trademarks is less compelling than for patents – since functionality is a bar to trademark protection – the impact of any remaining distortion caused by market power is less severe. That is, there is still the potential for static economic dead weight loss, but the alternative moral claims about output effects are mitigated.

188. *See, e.g.*, Stan J. Liebowitz & Stephen E. Margolis, Winners, Losers & Microsoft: Competition and Antitrust in High Technology (Oakland, CA: The Independent Institute 1999) (showing that truly inefficient outcomes are extrelemy rare instead that even situations of serial monopoly may be the best available in reality).

189. *See, e.g.* Rai, *supra* note 40 (arguing that IP rights impose greater transaction costs than the basic scientific norms in the open 'commons' of academics); Rebecca S. Eisenberg, *Proprietary Rights and the Norms of Science in Biotechnology Research*, 97 Yale. L.J. 177 (1987) (exploring the potential negative impact of patent rights on scientific norms in the field of basic biological research); *see also, e.g.*, Rebecca S. Eisenberg, *Patents and the Progress of Science: Exclusive Rights and Experimental Use*, 56 U. Chi. L. Rev. 1017 (1989) (exploring an experimental use exemption from patent infringement as a device for alleviating potential negative impact of patent rights on scientific norms in the field of basic biological research); Rebecca S. Eisenberg, *Public Research and Private Development: Patents and Technology Transfer in Government-Sponsored Research*, 82 Va. L. Rev. 1663 (1996) (offering preliminary observations about the empirical record of the use of patents in the field of basic biological research and recommending a retreat from present government policies of promoting patents in that field); Heller & Eisenberg, *Anticommons*, *supra* note 127 (arguing that patents can deter innovation in the field of basic biological research).

190. Kieff, *supra* note 41.

191. *Id.*

192. *See supra* notes 72–73, and accompanying text.

193. Kieff, *supra* note 41, at 703–4.

194. *See supra* note 131, and accompanying text.

195. John P. Walsh, et al. *View from the Bench: Patents and Material Transfers*, 309 Science

2002 (2005). Based on the reasoning explored here and in prior work by the present author, this result was expected. Kieff, *supra* note 41 at 704–5.

196. *See, e.g.*, Lemley *supra* note 155, at 133 (citing Mark A. Lemley, *The Economics of Improvement in Intellectual Property Law*, 75 Tex. L. Rev. 989, 1048–72 (1997) (discussing implications of relaxing rationality assumption for IP); Wendy J. Gordon, *Asymmetric Market Failure and Prisoner's Dilemma in Intellectual Property*, 17 U. Dayton L. Rev. 853, 857 (1992) (pointing out costs of rationality assumption).
197. *See supra* p. 290 (reviewing behavioralism problems).
198. *See* Paredes, *supra* note 86 (pointing out countervailing behavioralism problems for government actors, as well as public choice problems, and collecting sources).
199. *See, e.g.*, Heller & Eisenberg, *Anticommons*, *supra* note 189 (initiating literature on anticommons for IP); F. Scott Kieff, *Contrived Conflicts: The Supreme Court vs. The Basics of Intellectual Property Law*, 30 William Mitchell L. Rev. 1717, 1730 (2004).
200. That is, the IP owner may either actively license the IP to someone else who will in turn sell the subject matter protected by the IP, or the IP owner itself may sell the subject matter protected by the IP, which sale would include an implied license to the IP for its buyers. *See* Kieff *supra* note 181.
201. *See supra* notes 129–134, and accompanying text.
202. *See* Kieff, *Commercializing Inventions*, *supra* note 2 at 726 (commercialization risk and potential for future development provides incentives to license broadly).
203. *See* Kieff, *Perusing Property Rights in DNA*, *supra* note 200.
204. Ronald J. Mann, *Do Patents Facilitate Financing in the Software Industry?*, 83 Texas Law Review 961 (2005).

PART V

Innovation governance

Introduction

Tania Bubela

The chapters in this part discuss the wider social context in which scientists and industry create and distribute innovation. They explore policy options for intellectual property rights (IPRs) selected by public institutions and governments to manage innovation in biotechnology at institutional, national and international levels and the economic, social and ethical goals of IPRs as a component of innovation governance. The chapters consider whether the stated goals are being met, using empirical evidence, where possible, and potential IPR policy formulations that may maximize economic and social benefits flowing from biotechnology innovation.

Public opinion, our ability to regulate the use of technology and other concerns all play a role in shaping what innovation occurs and how it is introduced. Tim Caulfield's chapter analyses the problems surrounding public trust both in biotechnology research and in the products of biotechnology. Maintaining public trust is crucial for any kind of research, especially biomedical research, because of the need for dedicated public funds over the long term. Rightly or wrongly, biotechnology patents have become a flashpoint for the public as well as being the focus of policy debates as a proxy for more general concerns about commercialization of biomedical research. If we probe this generalized public angst around commodification of biomedical research, public survey data suggest that people are concerned about patenting life and the issue of access to technologies. There is a plethora of evidence that public trust in biomedical research is very fragile and that the research is carried out without taking into consideration the public's interests and values.

There are significant issues around the credibility of researchers and policy makers/politicians in the field of biotechnology. Publicly funded scientists are considered credible sources of information while industry researchers and publicly funded researchers who receive private funding are not. What is needed are governance structures for commercialization of biomedical research that maintain or enhance public trust. Caulfield suggests that patent pools may be one such governance structure. The benefits of patent pooling are risk spreading, decrease in transaction costs, and increase in investment possibilities. The criticisms are that it takes away incentives, increases monopoly control, and may be difficult to negotiate, especially in academic publicly funded research. If patent pools are structured so that they have an independent governance scheme, however, they may address the public's concerns about commodification, conflicts of interest and trust in the independence of the research community.

The chapter by Jasper Bovenberg focuses on the accessibility of biological data. The study of common complex disorders requires links to be made between genotype and phenotype; between abstract genomic data and concrete patient medical records. This, in turn, requires that these data be made accessible to those who were not necessarily the primary producers of the data. There is a spectrum of accessibility to genomic and phenotypic data from large-scale collections of abstract genomic data produced by publicly funded scientists (for example, the Human Genome Project) to small-scale collections of phenotype data collected by individual scientists, typically in a hybrid clinical/research setting. These collections tend to be only conditionally accessible, as the primary producers often feel that they have earned an exclusive right to use 'their' collection.

Two unrelated developments may impact on the degree of accessibility of these data. First, large-scale projects are moving from assembling raw genomic data with little or no 'utility' in the patent sense, to producing 'functional data' with increasing utility. The increased 'patentability' of these novel data may undermine the data release policies of these projects. Second, data in the small-scale collections are increasingly being assembled by the research participants themselves. Consequently, traditional proprietary claims by individual researchers may come under pressure from claims by research participants.

The model of the Human Genome Project, where data are governed by the Bermuda Principles, is based on an ethos of academic sharing of research results and publication related data may be more broadly applicable. The Project's ethos is reinforced by the prospect of reciprocity. There is a failure to address the problem of free-riding by other scientists and commercial researchers, however, which may be addressed, in part, through creative licensing provisions. In addition, the model may not translate well to the increasingly prevalent participation of patient groups as key players in the promotion, facilitation and acceleration of studies of the causal role of genetics in diseases. These groups create and maintain databases with epidemiological, medical and other information about the relevant families, and biomedical researchers have become heavily dependent on the long-term participation and co-operation of cell and tissue donors.

A strong case may now be made that the active participation and substantive contributions of tissue/cell donors justify the retention of title to the amassed collection. Bovenberg argues that the European database right could serve as a model. While traditional IP plays no role as regards the protection of collections of data, the recognition of a database right in large-scale collections of abstract (post-)genomic data could provide a meaningful remedy. It would be enforceable against all users who misappropriate or threaten to restrict the use to be made by others of the data collection, protecting the interests of researchers and patient groups who have invested substantial resources in the collection of valuable data and deserve a meaningful and enforceable positive right to exploit or control the collection.

The chapter by Sachin Chaturvedi addresses emerging trends in patent regime associated with agricultural biotechnology in developing countries. Plant variety protection (PVP) and patents have emerged as two important forms of IPRs. Both patents and PVP provide exclusive monopoly rights over a creation for commercial purposes for a limited period of time. Though the criteria for a patent are defined as novelty, inventiveness (non-obviousness), utility and reproducibility, along with provisions for compulsory licenses (CL), patent offices now grant biotechnology patents on microorganisms and, in some countries, on all life forms with no provisions for CL. In contrast, the intellectual property regime for plant variety protection emerged with a strong commitment to the public interest, including provision for CL.

Plant variety protection has worked well as a mechanism to promote the interests of plant breeders for developing new varieties through giving them proprietary rights, on the one hand, and treating plant breeders as custodians of public rights of access and use of genetic material, on the other hand. PVP encourages cross-licensing between a holder of plant variety rights (PVR) and a holder of a patent. Under the breeders' exemption of plant variety rights anyone may use protected material for breeding purposes. The patent regime does not reciprocate. In the patent regime the interpretation of research exemption is much narrower than that of the breeders' exemption in PVR. Unfortunately, the number of utility patents issued has grown very rapidly in the US. By December 1994, 324 utility patents had been issued for new plants or plant parts and 38 were issued for animals. As with Plant Variety Protection Certificates, most utility patents were awarded to the private sector.

The experience of the Indian seed industry, measured by the Research and Information System for developing Countries Seed Industry Study, shows the following results: (1) indigenous seed firms in India find it difficult to access relevant genes for development of new varieties in their biotechnological research as their sequence has already been patented by just one trans national company; (2) introduction of Bt (Bacillus thuringiensis) characteristic without license even in other plants is completely impossible; (3) license fees are not regulated, are arbitrary and are incredibly expensive, and (4) consolidation of the industry is driving out smaller firms.

The discussion of these chapters focused on the presentations of Caulfield and Bovenberg at the workshop in Florence. There was a general consensus that patent pools, especially those with an independent governance structure, would be a useful mechanism for improving public perceptions of biotechnology. This may be further facilitated if patent pools are first established in areas of biotechnology perceived as having high social value, such as forest biotechnology and the preservation of heritage trees.

In Europe, however, it may be difficult to separate public concerns over GMOs *per se* and biotechnology patenting. Some examples highlight the distinction. There was general public outrage when Monsanto attempted to patent terminator technology and this played out in the media as a patenting issue. Ironically, however, self-help strategies such as terminator technology and trade secrets are chosen by innovators when patent protection is weak and there is a fear that patents will not be enforced.

It is also interesting that the profit motive is so distrusted by the public when many studies have indicated that other motivations have a much greater effect on scientific research, for example, self-aggrandizement and the building of large institutions. It seems strange that the public mistrusts a private biotechnology company with $100,000 in assets but trusts a multi-billion dollar institution such as Harvard that styles itself as a not-for-profit educational foundation. The Walsh study found that individual desire to move forward academically had a more profound impact on data sharing than patents. Patents were not a significant factor in non-disclosure and the largest factor was career advancement. Other studies have shown race and gender effects.

Bovenberg suggested that without access to data there could be no scientific revolutions. This was explored from a historic perspective. Did this statement come from a view of the past where there was open access to data between members of the scientific research community that in fact never existed? From a historical perspective, there was a massive growth in mathematical science, chemistry and mechanics and the patent system/property rights from 1870–1914. At that point there were few mechanisms for technology transfer. In this respect, the illustration of patents and Watt's invention of the steam engine is actually more interesting from the perspective of what was not patented. Watt did not patent the efficiency measuring mechanisms he had in his works. Instead, he kept them a trade secret because they enabled him to gauge and understand what he was doing. It is likely that if the steam gauge had been made public, it would have done most of the work of diffusing the invention rather than stifling it for, some people argue, as long as 31 years. Thus, the issue about how technology moves forward is not simply about patenting but also the access or non-access to the means that allow evaluation of the technology or science.

From an economic perspective, the term during which an invention is useful is the important subplot in innovation because the useful life of an invention is shorter and shorter. In many industries, valuation is not possible for longer than five years of market value because the discount rate drops to nil. Thus, patent duration is of decreasing importance. The question of who should own academic data collections has many alternative answers including the public, scientists, funding agencies, government, research projects and institutions. Problems may arise when these different actors have different rules and cultures around managing IPRs. The better question may be – Should anyone own it? – It may be better to set up a system of attribution through meta-tagging. One could argue that authorship is a way of ensuring both attribution and certification. Communities could develop standards such that the meta-tagging would be accommodated within protocols for communication of data.

At our workshop in Florence, it occurred to us that as we discussed the papers that were presented in session, the discussion had moved subtly from the issue of IPRs in innovation to their role in commercialization. To a lessening extent IPRs were viewed as an incentive for innovation, but IPRs were seen as a facilitator for the movement of information from place to place. Viewed in this way, the question to be faced is the extent to which patents are necessary to move from innovation to commercialization, and to what extent there are non-IPR dependent methods yet to be explored.

The central question is the compatibility of patent and non-patent systems for different forms of information. Do patents overwhelm other methods once they are introduced? In terms of the question asked at this workshop – What is the role of IPRs in innovation? — the answer is likely that IPRs do not have much of a role at the inventive stage. At the same time, it is evident that once value is added to an invention, patents have a significant role in creating and limiting social access to innovation. How we tie this well-known phenomenon to the impact of IPRs on commercialization is an area in which future research ought to be conducted.

That said, in the business context, patents protect investment. The private sector does not want to bring out a product that can be easily imitated. Industry pushes for innovation that it can take to the market, and patenting is the mechanism that gives assurance for that investment. From this standpoint, we could argue that if patents do not encourage business-driven innovation, they certainly play a necessary facilitating role. Other forms or areas where society wants innovation are a separate issue – patenting is helpful for commercializing ideas, and at that level they spur innovation. There may therefore be a distinction in the role of patents in industry versus public science. In the private sector, patents play a role in providing a conduit for information to move into a commercial setting. Patents may provide an incentive to industry to move information around. Do patents have a negative effect on distributing other forms of knowledge, are these systems in parallel, do they conflict, or is there any relationship at all?

13. Accessibility of biological data: a role for the European database right?

Jasper A. Bovenberg

INTRODUCTION

One of the challenges in the post-genomics era is the study of common complex disorders (Collins, Green, Gutmacher and Guyer, 2003, 840). This type of research involves the study of links between genotype and phenotype; between abstract genomic data and concrete patient medical records. This, in turn, requires that collections of either type of data are accessible for those who were not its primary producers.

The accessibility of genomic and phenotype data runs a spectrum. At one end are large-scale collections of abstract genomic data produced by publicly funded scientists. These data are widely accessible as funders require immediate, unrestricted and even pre-publication release. At the other extreme are small-scale collections of phenotype data made by individual scientists, typically in a hybrid clinical/research setting. These collections tend to be only conditionally accessible, as the primary producers often feel that they have earned an exclusive right to use 'their' collection of data concerning 'their' patients or subjects.

Two unrelated developments might impact on the current degree of accessibility of the collections at either end of the spectrum. First, large-scale projects are moving from assembling raw genomic data with little or no 'utility' in the patent sense of the law, to producing 'functional data' with increasing utility. The increased 'patentability' of these novel data may undermine the data release policies of these projects. Second, data in small-scale collections are increasingly being assembled by the research participants themselves. Consequently, traditional proprietary claims by individual researchers may come under pressure from claims by these research participants.

This chapter will examine the current state of accessibility to different types of data at either end of the spectrum, how this state of affairs may be impacted by the above developments and how these two developments, while unrelated, may both be addressed by recognizing a novel intellectual property (IP) in collections of data.

ACCESSIBILITY OF ABSTRACT GENOMIC DATA: CURRENT STANDARD

The Bermuda Principles

The prime example of a large-scale collection in the biological data spectrum is the collection produced in the course of the Human Genome Project (HGP). Started in 1990 by the National Institutes of Health (NIH) of the United States (Cook-Deegan, 1994, 314), the goal of the HGP was to determine the sequence of the three billion chemical base pairs that make up the human genome, and to store this information in a database. From the start, a key issue was the promotion and encouragement of rapid sharing of the data generated by the project. On the one hand, sharing was considered essential to foster the project, to avoid duplication and to expedite research in other areas (National Genome Research Institute, 1991). Also, the academic sharing ethos mandates that scientists share the results of their findings and publication related data. This ethos is reinforced by the prospect of reciprocity: *do ut des*. On the other hand, rapid data release conflicts with the fundamental scientific incentive to be the first to publish an analysis of one's data. As biomedical research is increasingly data- rather than hypothesis-driven, a scientist could use his data collection more than once in the competition for new publications and new grants. It was considered essential to give the producers time to verify and validate their data and to gain some scientific advantage from the effort they invested (National Genome Research Institute, 1991). In addition, it was recognized that moving the results of genomic research from the bench to the marketplace would require intellectual property protection for some of the data. From 1995 to1996, when the human sequencing really began to take off Sir John Sulston, one of the leading scientists behind the international sequencing effort, felt the need to get some kind of commitment from the international sequencing community that genomic information would be made publicly available. At that time, Sulston notes in his personal account of the HGP, a gold rush mentality had evolved towards the genome, fostered by a new breed of genome-based private companies. His concern was highlighted by the controversy surrounding the patenting by Myriad Genetics of the BRAC2-gene.

Apart from pressure from the commercial sector, however, there was also the concern that large-scale sequencing centres, funded for the public good, would establish a privileged position in the exploitation and control of human sequence information. In view of these developments, Sulston and his American counterpart Bob Waterston organized a meeting to address the issue. The meeting was sponsored by the Wellcome Trust, a major supporter of the UK sequencing efforts, and took place in Bermuda. The attendants included sequencers – including Craig Venter, who by then had not yet started his privately financed venture to sequence the human genome – and representatives from the Wellcome Trust, the UK Medical Research Council, the NIH NCHGR (National Center for Human Genome Research), the DOE (US Department of Energy), the German Human Genome Programme, the European Commission, HUGO (Human Genome Organization) and the Human Genome Project of Japan. As Sulston observes, it was at that meeting that the attendants began to work out the *etiquette of sharing*. Feeling strongly that the principle of free data release had to be accepted or 'nobody would trust anyone else', Sulston scribbled on a white board:

> Human genomic sequence generated at large-scale center:
> *RELEASE*:
> - Automatic release of sequence assemblies >1 kb (preferably daily)
> - Immediate submission of finished annotated sequence
> - Aim to have all sequence freely available and in public domain for both research and development, in order to maximize its benefit to society
>
> *POLICY:*
> - The funding agencies are urged to foster these policies

To Sulston's amazement, all attendants agreed to what be known as the Bermuda Principles (The Human Genome Organization).

Extension to Community Resource Projects

Since their adoption, the Bermuda Principles have served as a point of reference for publicly funded large-scale sequencing projects. The principles have been reinforced, with some modifications to address technical issues, by a number of funding agencies. To date, these policies are claimed to have secured open access to at least 548 public genetic databases worldwide (Tyshenko and Leiss, 2004), available on the internet, including the large international nucleotide databases – EMBL and GenBank (National Genome Research Institute, 2003). At a meeting in 2003, convened by the Wellcome Trust, an international group of data producers, users, database personnel, journal editors and funding agency representatives unanimously agreed that pre-publication release of large-scale genome sequence data had been of tremendous benefit to the scientific research community (National Genome Research Institute, 2003). They further agreed that it was very important to ensure that such rapid release of sequence data continued. The group therefore reaffirmed the Bermuda Principles and recommended that they be extended to all types of sequence data. Furthermore, the attendants recognized that other large efforts, designated as 'community resource projects', would increasingly be generating data and other resources that should also be rapidly released to the community in an unrestricted manner. A 'community resource project' was defined as 'a research project specifically devised and implemented to create a set of data, reagents or other material whose primary utility will be as a resource for the broad scientific community' (National Genome Research Institute, 2003).

Projects that have been identified as community resource projects are the aforementioned International Human Genome Sequencing Consortium, the SNPs Consortium, the International HapMap Project and, recently, the Encyclopaedia of DNA Elements (ENCODE) Project carried out by the National Genome Research Institute. The attendants noted that these projects have become increasingly important as 'drivers of progress in biomedical research' and that immediate pre-publication release of the project data would serve the scientific community best. The attendants, however, also realized that the pre-publication data release model could jeopardize the standard scientific practice that the producers of primary data should have both the right and responsibility to publish the work in a peer-reviewed journal (National Genome Research Institute, 2003). Under a policy of rapid, unrestricted data release, second comers who have not produced any data can grab the data sets posted on the web, analyse them and publish 'their' results. As a result, the primary producers of the data can no longer publish these analyses themselves. In fact, as National Genome Research Institute Director Francis Collins told *The Scientist*:

> A paper would appear describing the sequence of an organism where the authors had not produced a single base pair of that sequence, but had basically analyzed what they could find for free on the Web. In a few instances they didn't even acknowledge where they got the information. That's obviously a fairly egregious breach of politeness [Powledge, 2003].

The obvious solution would have been the introduction of a publication exception. The scientific community would then have the permission to use the unpublished data for all purposes, except for the purpose of publishing the results of a complete genome sequence assembly or other

Table 13.1 Guidelines for a community resource system

Community Resource Project (CRP)		
Funding Agencies should:	*Resource Producers should:*	*Resource Users should:*
Designate appropriate efforts as CRP Require and enforce, as a condition of funding, free and unrestricted data release to appropriate searchable public databases Ensure that producers have sufficient support for curation, maintenance and distribution	Publish a Project Description of the CRP to inform the scientific community and to provide a reference Release data immediately, freely and without restrictions Recognize risk of abuse by Resource Users	Cite and acknowledge CRP appropriately Recognize interest of Resource Producers in publishing reports analysing the CRP Respect the producer's legitimate interest, while being free to use data in any creative way

large-scale analyses prior to the sequence producer's initial publication (Powledge, 2003). The attendants, however, concluded that the risk that sequence data were occasionally used in ways that violate normal standards of scientific etiquette was a *necessary risk*, to be accepted in view of the considerable benefits of immediate release (National Genome Research Institute, 2003). Instead of a publication exception, they encouraged producers to recognize this risk. In turn, sequence users were reminded that they 'are expected to acknowledge the source of the sequence data through the use of appropriate citations' and urged to recognize 'that producers have a legitimate interest in publishing their own data' (Powledge, 2003, fn.11).

A Tripartite System of Responsibility

To optimize the effectiveness of the system of rapid, pre-publication release of data from community resource projects, the attendees of the Wellcome Trust meeting proposed a system that addresses the three constituencies involved in the generation of such projects: data producers, data users and funding agencies. To ensure the development of a community resource system, each of them is encouraged to adhere to, *inter alia*, the guidelines in Table 13.1.

Extension of Bermuda principles to Phenotype Data?

The attendees of the Wellcome Trust meeting considered that beyond those large-scale 'community resource projects' many valuable small-scale data sets could come from other sources. Since those resources emerge from research efforts the primary goal of which is not resource generation, the contribution of their data to the public domain is more a voluntary matter. However, there is a clear benefit to be derived from converting these 'small-scale' data sets into community resources as rapidly as possible and, ideally, the producers of such data should release them into the public domain voluntarily. Whether that is the case is the subject matter of the next section of this chapter, which will examine the degree of data-accessibility of the small-scale collections of phenotype data at the other extreme of the biological data spectrum.

ACCESSIBILITY OF SMALL-SCALE PHENOTYPE DATA: CURRENT STANDARD

Survey

Significantly, in spite of the urgency of the issue, little empirical research into the practice of data sharing or data withholding within the biomedical research community has been done so far (Campbell et al., 2002). This section of the chapter presents part of the outcome of one such survey, albeit a limited one. The worldwide survey was carried out among the members of the Human Genome Organization, the American Society of Human Genetics and the European Society of Human Genetics in October and November 2003 (Bovenberg, 2006, 168–169). A questionnaire was completed by 118 human geneticists from over 15 different countries, but most of whom were from the United States, representing universities, hospitals and research institutes, as well as a few scientists employed in the commercial sector. While this is certainly not representative, the survey sketches a rough picture of how the responding human geneticists deal with the issue of access to their data. The results of the survey that are relevant for the present discussion are as follows.

Origin and secondary use of data

Most respondents collect their data both in connection with their treatment of patients and as part of research on healthy volunteers. The research shows the majority (more than 60 per cent) includes not only genomic, but also medical and genealogical data, as well as biological material. The databases they produce serve both clinical and research purposes. The research is not limited to the primary goal of the study involved, but also includes 'secondary use' of the collected data and tissue; most respondents (81 per cent) ask their patients or research subjects for their consent to use their data and material not only for the initial diagnosis, treatment or research, but also for future, unspecified research purposes.

Need for access

A small majority of the respondents (54 per cent) disagrees that unlimited access to third party databases is absolutely necessary for their research, whereas a large minority (42 per cent) claims that unlimited access to other databases is absolutely necessary for their research. A majority of the respondents (62 per cent) held that they actually have sufficient access to all databases that they need for their research. Some 34 per cent held that they do not have access to all the databases they need. Four per cent did not respond to either question.

Funding of databases

Notably, research funding for the generation of databases for the majority of the respondents (80 per cent) is entirely public. Only a small minority (12 per cent) are funded by both public and commercial sponsors. The research of most respondents (30 per cent) is funded by their national governments. Roughly one quarter are funded by their university departments and one fifth of the respondents receive funding from multiple public funding agencies. In more than 75 per cent of cases the funding comes from more than one source.

Ownership of databases

Notably, in view of the predominantly public nature of database funding, 43 per cent of the respondents claim ownership of the databases they build in the course of their research (see Figure 13.1).

Interestingly, 42 per cent of universities claim ownership of the databases created by their researchers (see Figure 13.2), but universities are reported to fund only 22 per cent of the research.

Furthermore, according to 83 per cent of the respondents, the research funding agencies funding did not claim ownership of the data (see Figures 13.3, 13.4 and 13.5).

Granting access

Notably, in view of the expressed need for unlimited access to third party databases, a majority of the respondents (51 per cent) say they do not grant access to their own databases to non-commercial institutions (see

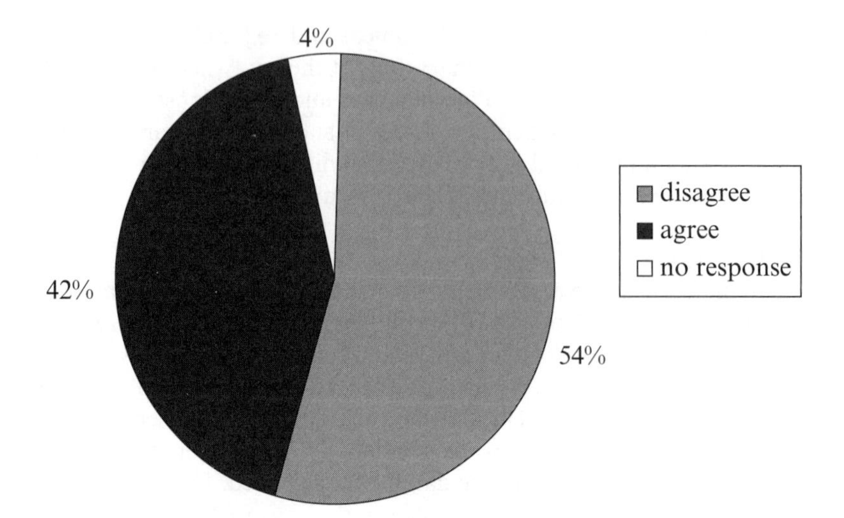

Figure 13.1 *Access to database. Unlimited access to third parties' databases necessary?*

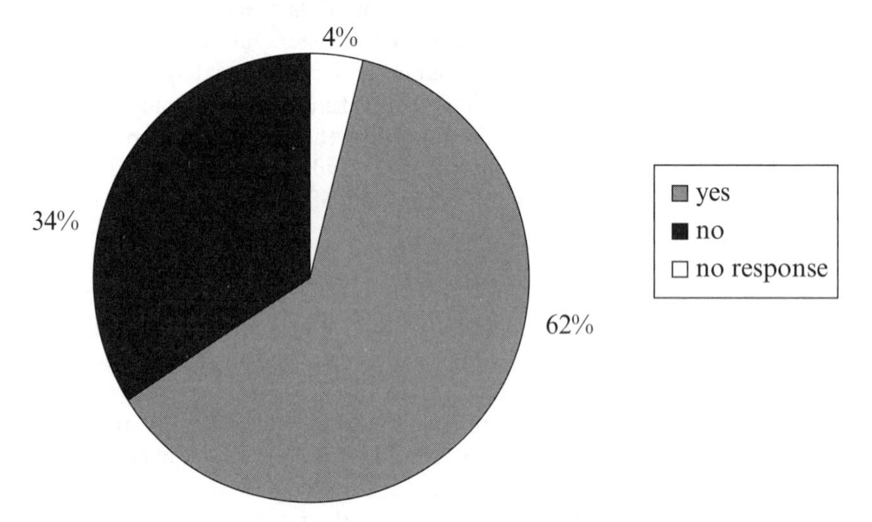

Figure 13.2 *Access to database. Do you have sufficient access to all databases needed?*

Figure 13.6 and 13.7). Surprisingly, in view of the predominantly public funding of the databases, only one third of respondents (35 per cent) reportedly grant access to their databases to non-commercial institutions for free.

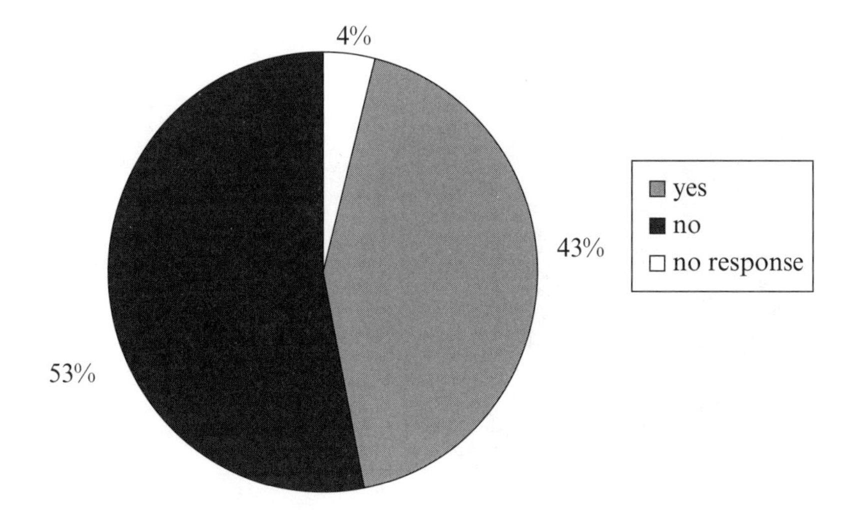

Figure 13.3 Ownership of database. Do you claim ownership?

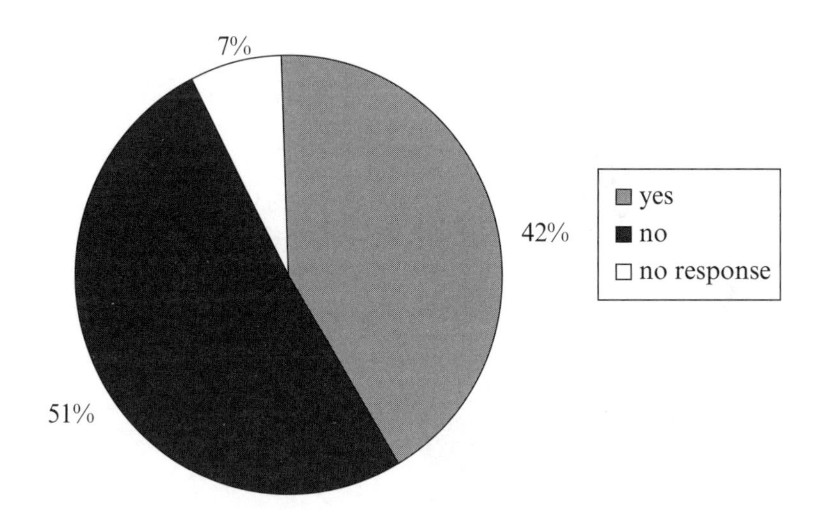

Figure 13.4 Ownership of database. Does your university claim ownership?

Although the survey was limited in terms of respondents, preliminary results suggest that the free flow of, access to and exchange of human genetic data and material assembled in clinical and scientific settings of human geneticists is not standard practice. This finding is supported by the study of Campbell et al. who conducted the first national survey on

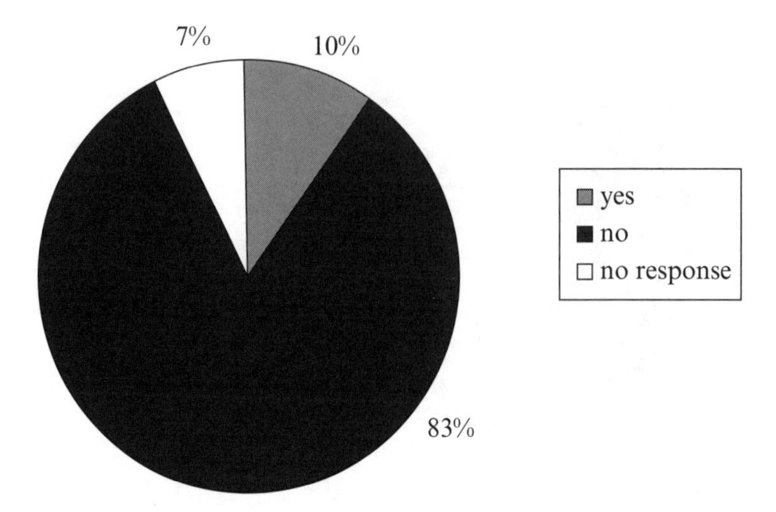

Figure 13.5 Ownership of database. Does any public funding agency claim ownership?

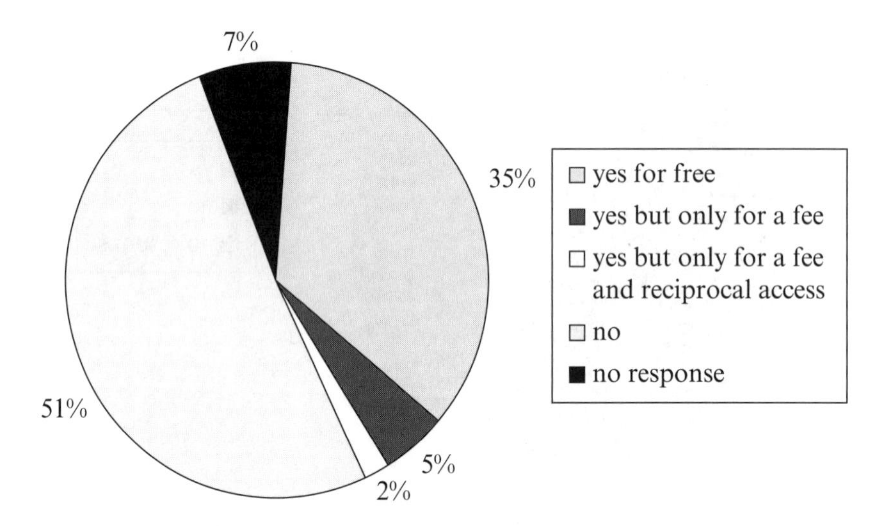

Figure 13.6 Do you grant access to your database to non-commercial institutions?

data-withholding among American geneticists. The Campbell study found that in a three year period, 47 per cent of the respondents had denied at least one request for additional information, data or material concerning published results.

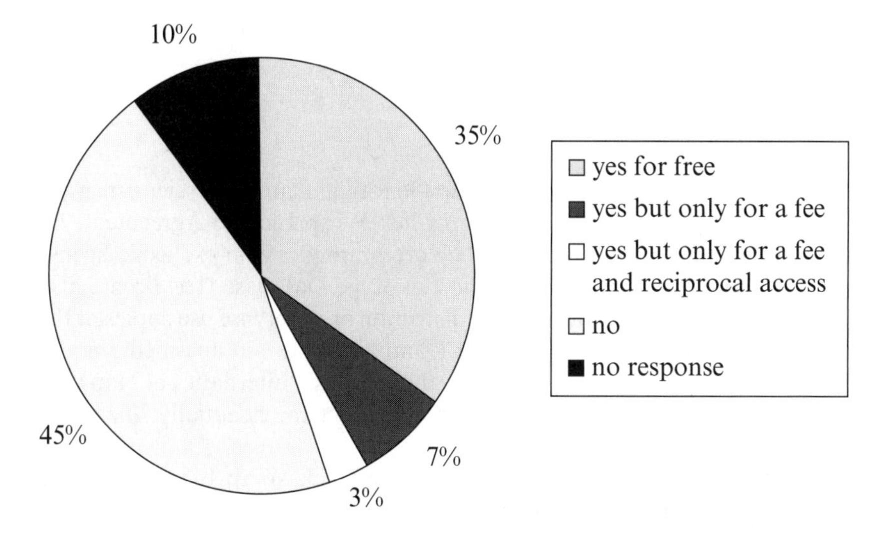

Figure 13.7 Do you grant access to your database to non-commercial institutions even if your database is unique?

ACCESSIBILITY OF ABSTRACT (POST-)GENOMIC DATA: DEVELOPMENTS

Parasitic Patenting

While the Bermuda Principles and the Wellcome Trust Tri-Partite Policy address the problem of free-riding by other scientists, they fail to address the problem of free-riding by commercial researchers. This became clear when the HapMap project was designed. The International HapMap Project is a multi-country effort to identify and catalog genetic similarities and differences in human beings (International HapMap Project 2004a). In principle, the project was to adhere to the data release principles developed by the Wellcome Trust meeting for Community Resource Projects (CRP). While the project satisfied the definition of a CRP, however, the way in which it was to generate its data made it vulnerable to 'parasitic patenting'. Specifically, the genotype data produced during the early stages of the project would make it possible for other parties to construct haplotypes by combining the project's data with their own. These other parties could then file for patents on those derived haplotypes, and in doing so potentially restrict others from using those haplotypes and underlying data. To prevent this from happening, the HapMap Project developed a licensing strategy in the form of the HapMap Click Wrap license (International HapMap Project 2004c).

The HapMap Click Wrap

To register for access to the HapMap Genotype Database, scientists must indicate acceptance of the HapMap Click Wrap License Agreement. By clicking on the 'Accept' button, they are granted a non-exclusive license to access and conduct queries of the Genotype Database. The license also includes the right to 'copy, extract, distribute or otherwise use copies of the whole or any part of the Genotype Database's data, in any medium and for all purposes, including commercial purposes' (International HapMap Project 2004c). This license however, is subject to, essentially, the terms and conditions set out in Box 13.1.

It follows from the rationale behind the Click Wrap License that the need for the license requirement will disappear once the HapMap Project has ensured that most of the data it has generated have been placed in the public domain. This was accomplished on December 10, 2004. When the project released all data publicly and abandoned the click wrap requirement for access.

The ENCODE Project

The HapMap complication seems to represent a trend whereby large-scale projects produce data of ever increasing patentability. Namely, a similar problem is presented in another successor programme of the HGP, the ENCODE Project. The ENCODE Project aims to identify all functional elements in the human genome sequence (International HapMap Project 2004b). The purpose of the ENCODE Project is to generate data that identify or define genomic DNA sequence elements that have biological function. Consequently, these data are likely to satisfy the utility threshold set forth in national patent laws. In fact, the danger of 'parasitic patenting' is even greater than in the case of the HapMap Project, as 'the HapMap data themselves do not have utility' (Ibid.). To maintain unrestricted data access, therefore, the agency funding the ENCODE project is considering making access to the data conditional on the terms of an agreement similar to the HapMap Click Wrap License Agreement (Ibid.).

The Limitations of a License

The question is, however, whether the strategy of a click wrap license is a meaningful remedy to ensure unrestricted access to the data produced by

BOX 13.1 INTERNATIONAL HAPMAP LICENSE (2004)

TERMS AND CONDITIONS FOR ACCESS TO AND USE OF THE GENOTYPE DATABASE ('LICENSE TERMS') ARE AS FOLLOWS:

(. . .) by your actions (whether now or in the future), you shall not restrict the access to, or the use which may be made by others of, the Genotype Database or the data that it contains; in particular, but without limitation,

you shall not file any patent applications that contain claims to any composition of matter of any single nucleotide polymorphism ('SNP'), genotype or haplotype data obtained from the Genotype Database or any SNP, haplotype or haplotype block based on data obtained from the Genotype Database; and

you shall not file any patent applications that contain claims to particular uses of any SNP, genotype or haplotype data obtained from the Genotype Database or any SNP, haplotype or haplotype block based on data obtained from, the Genotype Database, unless such claims do not restrict, or are licensed on such terms that that they do not restrict, the ability of others to use at no cost the Genotype Database or the data that it contains for other purposes; and

you disclose data obtained as a result of your access to and use of the Genotype Database only to other parties who have first confirmed to you in writing that they too are licensees under the terms of the International HapMap Project Public Access License and so are bound by equivalent terms and conditions to those that you have accepted under this License.

(License Version 1.1, August 2003, Copyright © Cold Spring Harbor Laboratory

Full acknowledgements to the GNU General Public License Copyright © 1989, 1991 Free Software Foundation, Inc.

59 Temple Place – Suite 330, Boston, MA 02111-1307, USA)

Community Resource Projects. First, by definition, the license itself, and thus the restrictions it contains, only operates between the parties to the license agreement, that is the licensor (the CRP consortium) and the licensee. As a result, the patent prohibition and the disclosure limitations set forth in the click wrap or any other license agreement are not enforceable against anyone who is not a party to the contract. Second, even between the licensor and the licensee, it is unclear whether the condition of the license could be effectively enforced in case of a violation of the terms of the license. To be sure, if the licensee breaches the agreement by filing a patent application, he will face an action for breach of contract. However, it is unlikely that such a breach would also invalidate the patent application or the patent issued (Opderbeck, 2004, 191). Clearly, given the interests at stake, the agencies funding these types of projects are in need of a more robust, positive right to protect (access to) the databases they produce.

ACCESSIBILITY OF PATIENT RELATED DATA: DEVELOPMENTS

The Collaborative Research Enterprise

Increasingly, patient initiatives bundle forces, collect material and data of their peers worldwide and make these available for the research community (Bovenberg, 2004). In some research areas, patients' contributions are no longer minimal, let alone undetectable. Patient groups have thus become key players in the promotion, facilitation and acceleration of studies of the causal role of genetics in diseases (Merz et al., 2002). They create and maintain databases with epidemiological, medical and other information about the relevant families. Research participants have even been reported to develop scientific expertise that can make substantive contributions to the direction and performance of research (Malakoff, 2004). Biomedical researchers depend heavily on the long-term participation and co-operation of cell and tissue donors. Thus there is a strong case that the active participation and substantive contributions of tissue/cell donors justify that they retain title to the collection they have amassed. *De facto*, they already do. For example, frustrated by the unwillingness of the researchers to share their data, families of patients diagnosed with autism created their own database, the Autism Genetic Resource Exchange. This database is accessible for every researcher and has so far yielded some 18 publications (Zitner, 2003). As patient groups play an ever larger role in the collaborative research enterprise (Merz et al., 2002), their substantial investments in the collection of data seem to justify the recognition of a property right in such a collection. Notably, however, such a right has been denied in the only published court case so far squarely to address the issue of whether patients retain a property right in data they have collected to assist researchers: *Greenberg v. Miami Children's Hospital Research Institute* (1066).

Patient Data in Court

As District Judge Moreno put it, the facts in *Greenberg v. Miami Children's Hospital* presented 'an unfortunate dilemma set against the backdrop of a historic breakthrough in the treatment of a previously intractable genetic disorder' (1066). A group of individuals and non-profit patient support groups had encouraged a scientist to expedite research into Canavan disease, a fatal genetic disorder with no known cure. The group provided the researcher with financial resources, blood, urine and tissue samples, autopsies and confidential medical information. Upon isolation of the gene, a patent was issued for the genetic sequence and related applications (US patent No. 5,679,635). The research institution then began negotiating exclusive licensing agreements and charging royalty fees across the board, thus restricting public accessibility of Canavan disease testing (*Greenberg v. Miami Children's Hospital*, at paragraph 30). The use restrictions prompted two parents of children affected by Canavan disease, Dan and Debbie Greenberg, together with several other parties, to file suit against the institution.

In one of their complaints, the plaintiffs alleged that they owned the Canavan registry which contained contact information, pedigree information and family information for Canavan families worldwide. They claimed that the hospital and Matalon had converted their names and the genetic information by utilizing them for the hospitals' 'exclusive economic benefit' (*Greenberg v. Miami Children's Hospital*, at paragraph 65). The Florida court, however, found no property interest in the genetic information voluntarily given to the researchers, because these were considered donations to research without any contemporaneous expectations of return (*Greenberg v. Miami Children's Hospital*, at 1074–1075). The court did permit, however, a cause of action for unjust enrichment, recognizing 'a continuing research collaboration that involved Plaintiffs also investing time and significant resources in the race to isolate the Canavan gene'.[1] That approach, however, as one commentator remarked, is too *ad hoc* to be relied upon by future research participants (Gitter, 2004, 257, Shields, 2004, 338) and a more meaningful remedy should be available.

ACCESSIBILITY OF BIOLOGICAL DATA: A ROLE FOR THE EU DATABASE RIGHT?

Accessibility of Biological Data: a Role for IP?

Given the substantial intellectual investment that goes into producing academic data collections, one would expect that accessibility of data is, at least partly, governed by intellectual property rights, if only by allocating the ownership rights in the collection of data. Traditional IP such as patents and copyrights, however, protect inventions. This is to say that they protect *creations* and not *collections of data*. In fact, the prevailing thought in legal discourse over IP and data is that data *per se* should not be protected. The general assumption is that academic research data form part of the public domain and that their accessibility is governed by the sharing ethos within the scientific community. However, while the concept of a public domain and the sharing ethos are noble principles, the developments sketched above seem to call, at either end of the academic biological data spectrum, for a robust intellectual property right in collections of data. As for large-scale collections of abstract (post-)genomic data, the recognition of such an exclusive right could provide a meaningful remedy, enforceable against all users who misappropriate or threaten to restrict the use to be made by others of the data collection. At the small-scale end of the spectrum, such a database right would provide those who have, in the words of Judge Moreno, invested time and significant resources in the collection of valuable data, with a meaningful and enforceable positive right to exploit or control the collection they have substantially invested in. In any event, such a right would provide better, upfront protection of substantial investments than an *ex post facto* cause of action for unjust enrichment.

The European Database Right

Obviously, when it comes to designing a right in collections of biological data, the European database right could serve as a model. The European database right was introduced in 1996 by the European Union. The right vests an exclusive right in the producer of a database to grant permission to extract and re-utilize the contents of the database (National Genome Research Institute, 1991). The database right aims to protect collections of data that meet the statutory definition of a database:

> A compilation of works, data or other elements, systematically or methodically arranged and independently accessible by electronic means or otherwise and of which the creation, control or presentation of the contents demonstrates in quantitative or qualitative respects a substantial investment.[2]

Notably, precisely because of its adverse implications for science, the right has been criticized by many observers of scientific research (International Council of Scientific Unions, 1997). Indeed, as this *sui generis* right was obviously not designed with the protection of scientific databases in mind, straightforward application to such databases would give rise to some awkward complications (Bovenberg, 2006). Some of the criticism, however, seems based on a misapprehension of a number of key elements of the database right.

First, the database right cannot give rise to the creation of a new right on data proper (National Genome Research Institute, 1991; Database Directive, Recitals 45 and 46). It only confers an exclusive right in a *collection* of data. It does not bar anyone from collecting the same data to create his or her own database. Second, if owned by an individual scientist or group of scientists, the database right could indeed be used as an instrument to legitimize and enforce data exclusivity. However, the database right typically will not vest in a single scientist or even a group of scientists. The database right vests in the *producer* of the database. Under the European database right the producer is the person who bears the *financial risk* of the investment in the database. Absent a substantial financial investment on their part, employees and contractors (National Genome Research Institute, 1991; Database Directive, Recital 41), while they may be the actual collectors of the data, will not qualify as the producers of the database and thus cannot be the owners of the database right. In most cases, and in any event in the case of the community resource projects, it will be the funding agency which will qualify as the investor. Third, while the database right is said to undermine the public domain, the Directive provides that the database right does not prejudice existing rights to access public documents. The rationale behind this provision is that public works produced by the public authorities should in principle be part of the public domain (Verkade and Visser, 140 fn.115). 'Public authorities' include public institutions, to the extent they are operating within the scope of their public remit and/or within their public competences. Obviously, public research funding bodies will usually satisfy this criterion. Fourth, the Directive provides for a research exemption, allowing the lawful user to use the database for research purposes without the consent of the owner of the database. This exemption has been criticized for being too narrow (International Council of Scientific Unions, 1997). Specifically, it is claimed that the exemption covers only the *use* and not the re-utilization of the database. On closer examination, this restriction is not necessarily an impediment to the conduct of research. Research use typically requires one or more of the following activities: accessing data, linking data with other data, selecting data as case controls for other data or pooling data with other data for meta-analysis (Lowrance, 2005, 15). None of these data involves the 're-utilisation' of data that is defined in the database-right law as the act of making the contents of a database available to the public. In brief, while the application of the database right to academic collections of biological data is certainly an awkward fit, it may not be the death knell for accessibility of academic collections of biological data. On the contrary, it could play a role in keeping publicly funded data collections accessible and in promoting the collection of data by patient advocacy groups by providing them with a meaningful and enforceable right.

CONCLUSIONS

Increasingly, innovation in biotechnology involves research relating abstract genomic data to concrete patient medical records. This calls for wide accessibility and wide availability of both types of data. On the one extreme of the data-accessibility spectrum the policy of immediate, pre-publication unrestricted release is under pressure as the large-scale projects increasingly produce data of 'patentable' utility. On the other side of the spectrum, traditional ownership claims by individual researchers in their small-scale clinical collections deprive the patient support groups which substantially invest in the data collection of the opportunity to exploit and control their data. As a result, such initiatives may be discouraged. While traditional IP plays no role as regards the protection of collections of data, the recognition of a database right in large-scale collections of abstract (post-)genomic data could provide a meaningful remedy, enforceable against all users who misappropriate or threaten to restrict the use to be made by others of the data collection. On the other side of the spectrum, such a database right would provide those who have invested substantial resources in the collection of valuable data, with a meaningful and enforceable positive right to exploit or control the collection.

NOTES

1. According to a September 29, 2003 joint press release, the parties have reached a settlement which provides for continued royalty-based genetic testing by certain licensed laboratories and royalty-free research by institutions, doctors and scientists searching for a cure; available at http://www.canavanfoundation.org/news/09-03_miami.php.
2. Article 1 (1) (a) of the Act of 8 July 1999, implementing Directive 96/9/EC into Dutch law, [1999] *Stb.* 303, as amended by the Act of 6 July 2004, [2004] Stb. 336, entered into force 1 September 2004 (Act on Database-right).

REFERENCES

Committee for a study on promoting Access to Scientific and Technical Data for the Public Interest, National Research Council. 2004. *A Question of Balance:*

Private Rights and the Public Interest in Scientific and Technical Databases. Washington, DC: National Academies Press.
Barton, J.H. and K.E. Maskus. 2004. 'Economic Perspectives on a Multilateral Agreement on Open Access to Basic Science and Technology'. *SCRIPT-ED, Journal of Law and Technology* 1:4. Available at http://www.law.ed.ac.uk/ahrb/script-ed/.
Committee on Issues in the Transborder Flow of Scientific Data, National Research Council. 1997. *Bits of Power: Issues in Global Access to Scientific Data.* Washington, DC: National Academies Press.
Bovenberg, J.A. 2004. 'Inalienably Yours? The New Case for an Inalienable Property Right in Human Biological Material: Empowerment of Sample Donors or a Recipe for a Tragic Anti-Commons?' SCRIPT-ED, *Journal of Law and Technology* 1:4. Available at http://www.law.ed.ac.uk/ahrb/script-ed/.
Bovenberg, J.A. 2006. *Property Rights in Blood, Genes & Data: Naturally Yours?* Boston: Martinus Nijhoff Academic Publishers.
Campbell, E.G. et al. 2002. 'Data Withholding in Academic Genetics, Evidence from a National Survey'. *Journal of the American Medical Association* 287: 473–480.
Collins, F.S., E.D. Green, A.E. Gutmacher and M.S. Guyer. 2003. 'A Vision for the Future of Genomics Research'. *Nature* 422:835–47.
Cook-Deegan, R.M. 1994. *The Gene Wars: Science, Politics and the Human Genome.* New York: W.W. Norton & Company.
Database Directive 96/9/EC of the European Parliament and the Council, adopted March 11, 1996 concerning the protection of databases [1996] OJ L77/20.
Gitter, D.M. 2004. 'Ownership of Human Tissue: A Proposal for Federal Recognition of Human Research Participants' Property Rights in their Biological Material'. *Washington and Lee Law Review* 61: 257–364.
Greenberg v Miami Children's Hospital Research Institute, Inc. 2003. 264 Federal Supplement, 2d Series 1064. 1064–78.
International HapMap Project. 2004a. 'About the HapMap'. Available at http://www.hapmap.org/thehapmap.html.en.
International HapMap Project. 2004b. 'Accessing ENCODE Data'. Available at http://www.genome.gov/10005107#3.
International HapMap Project. (2004c). 'Registration to Access the HapMap Project Genotype Database'. Available at http://www.hapmap.org/cgi-perl/registration.
International Council of Scientific Unions. 1997. 'Position Paper on Access to Databases, prepared by the ICSU/CODATA group on Data and Information'. WIPO Information Meeting on Database Protection Geneva. 17-19 September. Available at www.codata.org/codata/data_access/index.html.
Lowrance, W.W. 2005. *Access to Collections of Data and Material for Health Research,* A report to the Medical Research Council and the Wellcome Trust. Available at http://www.wellcome.ac.uk/About-us/Publications/Reports/Biomedical-ethics/WTX030843.htm.
Merz, J. et al. 2002. *American Journal of Human Genetics* 79: 965–71.
National Genome Research Institute. 1991. 'NIH-DOE Guidelines for Access to Mapping and Sequencing Data and Material Resources'. Available at http://www.genome.gov/10000925.
National Genome Research Institute. 2001. 'Reaffirmation and Extension of NHGRI Rapid Data Release Policies: Large-scale Sequencing and Other Community Resource Projects'. Available at http://www.genome.gov/10506537.

National Genome Research Institute. 'The ENCODE Project: Encyclopedia of DNA Elements'. http://www.genome.gov/10005107.

Opderbeck, David W. 2004. 'The Penguins Genome, or Coase and Open Source Biotechnology'. *Harvard Journal of Law & Technology* 18(1): 168–98.

Powledge, J. 2003. 'Revisiting Bermuda: Proposed Revision of Data-sharing Principles Recognizes Changing Times and Technologies'. *The Scientist* 4(1): 30311–03. 11 March.

Reichmann, J.H. and P. Samuelson. 'Intellectual Property Rights in Data'. *Vanderlsilt Law Review*. 50: 52–166.

The Royal Society. 2003. *Keeping Science Open*: *The Effects of Intellectual Property Policy on the Conduct of Science*. Available at http://royalsociety.org/document.asp?id=1374.

Malakoff, D. 'Showdown Expected in Congress'. 2004. *Science* 305: 1226.

Shields, M.A. 2004. 'Annotation, Liability for Conversion and Misappropriation of Genetic Material, *American Law Reports*. 121: 315–65.

The Human Genome Organization. 1996. 'Summary of Principles Agreed at the International Strategy Meeting on Human Genome Sequencing Bermuda'. Sponsored by the Wellcome Trust. 25–28 February. Available at http://www.gene.ucl.ac.uk/hugo/bermuda.htm.

Tyshenko, M.G. and W. Leiss. 2004. 'Genomics and the Current State of Public Databases'. GE3LS Symposium. Poster presentation with abstract. Vancouver, Canada.

US patent No. 5,679,635 issued to the hospital in October 1997.

Verkade, D.W.F. and D.J.G. Visser. 2009. *Inleiding en Parlementaire Geschiedenis Databankenwet.* Den Haag: Boom.

Zitner, A. 2003. 'Whose DNA Is It, Anyway? Many People, Hoping for Medical Advances, Give Genetic Material. But Some Researchers' Refusal to Share Samples has Donors Up in Arms'. *Los Angeles Times*, 18 July.

14. Biotechnology patents, public trust and patent pools: the need for governance?

Timothy Caulfield[1]

Over the past few decades, biomedical research has caught the public imagination like never before. The mapping of the human genome and stem cell research, for example, are topics that have received intense media attention. They are topics that the public cares about and, to a large degree, they are areas of research that the public supports. However, biomedical research has also produced a great deal of social controversy. The scandals associated with clinical trials, the recent Korean cloning fiasco and the continued public debate about the moral acceptability of stem cell research have also been the focus of the media limelight and public debate.

Such controversies have the potential to erode confidence in the research enterprise. In addition, the public seems to be growing increasingly skeptical about the impact of commercial pressures on research – this at a time when academic biomedical research has unprecedented ties with industry. Indeed, as we will see below, many of the most high profile biomedical research controversies are associated with industry influences (Lemmens, 2004; Morin, et al., 2002). In addition, the patenting of biological material, such as human genes and stem cell lines, has also created a level of social unease.

In this chapter I briefly review, by way of example, some of the social controversies associated with the involvement of industry in controversial biomedical research. I consider what the available data on public opinion tells us about how the public views 'life patents' and the commercialization process. I then consider the use of patent pools as a governance strategy to moderate the social concerns associated with the involvement of market forces in biomedical research. While hardly ideal, patent pools may assist in creating a research environment that maintains public trust while still facilitating the involvement of needed industry partners.

PATENTS AND SOCIAL CONTROVERSY

While this chapter is not the place to critique the benefits and risks associated with a close link between the private sector and biomedical research, it is hard to deny that over the past few decades the research environment has become increasingly focused on commercialization (Caulfield, 2003). Indeed, it has become common for major public funding agencies to have commercialization as an explicit goal. In my home country of Canada, for example, the Canadian Institutes of Health Research (CIHR), Canada's primary public supporter of biomedical research, has enabling legislation that states that the goals of the CIHR are to 'encourag[e] innovation, facilitat[e] the commercialization of health research in Canada and promot[e] economic development through health research in Canada' (2000).

Some of the biggest controversies in science policy have been a direct result of this growing relationship between biomedical science and commerce. In the US, for example, the death of a young man, Jesse Gelsinger, raised questions about the close affiliation of the researchers and an industry sponsor. Specifically, there was concern about whether financial consideration had an inappropriate impact on the running of the gene therapy research trial. The case generated a great deal of media coverage and led to some policy-making activity (Nelson and Weiss, 2000). The Canadian dispute between Nancy Olivieri and Toronto Hospital for Sick Children had a similar impact – generating magazine covers and numerous policy responses (CMA, 2001). In the Olivieri case, Apotex Inc., which was sponsoring some of Olivieri's work, threatened to bring legal action when Olivieri took steps to inform patients enrolled in clinical trials that the treatment drug might be causing life-threatening side-effects. The company claimed that such disclosure was a breach of the clinical research agreement between Dr Olivieri and Apotex Inc. Though the facts of the disagreement are complex and convoluted (Litman and Sheremeta, 2002), the case has nevertheless come to stand as a cautionary tail about potential harms and conflicts that come with industry involvement in biomedical research (Thompson, et al., 2001).

In Canada, there have also been several high profile public debates about the commercialization process and the way science is funded. Genome Canada (GC), a publicly funded non-profit organization which is one of the country's leading sponsors of large-scale genomic research, requires scientists to obtain co-funding for all projects. For some in the research community, this requirement has been interpreted as pressure to get industry funding. This led many scientists publicly to denounce Genome Canada and the perceived commercialization agenda (Wells,

2005). Indeed, it led to the creation of a web based petition, signed by over 1,400 Canadian researchers, and a highly publicized letter to the journal *Science* (CSBMCB, 2005; Tyers, et al., 2005).

The reality of GC funding is, in general, quite different from that portrayed in these public debates (in fact, only 11 per cent of the co-funding comes from industry and none of the current funding from ethical, legal and social issues projects receives funding from industry (Genome Canada, 2005)). Regardless, the controversy once again highlights the ease with which the mixing of private markets and biomedical research can engender public scandal. Many Canadian researchers felt, understandably, that pressure to obtain industry support would adversely impact the overall research agenda by allowing commercial goals to supersede scientific objectives.

Of course, a number of research commercialization stories flow directly from the patenting process – the Myriad Genetics controversy being the most obvious (Caulfield, 2005). Indeed, the decision by Myriad, a Utah based company, to enforce its patents over the BRCA1/2 genes was, arguably, the most significant recent catalyst of national policy making activity, media coverage and public outrage on the topic of gene patents (Foster, 2001; Benzie, 2001; Heath, 2005). In 2001, the company sent a 'cease and desist' letter to most Canadian provinces that used the test as part of the publicly funded health care system. The case was viewed, rightly or not, as a harbinger of the policy challenges created by gene patents (Williams-Jones, 2002).

Other high profile biotechnology patent stories, such as the Supreme Court of Canada's decisions in *Harvard College v. Canada (Commissioner of Patents)* [2002] 4 SCR 45 and *Monsanto Canada Inc. v. Schmeiser* [2004] 1 SCR 902, also engendered public interest. In these decisions, the Supreme Court of Canada was struggling with the patentability of higher life forms. In the *Harvard College* case, the Court concluded that the 'onco-mouse' was not a patentable innovation under Canadian patent law. The case makes Canada the only country with a High Court decision rejecting the patentability of higher life forms. Just two years later, however, in the *Monsanto* decision, the court seems to come to the opposite decision. The Supreme Court of Canada concluded that Monsanto had 'control' over a genetically modified canola plant because it contained a genetically modified element that was an appropriate subject of a patent. For the purposes of this chapter the details of the cases are not critical. However, the decisions do highlight the high degree of public and policy-making interest in biotechnology patents (for example, one study found that both the *Harvard College* and *Monsanto* cases got a tremendous amount of media coverage (Laing, 2004). Indeed, a survey taken before the SCC had overturned the decision found that 'about 50 per cent of Canadians said they were not comfortable with the Appeals Court decision' to allow the patenting of higher life forms.

Naturally, not all science controversies can be traced to the involvement of industry. The recent stem cell research hoax involving a group of Korean scientists seems to have been motivated by one of the longest standing sources of unethical behavior, ambition (Wade, 2005). In addition, the evidence about the actual impact of patents on the research environment remains unclear. For example, there is at least some evidence that patenting and commercialization have only a small impact on the data sharing behavior of basic researchers (Walsh, et al., 2005). Nevertheless, as we will see below, there are mounting data to support the conclusion that commercialization and patenting pressures can have an impact on public perceptions.

PUBLIC PERCEPTION AND PATENTS

Given the profile of these controversies, it is hardly surprising that the public may have growing concerns with both 'life patents' and the commercialization process more generally (Einsiedel, 2005; Caulfield, 2006). There is at least some evidence that the Canadian public is becoming increasingly uncomfortable with 'life patents' of all kinds. For example, a 2002 survey of Canadians on the topic of biotechnology patents found that '46% said there are likely more risks than benefits in allowing such patents, up from 37% in 2000' (Pollara, 2002). Similarly, a recent focus group study conducted on behalf of the Canadian government found that while most the public are supportive of the general concept of patents as a mechanism for encouraging innovation, they have reservations about biotechnology patents. In fact, the more they learned the more concerned they became. As noted by the authors of the study, 'as discussion introduced the patenting of genes, body parts and whole organisms (some of these raised by more knowledgeable participants), most people became quite startled and began to revisit first principles, and in most cases resulted in diminished support for patenting of higher life forms' (Earnscliff, 2003).

There is also evidence that commercial involvement with biomedical research has the potential negatively to affect the trust the public places in the research enterprise. For example, there are survey data which show that a perceived connection with commercial forces has an adverse impact on the perceived credibility of researchers. A recent survey of the Canadian and US public found that publicly funded university researchers are, in the context of biotechnology, one of the most trusted and credible voices.

Only the WHO and peer reviewed scientific journals were rated higher. However, scientists working for biotechnology companies and university researchers funded by industry were rated as one of the least credible voices (Canadian Biotechnology Secretariat, 2005, 55 (slide 110)). In the study, individuals were asked to rate credibility on a 1–5 scale. Fifty three per cent of Canadians gave university scientists funded by government a score of four or five. However, only 23 per cent gave researchers funded by industry the same score. A 2000 focus group study done on behalf of the Canadian government came to a similar conclusion, finding that 'many people say university scientists are much more credible than other scientists because it is assumed they are free from funding pressures and therefore, more "independent"' (Pollara, 2000, 13). If there is a perception that funding comes from less independent sources, credibility deteriorates.

In addition, it should not be forgotten that many areas of biotechnology research are morally contentious – as exemplified by the ongoing public debate about embryonic stem cell research. In these areas, public trust is undoubtedly fragile. The public may be setting aside moral concerns based on the belief that the research is being done in the public's interest. If the public comes to believe that commercial interests are dominating the research process, trust could easily be lost. And it should not be forgotten that many in the public are already suspicious of the motivations behind biotechnology innovation, especially in Canada. In the 2005 study only 49 per cent of the Canadians surveyed (compared with 57 per cent of Americans) thought that biotechnology was being developed with consideration to their 'interests, values and beliefs' (Canadian Biotechnology Secretariat, 2005, 56 (slide 111)).

There are, of course, numerous other policy concerns associated with the commercialization of biomedical research, including, its possible impact on the access to useful technologies, the skewing of the research agenda away from needed basis research and the potential to drive up the cost of useful health care technologies.

GOVERNANCE AND PUBLIC TRUST

The controversial nature of some areas of biotechnology research seems unlikely to go away. Stem cell research will always be morally problematic for some, and others will continue to believe that patenting will inevitably lead to the commercialization and commodification of 'life'. Nevertheless, in most jurisdictions, patenting will remain, unavoidably, a vital component of the research and innovation process, even in controversial areas. As noted by Resnik, 'without the incentives associated with patents, companies and researchers may do less research (or spend less money on research) or they may use other means of protecting their property interests, such as trade secrets' (2002, 131).

What, then, can be done to help maintain public confidence in the research enterprise? If we believe, as some of the existing data tell us, that close ties with commercial interests are part of the problem, we should strive to devise governance structures which temper the perceived connection between industry funding and the actual research. Such approaches seem particularly important to long-term projects which require a significant commitment from public funding agencies, as in stem cell research and population genetic studies.

One strategy would be to build a governance framework which seeks to ensure the research is focused on the public interest and, to some degree, removed from a commercialization agenda. This is not to say that the maintenance of public trust demands that private funds not be used. Rather, the actual and perceived adverse influence of private investment should, as much as possible, be moderated. There is a desirable and inevitable role for private funds in biomedical research. In most countries, industry must be involved in the translation and development of new technologies.

Below, I briefly outline how an independently governed patent pool might be used as an important element of a governance strategy aimed at protecting public trust while still facilitating an appropriate commercialization process.

AN INDEPENDENTLY GOVERNED PATENT POOL

A patent pool is an arrangement in which 'two or more patent owners agree to license certain of their patents to one another and/or third parties' (US Department of Justice, 1995; Goldstein, et al., 2005, 87). In other words, patent pools bring together numerous patent holders, usually in a specific or related area of innovation, to facilitate the efficient use and development of a technology. The patents are 'pooled' in the sense that the arrangement allows inventors in the pool to use all patented invention under favorable licensing terms. Any benefits that may materialize are then shared among the group. Patent pools are hardly new. At the beginning of the twentieth century, for example, Edison, Armat and others entered into an agreement on the early inventions relevant to the motion picture industry, thus facilitating the development and translation of emerging technology. More recently, patent pools have been used with aircraft, video and DVD technology (Clark et al., 2000).

Patent pools can be used to achieve a variety of policy and economic goals. For example, it has been suggested as one way of addressing the concerns associated with the 'anti-commons' dilemma (Heller and Eisenberg, 1998) – that is, the possibility that patents may hurt innovation by driving up research costs. The anti-commons dilemma is especially problematic in areas, such as biotechnology, which involve 'cumulative innovation', where each new innovation builds on or requires access to another innovation (Shapiro, 2001). Patent pools help to moderate this problem by promoting favorable licensing terms for all members of the pool and, if most of the relevant players are involved, avoiding blocking patents (Resnik, 2003).

However, patent pools could also be an important element of a governance scheme designed to build and maintain public trust. It has been noted that, ideally, an independent body should govern patent pools so as to avoid the potential that particular members of the pool are favored over others. Resnik, for example, has recommended as follows:

> In order to avoid bias or the appearance of bias, the patent pool should be an independent, non-profit corporation. Although influential companies and government agencies could play a key role in launching the patent pool, the patent pool should have its own charter, bylaws, trustees, and management. . . . An independent board established by pool would arbitrate and review disputes put forward by patent holders [Resnik, 2003].

Building on this suggestion, if the independent board were structured with the appropriate elements, the board might help to facilitate ongoing public confidence in the research activity. For example, the board's activities should be open to public scrutiny and should have several 'public' members. Decisions about commercialization, access and licensing would be left in the hands of the board. Individual researchers would not make decisions about the use of patents or licensing schemes. The explicit mandate of the board should be to promote the public good, thus, from the perspective of the public, shifting the focus away from profit as the sole motivator. The definition of 'public good', however, should be interpreted broadly and should reflect to realities of the relevant research environment and health care system. For example, when considering the licensing of a relevant technology to a public health care system, the board would need to balance the necessity of industry involvement, the interests of researchers and the desire to keep licensing terms reasonable to ensure that the public has access to valuable technologies (OECD, 2005).

By constructing the kind of independent governance framework suggested above, researchers are removed from direct involvement with industry and commercial forces. Likewise, there is a clear mandate to consider issues beyond those traditionally associated with commercial activities, namely the profit motive. Rightly or not, available data indicate that the public views such motives as compromising the integrity of the research process. An independent board may help to create a degree of independence.

Obviously, not all areas of research lend themselves to this kind of governance scheme, as in the case of a small, targeted project aimed at one or a few specific, and possibly patentable, products or innovations. However, such an approach would seem particularly valuable for a large, government funded, initiative in controversial areas – such as stem cells research (Ebersole, et al., 2005). In Canada, for example, there is a relatively small and well integrated, group of stem cell researchers. Most have some association with the federally funded Stem Cell Network.[2] In this context, an independently governed patent pool might be a workable strategy. Research and development in the area of stem cells will involve 'cumulative innovation'. Researchers cannot go it alone. They will need to rely on the innovative steps and products – steps and products that may be covered by a patent – of other researchers (Verbeure, et al., 2006).

Indeed, a well run, independently governed patent pool could be a win-win for all. It could help to create an efficient research network and, as a result, allow research to be done in a manner that is cheaper for both the taxpayers and the research funding agencies. From the licensee's perspective, patent pools can also be seen as simplifying the innovation and commercialization process – 'both for convenience of "one-stop shopping" and because a subset of the required patents may be of little or no value by themselves' (Shapiro, 2001, 17). In other words, patent pools can work to the advantage of industry by making the overall process more efficient.

There are challenges associated with patent pools. Though they provide the members of the patent pool with a degree of certainty and lower research transaction costs, they also reduce the likelihood of a big payoff. The risks are spread, but so are the potential profits. If an independent board governs the pool, as I have suggested, than these issues are magnified. The control of the patent portfolio is, to a large degree, removed from the researchers and profit is not meant to be the sole agenda. Given the resultant watered down financial incentive scheme, will biotechnology innovators, such as stem cell researchers and private investors, want to get involved? Some authors, however, have noted that for the vast number of biotechnology patents, including human gene and stem cell patents, the value of the patents remains unclear or, more particularly, the true value 'may be unascertainable' without the cooperative efforts of multiple patent holders (Sung, 1998). In this light, patent pools can be viewed as increasing the value of individual patents. In addition, one of the goals of the proposed patent pools is to build long-term confidence and comfort in the industry, thus, paradoxically, facilitating the commercialization process. Admittedly, the building of this kind of public 'good will' may not be enough to satisfy all investors. However, if, in the aggregate, it can be shown that the patent pools will contribute to the possibility of a long term and stable revenue stream, it should be sellable to most investors interested in the biotechnology sector. As noted by Verbeure, et al.:

> The major incentive for all parties is economic benefit. In order for a patent pool to be an effective solution, the right balance has to be achieved between the cost of creating a pool and the prospect of adequate revenue generated by royalties on the end-product [2006, 19].

Another problem with patent pools is that they have the potential to promote anti-competitive practices and, as a result, trigger anti-competition laws. In 1995, the US Department of Justice and the Federal Trade Commission (1995) provided guidelines on this issue. The report suggests that if the intention of the patent pool is to fix prices or exclude competitors from the area, then it may be found to be anti-competitive (Ebersole, 2005; Verbeure, et al., 2006). However, the goal of most biotechnology patent pools, and certainly the strategies recommended here, would be to facilitate research and the efficient use of emerging technologies. As such, anti-competition laws should not be a problem. Indeed, the 1995 report explicitly notes that patent pools 'may provide procompetitive benefits by integrating complementary technologies, reducing transaction costs, clearing blocking positions, and avoiding costly infringement litigation. By promoting the dissemination of technology, cross-licensing and pooling arrangements are often procompetitive' (1995, 5.5).

Finally, there would likely be a variety of practical issues associated with the creation of patent pools. An effective pool would require cooperation from a wide variety of stakeholders which may have an interest in the intellectual property associated with research activities such as those of funding agencies and universities. Negotiating intellectual property agreements with universities could be a particularly daunting task.

CONCLUSION

Despite the possible obstacles associated with the creation of patent pools, they remain a potentially valuable governance tool. It is axiomatic that public support is needed for the sustainable development of biomedical research. Public confidence is hard to win and easily lost. As a result, large-scale research projects, particularly those in controversial areas which involve the use of public funds, need to develop governance schemes that foster public confidence that the research is being conducted in the public good. Patent pools, while not without challenges, offer one possible approach. Most importantly, they provide the opportunity to govern the commercialization strategies in a manner which isolates the research activity from the influence of market pressures.

NOTES

1. I would like to thank Tania Bubela, Ubaka Ogbogu and the Workshop participants for their insightful comments. I would also like to thank Genome Alberta, the Stem Cell Network and SSHRC for the funding support.
2. (see www.thestemcellnetwork.com).

REFERENCES

Benzie, R. 2001. 'Ontario to defy U.S. Patents on cancer genes'. *The National Post*, September 20, sec. A15.

Canadian Biotechnology Secretariat 2005. *International Public Opinion Research on Emerging Technologies: Canada–U.S. Survey Results*. Ottawa: Canadian Biotechnology Secretariat Industry Canada, March 2005. Available at http://www.biostrategy.gc.ca/CMFiles/E-POR-ET_200549QZS-5202005-3081.pdf.

CIHR (*Canadian Institutes of Health Research Act*), S.C. 2000, c. 6.

Canadian Medical Association. 2001. 'Physicians and the Pharmaceutical Industry, Update 2001'. *CMAJ* 164(9): 1339–1344.

Caulfield T. 2003. 'Sustainability and the Balancing of the Health Care and Innovation Agendas: the Commercialization of Genetic Research'. *Saskatchewan Law Review* 66: 629–645.

Caulfield T. 2005. 'Policy Conflicts: Gene Patents and Health Care in Canada'. *Community Genetics* 8: 223–227.

Caulfield T. 2006. 'Stem Cell Patents and Social Controversy: A Speculative View From Canada'. *Medical Law International* 7: 219–232.

Canadian Society for Biochemistry and Molecular and Cellular Biology Bulletin (CSBMCB). 2005. 'Petition to the Canadian Federal Government'. *CSBMCB* Available at http://www.sciencefunding.ca/.

Clarke, J. et al. 2000. 'Patent Pools: A Solution to the Problem of Access in Biotechnology Patents?'. Washington, DC: United States Patent and Trademark Office. Available at http://www.uspto.gov/web/offices/pac/dapp/opla/patentpool.pdf.

Earnscliff Research and Communications. 2003. *Patenting of Higher Life Forms Research Findings*. Report prepared for Industry Canada, 5. Ottawa: Earnscliff Research Communications.

Ebersole T., R. Esmond and R. Schwartzman. 2005. 'Stem Cells – Patent Pools to the Rescue?'. Sterne, Kessler, Goldstein, Fox, P. L.L.C. Available at http://www.skgf.com/media/news /news.176.PDF.

Einsiedel, E. and J. Smith. 2005. 'Canadian Views on Patenting Biotechnology'. Canadian Biotechnology Advisory Committee (June 2005). Available at http://www.cbac-cccb.ca/epic/internet/incbac-cccb.nsf/vwapj/FINAL_inseidel_e.pdf/$FILE/FINAL_inseidel_e.pdf.

Foster, S. 2001. 'Gene Patent Fight Imperils Health Care System'. *The Edmonton Journal*, August 24, sec. A1.

Genome Canada. 2005. 'Co-Funding Update'. *Genome Canada* (August 22). Available at http://www.genomecanada.ca/xcorporate/about/coFunding.asp?l=e.

Goldstein J, et al. 2005. 'Patent Pools as a Solution to the Licensing Problem of Diagnostic Genetics'. *Drug Discovery World*, Spring, 86–90.

Heath, A. 2005. 'Preparing for the Genetic Revolution – the Effect of Gene Patents on Healthcare and Research and the Need for Reform'. *Canterbury Law Review* 11: 59–90.

Heller, M. and R. Eisenberg. 1998. 'Can Patents Deter Innovation? The Anticommons in Biomedical Research'. *Science* 280: 698–701.

Laing, A. 2004. 'A Report on News Media Effects and Public Opinion Formation Regarding Biotechnology Issues'. Unpublished manuscript. Available from the Canadian Biotechnology Secretariat, Ottawa.

Lemmens, T. 2004. 'Leopards in the Temple: Restoring Scientific Integrity to the Commercialized Research Scene'. *Journal of Law, Medicine and Ethics* 32: 641–657.

Litman M. and L. Sheremeta. 2002. 'The Report of the Committee of Inquiry on the Case Involving Dr. Nancy Olivieri: A Fiduciary Law Perspective'. *Health Law Review* 10: 3–13.

Morin, K. et al. 2002. 'Managing Conflicts of Interest in the Conduct of Clinical Trials'. *JAMA* 287: 78–84.

Nelson, D. and R. Weiss. 2000. 'Penn Researchers Sued in Gene Therapy Death'. *The Washington Post*, September 19, sec. A03.

OECD (Organization for Economic Cooperation and Development). 2005. 'Genetic Inventions, Intellectual Property Rights and Licensing Practices: Evidence and Policies'. Paris: OECD. Available at http://www.oecd.org/dataoecd/42/21/2491084.pdf.

Pollara (Pollara and Earnscliffe Research and Communications). 2000. 'Third Wave: Executive Summary (Prepared for and presented to the Biotechnology Assistant Deputy Minister Coordinating Committee (BACC), Government of Canada)'. *BioPortal* (December). Available at http://www.biostrategy.gc.ca/english/view.asp?x=551&all=true.

Pollara (Pollara and Earnscliffe Research and Communications). 2002. 'Seventh Wave Report: Executive Report (Prepared for and presented to the Biotechnology Assistant Deputy Minister Coordinating Committee (BACC), Government of Canada)'. *BioPortal* (December). Available at http://www.biostrategy.gc.ca/english/view.asp?x=547&all=true.

Resnik, D.B. 2002. 'The Commercialization of Human Stem Cells: Ethical and Policy Issues'. *Health Care Analysis* 10: 127.

Resnik, D.B. 2003. 'A Biotechnology Patent Pool: An Idea Whose Time Has Come?' *Journal of Philosophy, Science and Law,* 2003: 3. Available at http://www6.miami.edu/ethics/ jpsl/archives/papers/biotechPatent.html.

Shapiro, C. 2001. 'Navigating the Patent Thicket: Cross Licenses, Patent Pools and Standard-Setting', in A.B. Jaffe, J. Lerner, and S. Stern(eds), *Innovation*

Policy and the Economy, Vol. 1. Cambridge: MIT Press, at 119–150. Available at http://haas.berkeley.edu/~shapiro/thicket.pdf

Sung, Lawrence. 2002. 'Greater Predictability May Result in Patent Pools'. Washington, DC: Federal Trade Commission (April). Available at http://www.ftc.gov/opp/intellect/020417lawrencemsung1.pdf.

Thompson J., P. Baird and J. Downie. 2001. *Report of the Committee of Inquiry on the Case Involving Dr. Nancy Olivieri, The Hospital For Sick Children, The University Of Toronto and Apotex Inc.* Toronto: James Lorimer & Co. Ltd.

Tyers M., E. Brown and D.W. Andrews, et al. 2005. 'Problems with Co-Funding in Canada'. *Science* 308(5730): 1867.

US Department of Justice and the Federal Trade Commission. 1995. *Antitrust Guidelines for the Licensing of Intellectual Property*. 4 Trade Reg. Rep. (CCH) par. 13,132.

Verbeure, B., E. van Zimmeren, G. Matthijs and, G. Van Overwalle. 2006. 'Patent Pools and Diagnostic Testing'. *Trends in Biotechnology* 24: 114–120.

Wade, N. 2005. 'Scientist Faked Stem Cell Study, Associate Says'. *The New York Times,* December 15. Available at http://www.nytimes.com/2005/12/15/science/15cnd-clone.html.

Walsh, J., C. Cho and W. Cohen. 2005. 'View from the Bench: Patents and Material Transfers'. *Science* 309: 2002–2003.

Wells P. 2005. 'Our Mad Scientists: They're Getting Very Angry Over Funding', *MacLeans,* June 23. Available at http://www.macleans.ca/topstories/canada/article.jsp?content= 20050627_ 107830_107830.

Williams-Jones B. 2002. 'History of a Gene Patent: Tracing the Development and Application of Commercial BRCA Testing'. *Health Law Journal* 10: 123–146.

15. Agricultural biotechnology and trends in the intellectual property rights regime: emerging challenges for developing countries

Sachin Chaturvedi

INTRODUCTION

In the past few years, recombinant DNA technology, along with other key techniques in biotechnology, has become an important feature of modern agricultural research and development (R&D). These frontier technologies have also assumed importance in developing countries which face stagnation in agricultural productivity, in addition to confronting biotic and abiotic agricultural stresses which affect their crops. In these countries, productivity constraints in agriculture have become much more acute since the late 1980s, when Green Revolution crop varieties reached their maximum yield potential. These constraints are forcing even major agricultural producers, like India, to import basic grains – in 2006 India imported more than three million tonnes of wheat (The Economic Times, 2006). Advancements in biotechnology seem to offer a way out of this impasse by generating opportunities to attain higher productivity while reinstating features that ensure sustainable development in the agricultural sector (Linder, 1999; Chaturvedi, 1997, 1999).

Developments in biotechnology are, however, accompanied by a stronger intellectual property rights (IPR) regime. In fact, with the advancements in this technology, the instruments that are being used for its protection have become highly exclusionist in their approach. This may pose severe challenges for developing countries, as advances are largely occurring within the private sector, and these new trends in the IPR regime seem to foreclose the entry of latecomers into the technology race. This foreclosure is occurring despite the fact that, since 1995, a large number of developing countries have agreed to a relatively newer IPR regime, and now that the deadline of 2005 is over, these regimes are fully in place in almost all the member countries. In fact, IPR coverage of the agricultural sector is a recent phenomenon in these countries; in many leading economies in Asia, including India, plant variety protection is a rather new addition to IPR protection.

There is an urgent need to address some of these issues on a priority basis. The IPR regime as it is now unfolding may not be in a position to take note of this dynamic on its own, unless government policies are adjusted to create space for such absorption. These developments are all the more important in the context of developing countries, where agricultural biotechnology is being regarded as a major instrument for overcoming food security concerns (Cullet, 2003). Restrictions on the use of genetic technologies constitute a specific challenge in the sense that, when enforced, private companies will be able to ensure that farmers fully respect IPR rules (Byerlee and Fischer, 2002).

Apart from the impact of IPR on access to technology, there is also the question of innovation itself. A strong IPR regime in an upcoming area like agricultural biotechnology may stifle the innovative ability of the developing countries. This situation calls for better management of the huge landscape of the IPR regime. In this context, several issues pertaining to the role of government and the space for the public sector supported by R&D in agriculture – aside from public–private partnership – have been raised (Padolina, 2000; RIS, 2003). One idea has been that public sector R&D institutions should develop more strength and competence in the realm of this frontier technology. While the existence of a strong physical infrastructure is necessary for the development of an effective R&D system, the critical factors remain the institutional set-up which supports this system and the cohesion between the overall developmental objectives and the R&D endeavours in different streams. In fact, these factors play a far more significant role in frontier technologies, and in biotechnology in particular, than in traditional technologies.

This chapter examines the strong policy hurdle created by the IPR regime and faced by developing countries, in light of biotechnology's promise to rescue agricultural systems from the various technological challenges. The second section addresses these promises over the last decade, briefly presenting the broad contours of public sector constraints in agricultural research. Various facets of the IPR regime are discussed in the third section. The last section draws conclusions.

ADVANCES IN BIOTECHNOLOGY AND CONSTRAINTS WITH PUBLIC SECTOR RESEARCH

The potential of plant biotechnology for agriculture includes a diverse range of techniques, which appear to offer the scope needed to help solve some of the problems of developing countries, providing potential tools to solve agronomic problems. In fact, many developing countries, including India, launched a series of programmes to take advantage of this opportunity.[1] The 1992 Convention on Biological Diversity (CBD) defined biotechnology as 'any technological application that uses biological systems, living organisms or derivatives thereof, to make or modify products or processes for specific uses'. Biotechnology assists plant breeders in monitoring the outcomes of conventional crossings and selection, allows useful genes to be identified and cloned, and makes it possible for genes from the same species to be used more quickly and precisely than through traditional methods of plant breeding. Although the scope of plant biotechnology is very wide, the major part of research attention is currently focussed on transferring genes between different species, creating transgenic crops often referred to as GMOs (ODI, 1999). A vast body of literature is available which discusses the probable advantages of biotechnology (World Bank, 1991; RIS, 1988; Chaturvedi, 1997). Fransman (1994) has taken stock of many *ex ante* studies, although *ex post* evaluation – particularly of economic returns – makes the inferences clearer.

Specific Trait Improvements

There are many possible means by which biotechnology may improve the nutritional value of cereals by enhancing the presence of special nutrients or chemicals. A commercial example is the increase in the levels of biotin (vitamin H) for applications in animal and human nutrition.

Another example is biotechnologies targeted at rice to improve its Vitamin A content.[2] Vitamin A deficiency, which also interferes with the bio-availability of iron, affects 413 million children worldwide – that is about 7 per cent of the world population. Rice endosperm does not contain any pro-vitamin A. However, techniques targeted at transgenic plants carrying the genes that produce seeds with yellow endosperm have been developed. Biochemical analysis has confirmed that this yellow colour indicates the presence of pro-vitamin A (Massey and Hemming, 1999). Public sector breeders have also been looking into similar special purpose applications, such as inserting genes into rice so that vitamin A and iron become available through consumption (Massey and Hemming, 1999).

Potentially more important applications are those that seek to improve the quality of feed crops for specific markets. For example, new varieties of transgenic maize which contain higher oil levels to boost energy and improve feeding efficiency or have characteristics which reduce phosphorous in animal waste are currently under development (USDA, 1999). An interesting development that is certainly relevant to feed grains is a patent covering the insertion of a protein into plants which, when eaten, would facilitate control of animal parasites.

Biotechnology offers several ways by which average crop yields can be directly increased. For example, improvements in the 'architecture' of the plant can enable it to absorb more photosynthetic energy or to convert a larger portion of that energy into grain rather than stem or leaf. This was, in essence, the approach taken by the 'Green Revolution' – breeding dwarfing genes into plants so that the plants could make better use of fertiliser and water, and therefore produce more grain. This approach is being pursued anew in studies on rice architecture by the International Rice Research Institute (IRRI), as well as by some private sector interests undertaking research into the fundamental mechanisms that control plant architecture. For certain climates, another useful approach is to modify the plant for a shorter growing season by enhancing the efficiency of its use of fertiliser, pesticides and water. Molecular hybridisation has also been demonstrated to increase the productivity of several crops, including rice and wheat, by 15 to 20 per cent (James and Krattiger, 1999). But it must be noted that the on-farm yield improvements observed thus far have been for transgenic varieties developed to reduce on-farm production costs, rather than for the purpose of increasing crop yields.

It is not yet clear, however, whether increases in crop yield experiences recorded thus far reflect a one-time advance, or the first stage of continuing yield increases. Considering that there are many new technologies that will, over time, be applicable for plant improvements and/or be integrated into plants, the most reasonable conjecture is that the new technologies will continue to provide yield increases, that these will be introduced on a regular basis, and that each of the associated increases will somewhat surpass historical trends (Toenneissen, 1991).

There are several new trends being added with the advancement in the technology. Biotechnology in food grains has addressed only the development of single traits – mostly for herbicide and pesticide tolerance. However, recently some companies such as Garst Seeds, a subsidiary of Advanta, have developed maize hybrids which can tolerate two different classes of chemical herbicides (Spinney, 1998). In the United States, currently about 20 per cent of the maize production is destined for such markets, given that the largest number of industrial applications is the production of high-fructose corn syrup and of alcohol (US International Trade Commission, 1998, A-6). Maize and sorghum are among the crops that produce a high yield of starch/energy per hectare, and are the leading temperate zone crops for this purpose. In essence, it has become possible to vary widely the feed or starch production characteristic of important crop plants, making it possible to use almost any starch producing plant for many industrial purposes.

There are other non-traditional uses of cereal crops, the most important example of which is cellulose. Although cellulose is clearly available from other sources, it may be usefully produced through grain cultivation under certain circumstances. These developments may have significance for rice and other cereals, which are more widely grown in developing countries. To the extent that imported cereals are higher priced than those domestically grown, using starch and other traits from domestically produced bioengineered cereals could lead to cost savings and boost farm incomes in developing countries. Another important possibility in genetically altering crop plants is the production of proteins of pharmacological significance. Some of the patents in this area have wide applicability to different products, including, for example, the production of maize. One patent has very broad claims, but its examples emphasise the production in rice. Several of the patents mention specific products, not all of which are therapeutic. However, commercial applications of these technologies are not yet widely available.

Constraints on Public Research

The agricultural sector in developing countries is passing through a difficult phase. The challenges range from the post-Green Revolution stagnation in primary agricultural crops to large-scale malnutrition and declining R&D allocations. Almost all developing countries grapple with at least some of these constraints.

One of the major constraints in agriculture relates to farm productivity. The Green Revolution contributed to achieving higher yields. The semi-dwarf high-yielding varieties of wheat and rice planted in the late 1960s in South and Southeast Asia revolutionised growth productivity. Higher growth rates were also achieved through the development and adoption of improved technologies, and appropriate government policies and programmes for widely disseminating enhanced varieties. Productivity in most of the food crops, however, has been stagnating since early 1990s, although the issue of whether crop yield is approaching a plateau has become increasingly controversial (Ruttan, 1999; Reilly and Fuglie, 1998). Lower agricultural growth productivity would have a bearing on *per capita* food availability, especially for the growing population in developing Asia, where the population is expected to grow to 3,726 million by the year 2010 (FAO, 1999).

Another major constraint is the reduction of R&D allocations to the

Table 15.1 Average annual agricultural research expenditure

Countries	Agricultural research expenditures					Annual growth	
	1971–75	1976–80	1981–85	1986–90	1995	1971–80	1981–93
	(million 1985 international dollars)[a]					*(percentage)*	
Bangladesh	51.7	68.8	111.2	131.0	123.8[c]	**6.8**	**2.7**
China	576.9	842.5	1165.3	1460.0	1867.6[b]	**8.4**	**4.8**
India	404.4	657.6	874.6	1296.5	1561.8[b]	**9.9**	**7.5**
Indonesia	61.6	108.0	147.2	202.4	208.2[b]	**9.5**	**6.2**
Pakistan	74.6	111.6	165.7	201.8	198.3[c]	**8.5**	**3.5**
Sri Lanka	19.4	31.8	37.3	31.3	35.5[b]	**9.6**	**−1.3**
Malaysia	42.7	91.2	124.5	151.0	170.5[c]	**16.1**	**3.6**
Thailand	119.4	143.8	196.9	245.6	428.0[b]	**3.9**	**8.3**

Notes:
a. To obtain an internationally comparable measure of the volume of resources used for research, research expenditures were compiled in local currency units, then deflated to base year 1985 with a local GDP deflator (World Bank, 1995), and finally converted to 1985 international dollars using 1985 purchasing power parities indexes (PPPs) (Summers and Heston, 1991).
b. 1990 figure.
c. 1992 figure.

Source: Tabor (ed.) 1998, ISNAR.

agricultural sector in developing countries (see Table 15.1). In light of capital-intensive biotechnologies, this has become a major concern. In the case of Asian developing countries, the growth rate had been much higher between 1970 and 1980, but it slowed down in the late 1980s. The sharp decline in allocations for agriculture R&D in Sri Lanka is very intriguing. Similarly, the growth rate in Bangladesh has been on the decline, from 6.8 per cent during the 1970s to 2.7 per cent after 1980. In addition, there is a shift in the financing of agricultural research from the public to the private sector in these countries, carrying on a trend already established in the developed countries, where the basic R&D, which has always been seen as an exclusive domain of public research, has attracted considerable private interest in the realm of plant genomics and frontier biotechnologies. Now more than US$300 million is being spent by private firms on sequencing the genes of different plants. The explanations for these developments may be attempted at two levels – in the wider international setting within which newer technologies are coming up, and in the public sector of developing countries, examining a purely endogenous factor reflecting on their capability *per se*. Kalaitzandonakes (1999) explains that private investment in knowledge generation and transfer has increased because knowledge assets are gradually becoming less public in nature. The changing nature of the economy from 'materials based' to 'knowledge based' – with an emphasis on a stronger intellectual property regime covering biological systems – has largely brought about this transformation. On the other hand Kumar and Sidharthan (1997) and CIPR (2002) explain that the ongoing structural adjustment programmes have severely affected the ability of developing countries to support public R&D budgets, while Tabor (1998) observes that the decline in public spending reflects a general lack of confidence in the ability of the public research system to play a meaningful role in agricultural development.

The relevance of this technology for developing countries has to be seen in the light of two factors. The first pertains to the priorities which agro-biotech research has seen thus far; the second relates to the possibilities small farmers have to access this technology. It is important to establish the integration of biotechnology into the overall agricultural research context. As plant breeding has been the major tool for agriculture R&D since the early 1960s, the challenge is in finding the factors that will maximise the mixture of biotechnology and traditional methods in future plant breeding programmes. The first determinant emanates from the very perception of the usefulness of genes accessible from incompatible species. For instance, the Round-up tolerance and Bt genes from bacteria have benefited crop production in many ways. The second determinant is the relative cost of using biotechnology and traditional plant breeding methods for cultivar development. The factor of relative cost becomes more important when both traditional techniques and biotechnology can assist in cultivar development.

The direction of public sector research was in keeping with the basic objective of achieving larger access to advanced technologies, insuring food security through both food and commercial crops. Therefore, public sector research took a balanced view, giving full attention to the crops linked with the food security of a developing country. This approach is particularly important in most developing countries, given that small and marginal farmers form a large proportion of the farming communities in these countries. They work within a typical 'low external input sustainable agriculture' (LEISA) production model. This model by its own nature is sustainable; however, the challenge before modern technology is to enhance the productivity of such farms without adversely affecting their sustainability. This location-specific dimension demands that any strategy for R&D support should involve local resources – local knowledge and biodiversity. Only this approach will provide for a sustainable agricultural system.

Thus the emergence of biotechnology has given a lot of hope to the developing countries. However, the international environment in which these technologies are being developed is considerably different from the one that saw the adoption of the Green Revolution (Brenner, 1998). The most significant difference is that unlike developments of Green Revolution plant varieties, which were primarily developed in the publicly funded organisations, developments in biotechnology are being spearheaded by commercial companies, as we shall see in the following section. Thus far, although in most developing countries the level of research, development and use of biotechnology in the public sector has been much higher than in the private sector, this picture is changing rapidly.

FACETS OF THE TRIPS REGIME

The conclusion of the Uruguay Round of GATT Negotiations which included in an Agreement on Trade related Intellectual Property Rights (TRIPs) was a major step in terms of establishing a legally binding international intellectual property protection regime. During the negotiations there was an interesting debate about the scope of patenting being extended to life forms. The sharp differences between the European Union and the United States led to the inclusion of a provision enabling a review of Article 27.3(b) four years after TRIPs came into force – that is by 1999. This Article basically allows national governments to exclude certain inventions from their patent regimes, especially ones based on plants, animals and 'essential biological processes' – including microorganisms and non-biological and micro-biological processes. During the review process the scope of the debate expanded considerably. At the Doha Ministerial the developing countries also joined the debate. As a result, the issues related to indigenous knowledge systems (IKS) and access and benefit sharing (ABS) were also included. As a result the Doha Development Agenda (DDA) provides for a relationship between the UN Convention on Biological Diversity (CBD) and the TRIPs Agreement, and explicitly acknowledges IKS and ABS (Paragraph 19 of TRIPS).

In this debate several developing countries contributed their submissions on issues of biodiversity and indigenous knowledge systems, and thus have expanded the scope of the debate. The WIPO General Assembly also established an Inter-governmental Committee on Intellectual Property and Genetic Resources, Traditional Knowledge and Folklore. The conclusion of the International Treaty on Plant Genetic Resources for Food and Agriculture (ITPGRFA) under the aegis of the FAO further encouraged the protection and promotion of farmers' rights and indigenous knowledge systems.

However, one finds that the advances in agricultural biotechnology to some degree are paralleled by enhancement in the IPRs under TRIPs, at times by the addition of TRIPs. This has brought the optimal patent scope and its coverage under sharp focus, and which are now being discussed at length in the TRIPs Committee. The debate largely centres on the patentability and non-patentability of plant and animal inventions and the protection of plant varieties. The checklist of issues submitted to the WTO by Brazil, Cuba, India and Peru – among others – is a recent example of the kind of documents that are needed at this stage to delineate the concept of disclosure requirements and related issues, drawing the negotiating process away from the current status quo. Switzerland's proposal to make the disclosure requirement optional under the national legislation, rather than a mandatory provision at the international level, may be explored further to facilitate movement in the debate. The African group wants the TRIPs agreement to prohibit patenting of all life forms, including microorganisms, and wants *sui generis* protection for plant varieties to preserve farmers' rights. The US position is closer to that of Switzerland in the sense that the US proposal argues for national legislation to address CBD objectives on access to resources, traditional knowledge, and benefit sharing.

In the case of India, at least some of these objectives have been included in the national legislation and guidelines. However, given the emerging trend of free trade agreements across various countries and the IPR provisions they contain, it is important that one look into the IPR regime beyond the TRIPs framework as well. One may find this trend completely opposed to the spirit and objective of Article 7 of TRIPs. The objective of TRIPs as stated is 'the protection and enforcement of intellectual property rights should contribute to the promotion of technological innovation and to the transfer and dissemination of technology, to the mutual advantage of producers and users of technological knowledge and in a manner conducive to social and economic welfare, and to a balance of rights and obligations.' This section briefly explores three broad trends in the patent regime important from the standpoint of the developing countries in the context of their access to biotechnology.

Over the years, TRIPs has emerged as one of the most widely debated WTO agreements. The development of biotechnology has further intensified the debate. The writings have largely been focussed on Article 27.3(b) of the WTO agreement on TRIPs, which requires developing countries to provide either patents or an 'effective *sui generis*' protection for the ownership of plant varieties by the year 2000. For the least developed countries, the deadline was extended to 2005.

Although the US and other developed countries proposed a formal review of this Article, resistance from other trading nations prevented this action at the WTO. This is of course a temporary relief. If this review is pushed back on the formal agenda, it would not come up before the next round of negotiations, despite the Hong Kong WTO Ministerial. There is a large body of literature defining the concept *sui generis,* and discussing whether the UPOV of 1978 or 1991 is of greater relevance for the developing countries (Dhar and Chaturvedi, 1998; Shiva, 1996; Sahai, 1996), a detailed discussion of which is outside the scope of this chapter. But the pertinent questions are: what agenda should developing countries pursue, and to what extent is it technologically tenable? An answer to this should guide the future course of action for the developing countries, not only with respect to TRIPs negotiations, but also their IPR policy in general. In this context, it needs to be mentioned at the outset that the increasing importance of genetic engineering in agricultural research across the world, a continued increase in genetically modified organisms (GMOs) and the ability to patent plants have actually reduced the importance of plant variety protection and consequently of the *sui generis* system within the IPR regime.

At this point it would be interesting to take a stock of the broad trends in the intellectual property regime at the level of individual developed countries. Until recently, life forms used to be exempted from patenting. However, developments in biotechnology have compelled a revision of this approach towards the intellectual property regime. These policy changes have largely been taking place in the US, but now the European Union and Japan are also all set to follow closely in this race. Accordingly, various national governments are bringing in changes in their national laws in order to protect and encourage investments in biotechnology. These policy changes have further widened the scope of the ongoing debate to include a wide range of issues, such as the range of product patents and the patentability of genes, gene-sequences and parts of gene-sequences derived from humans, animals, plants or microorganisms. The additional aspect to consider is the relationship between the patent system and the plant variety system. Moreover, patenting, especially of human body parts, has posed an ethical limit for biotechnology.

Advancements towards Plant Patents

In the recent past, plant variety protection (PVP) and patents have emerged as two important forms of intellectual property rights. In the context of developing countries, while PVP has been established for some time, patents for plants are a recent phenomenon. Both patents and PVP provide exclusive monopoly rights over a creation for commercial purposes over a period of time (Table 15.2). A patent is a right granted to an inventor to prevent all others from making, using and/or selling the patented invention for 15 to 20 years. The criteria for a patent are novelty, inventiveness (non-obviousness), utility and reproducibility. Although patents were designed for industrial applications, with the advent of biotechnology patent offices now grant patents on microorganisms and, in some countries, life forms.

The intellectual property regime for plant variety protection emerged with a strong commitment to the public interest in mind. The provision for compulsory licensing was introduced with this intention. Through compulsory licensing, a holder of plant breeders' rights can neither refuse nor offer unreasonable terms of use to any applicant. Plant variety protection has worked well as a mechanism to promote the interests of the plant breeders developing new varieties by giving them proprietary rights on the one hand, and establishing them as a custodian of public rights of access and use of genetic material on the other hand. PVP gives patent-like rights to plant breeders, protecting the genetic makeup of a specific plant variety. The criteria for protection are different: novelty, distinctness, uniformity and stability. PVP laws can provide exemptions for breeders, allowing them to use protected varieties for further breeding, and for farmers, allowing them to save seeds from their harvest. Thus, in plant breeding, PVP is the weaker sister of patenting, mainly because of these exemptions. PVP also encourages cross-licensing between the holder of PVP and the holder of a patent. Under the breeders' exemption of plant variety rights anyone may use protected material for breeding purposes. However, the patent regime does not reciprocate this exception.

In the patent regime, the interpretation of the research exemption is much narrower than that of the breeders' exemption in PVP (Table 15.2). Thus, for instance, if a breeder wants to produce a new variety and needs a compulsory cross-licence from a patent holder, the breeder has to demonstrate that the breeding programme will produce technical progress, but all results of a breeding programme take a long span of research and development effort. How, then, can we demonstrate technical progress right in the beginning? Cross-licensing for a plant breeder, is hardly of value (*The Economist*, 1997).[3] Thus for all practical purposes, PVP ends up protecting small advances in the breeding process, while the patent regime actually leads to the protection of bigger leaps in technological achievements.

In Europe, animal and plant varieties have always been excluded from patentability under Article 53(b) of the European Patent Convention (EPC). This convention was signed in 1973. The term 'variety' was not defined in the EPC as plant varieties could be protected either through

Table 15.2 A comparison of Trade Related Aspects of Intellectual Property (TRIPs), US utility patent protection, the European Patent Convention (EPC) and the International Convention on the Protection of New Varieties of Plants (UPOV)

	TRIPs Agreement	US Utility Patent	European Patent Convention	UPOV Convention 1991
Granting Criteria	Novelty, Inventive Step and Industrial Applicability	Novelty, Non-obviousness, Utility	Novelty, Inventive Step, Industrial Application	New, Distinct, Uniform and Stable
Industrial Applicability/ Utility	Not defined	Advantage over the Prior art	The invention must be capable of industrial application–this includes agricultural use but does not include methods of human treatment	Not a requirement
Distinctness	Not defined even as a requirement for the *sui generis* system of protection mandated for plant varieties under Article 27(3)(b)	Not a requirement	Not a requirement	The variety must be clearly distinguishable in its essential characteristics from other varieties, which are a matter of common knowledge (e.g. protected by a plant variety right at the time of application).
Extent of Protection	a) Where the subject matter of the patent is a product the right allows the holder to prevent third parties	a) Right to prevent all others from using the invention	a) Right to prevent all others from using the invention	a) Right to produce, reproduce, sell or stock any plant variety

	b) Patent holder can deny usage of the process he has developed or even the sale of product of that process	b) Protection extends to all biological materials, genes to genotype	b) Broad claims are not permitted c) Protection extends to all biological materials, genes to genotype and includes plant groupings but not plant variety	Right extends to harvested material and other products obtained from material of the variety provided
Farmers' Privilege	Not specific, but possibly permitted via Article 30	Not permitted	Not permitted	Optional Contracting Parties may, within reasonable limits and subject to the safeguarding of the legitimate interests of the breeder, restrict the breeder's right in relation to any variety in order to permit farmers to use for propagating purposes
Breeders/Research Exemption	Not specific–but possibly permitted via Article 30	Free use of protected material for research purposes is permitted but only where it is for non-commercial purposes	No – but such an exemption is usually provided in the national patent laws of Member States of the EPC	Yes – non-infringing acts include acts done privately and for non-commercial purposes acts done for experimental purpose and for breeding

Table 15.2 (continued)

	TRIPs Agreement	US Utility Patent	European Patent Convention	UPOV Convention 1991
Compulsory Licences	Yes, but only where a) the applicant has requested and been refused a licence from the patent holder b) the use for which the applicant wishes to use the protected invention is non-exclusive (c) the use is predominantly within the domestic market (d) the licence holder pays adequate remuneration Where the licence is needed in order to exploit a second patented invention which is dependent then a licence will be granted only where (1) the invention claimed in the second patent involves an important technical advance of considerable economic significance in relation to the invention claimed in the first patent;	No, although the ability of the patent holder (the licensor) to dictate the terms of any licence s/he chooses to grant subject to extensive restrictions via the common law doctrine of patent misuse and anti-trust laws		Not mentioned as such Article 17 states that (1) Except where expressly provided in this Convention, a non Contracting Party may restrict the free exercise of a breeders' right for reasons other than of public interest (2) When any such restriction has the effect of authorising a third party to perform any act for which the breeder's authorisation is required, the Contracting Party concerned shall take all measures necessary to ensure that the breeder receives equitable remuneration

	(2) the owner of the first patent is entitled to a cross-licence on reasonable terms to use the invention claimed in the second patent; and (3) the use in respect of the first patent is non-assignable except with the assignment of the second patent. Each case is assessed on its individual merits, it is non-assignable, it is subject to termination when the circumstances change and any decision is subject to judicial review			
Duration of Protection	20 years from the date of filing	20 years from the date of filing	20 years from the date of filing	30 years for trees and vines, 25 years for all other varieties (Article 19)

Source: US Department of Agriculture Economics Research Service (1998).

the existing national laws (plant breeders' right) or through the UPOV Convention. The European Patent Office started establishing plant varieties under the jurisdiction of the patent regime. As noted in *The Economist*, one of the major reasons cited for slow growth in the biotechnology industry in Europe is the lack of certainty concerning intellectual property protection for biotechnology inventions. The proponents of biotechnology suggest that the conflicts between the ethical aspects of technology development *vis-à-vis* commercial gains from technology have not allowed the growth of this industry. The Novartis decision (Decision G01/98) seems to confirm this conclusion. The decision suggests that plant varieties are not patentable but patent on a genus is possible. The genus is made up of species and sub-species and varieties. This means that patent control of varieties is acquired through the proprietary control of a genus. In the Novartis case, at issue was the claim to plants containing a gene conferring resistance to plant pathogens. The Technical Board of Appeal referred the question to the Enlarged Board of Appeal (Blochlinger, 2000). The Board found that a claim in which plant varieties are not cited is not excluded from patentability under Article 53(b), even though it may embrace plant varieties. The Board further concluded that inventions ineligible for protection under the plant breeders' rights system were intended to be patentable under the European Patent Convention if they met all the other requirements of patentability.

However, a directive from the European Community on the protection of biotechnology inventions (Directive 98/44EC) contains specific provisions on the patentability of genetically engineered biological material including plants and animals. This marks a major departure in the European Union (EU). The Directive was adopted by the EU on 6 July 1998 and all the necessary amendments were ratified on 1 September 1999. The most significant feature of the Directive is the provision pertaining to the patentability of the biological material, including inventions relating to plant and animal varieties, the human body and sequences or partial sequences of genes (Erratt and Sechley, 2000). The individual Member States of the EU have two years to amend their national laws to bring them into conformity with the Directive, and the Directive has already been challenged in the European Court of Justice. The explanatory notice published in the EPO Official Journal records that since the early 1980s, the EPO has received about 15,000 applications in the field of biotechnology, of which about 3,000 patents have been granted: 15,00 applications relate to transgenic plants, 600 to transgenic animals, and 2,000 to DNA sequences. The Biotech Directive had to be implemented into national law by 30 July 2000.

Germany has asserted that the Directive is inadequate for the promotion of biotechnology and that in order to retain the competitiveness of the European biotechnology industry the Directive should be further strengthened. Most of the implementation was done by the UK through Statutory Instrument (SI) 2000/2037. UK law is already largely compatible with the Biotech Directive. The UK is attempting a balancing act between PVP and patents. In one or two areas, however, changes are required. In particular, derogations, equivalent to those in plant varieties legislation, need to be introduced into patent law to enable farmers to save and use seed. A cross-licensing provision also needs to be introduced to acts for both plant variety rights and patent rights, to account for instances where the invention or plant variety constitutes 'significant technical progress' and could not be exploited without infringing the protection conferred on the other rights holder. 'Significant technical progress' is a high hurdle to overcome. This part of the Directive is still to be implemented. Although these compulsory licence provisions are proposed to compensate the other rights holders, they also require applicants to cross-license their own rights with the person from whom they are seeking the compulsory licence (Eversheds National Bioscience Group, 2001). The UK SI has produced new schedules in order to ensure uniformity between UK patent legislation and the Biotech Directive, the Patents Act and rules to set out clearly the provisions of the Biotech Directive in national legislation. The government is taking this opportunity for minor tinkering with the Patents Act to ensure compliance with TRIPs (including Trade in Counterfeit Goods–part of the last round of the General Agreement on Tariffs and Trade) and the recent changes to the EPC.

PVP and Utility Patents

In the US, the extension of IPRs to new plant varieties and biological inventions, including the development of biotechnologies, has stimulated private companies to invest in plant breeding. The Plant Patent Act of 1930 and the Plant Variety Protection Act (PVPA) of 1970 established plant breeders' rights for new plants and plant varieties. In 1980, a Supreme Court decision, *Diamond v. Chakrabarty,* authorised the use of patents for biological inventions, specifically microorganisms. Several recent decisions by the Patent and Trademark Office have broadened the use of patents for plants and created space for utility patents (*ex parte Hibberd* in 1985) and animals (*ex parte Allen* in 1987). Utility patents are for any 'new and useful process machine, manufacture, composition of matter or any new and useful improvement therefore', and can protect all parts of plants including genes, seeds' physiological and physical traits. Utility patents have a larger coverage than PVPs in the sense that they

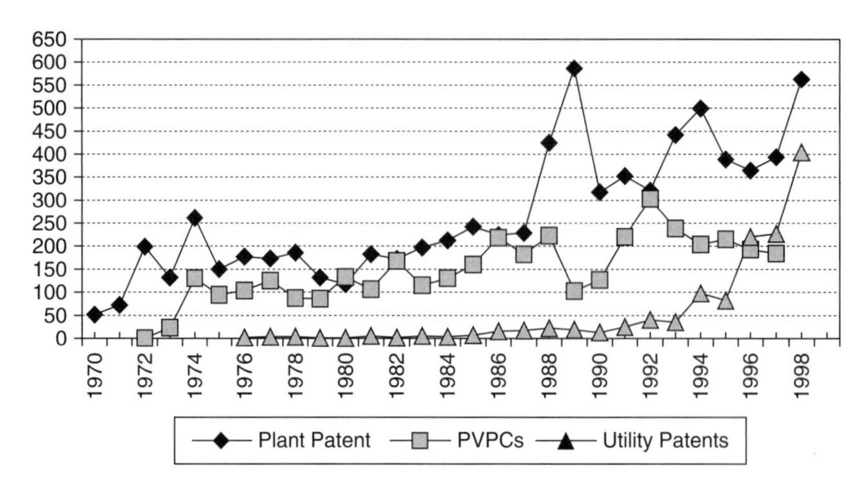

Figure 15.1 Trends in IPR in US

cover not just a single variety, but also all other varieties with similar traits and functional properties (Table 15.2). Further, utility patents provide protection for multiple claims covering plant parts including flowers, fruits and cuttings. Protection is not dependent on whether the plant is sexually or asexually produced.

As a result, private sector research expenditures for plant breeding have increased from $6 million in 1960 to $400 million in 1992 (Klotz, Fugile and Pray, 1995; Fugile, Klotz and Gill, 1995). Nearly 70 per cent of private sector plant breeding research expenditures in 1989 was for corn, vegetables, and soyabeans. Private firms have also reacted to changes in IPRs by investing heavily in biotechnology techniques.

The number of plant patents, Plant Variety Protection Certificates (PVPCs) and utility patents issued over the last 25 years has risen (Figure 15.1). The PVPA stimulated the development of new field crop varieties. In total, by the end of 1994, 3,306 PVPCs had been issued for new crop varieties, approximately 87 per cent of which were issued to the private sector. The number of PVPCs issued specifically for new varieties of field crops, grasses and vegetables climbed from 153 from 1971 to 1974, to 992 from 1991 to 1994. New soyabean, corn and vegetable varieties accounted for 56 per cent of the total number of PVPCs awarded. Oats was the only crop in which the public sector held a higher share of PVPCs. Utility patents are the most difficult to obtain, and have been awarded primarily to new biotechnology innovations, such as genetically engineered plant varieties. The number of utility patents issued has grown very rapidly in the US – by December 1994, 324 utility patents had been issued for new plants or plant parts and 38 issued for animals. As with PVPCs, most utility patents were awarded to the private sector (Fugile, Klotz and Gill, 1995). Thus, IPR has encouraged the private sector to develop new agricultural technologies by enabling firms to capture a greater share of the commercial value of their inventions.

Another emerging trend is the breadth of claims, which at times extend so far as to encompass research tools necessary for further research and development. Some of the research tools that have attracted patenting attention include expressed sequence tags (ESTs), restriction enzymes, screening systems, techniques related to DNA sequencing, and single nucleotide polymorphisms (SNPs) (Hirai, 2001). As these research tools by definition have the power to control the downstream research into pharmaceuticals, they can wield an extremely large influence when patented. The problem of broad patenting has actually grown over the years. Good examples include the Agracetus patent on all transgenic cotton (US patent 5,159,135) and similar patents on all transgenic soyabean. Some of these patents are subject to re-examination or litigation to determine their validity. Furthermore, a new US patent awarded to Monsanto in 2001 gave an exclusive monopoly right on a crucial method for identifying modified plant cells in the laboratory. US Patent No. 6,174,724 covers all practical methods of making transformed plants that employ antibiotic resistance markers. The technique has been used in virtually all commercial genetically modified crops. Moreover, an earlier patent granted to another major US firm, Syngenta, covers a marker that enables plant cell transformation and selection without the use of antibiotic resistance marker. This technology was first developed in a very small firm, Danisco, in Denmark. This company sold the patent in 1998 to Sandoz, which later became Novartis, which became Syngenta in 2000 (Anonymous, 2001).

These issues may give rise to several policy challenges when reviewed in the context of developing countries. These developments may just foreclose the entry of the late comers in the technology race, which would eventually affect public sector research endeavours, mostly practised in the developing countries. In the larger context of investment in research for developing such techniques, patents seem to be the only way to recover the investment. However, given the need to analyse the research trends in the overall context of social requirement – as with individual technological features of research tools – the scope of the technology and its contribution to society may differ. For instance, in developing countries, ensuring a higher crop yield for adequate food supply would always be a priority over the necessity of developing a drug required for lifestyle diseases.

Severe implications of EST patents for the future progress of genomic industries are also being seriously analysed in Japan. The opinion that is emerging there suggests that the utility of ESTs will not be recognised through the mere disclosure of a general function – that is, capability of use as a research tool – and such EST inventions will not be patented (Hiraki, 2000). However, the utility of EST can be recognised if it can be used as a probe for a gene encoding a specific useful protein or as a tool to diagnose a specific disease. In this context, the agricultural sector also provides some interesting examples.

BY WAY OF CONCLUSION

As is evident in this chapter, varietial protection is being attempted through a much stronger patent regime, which does not allow a farmers' or even research exemption, and is much narrower in its scope than the plant patents or plant variety protection. It is worth recalling that at the WTO, the debate is still stuck at Article 27.3(b), which refers to the patent regime or an effective *sui generis* system for the protection of plant varieties. Meanwhile, the developing countries are engaged in debating the various aspects of *sui generis* systems, in particular as it has become very clear that there are forums, other than the WTO, where IPR related developments are defining new contours of intellectual property protection that are eventually legitimised through FTAs. The continuous urge in the US toward a stronger IPR regime – as is discernible through a sharp growth in the utility patents – and similar mechanisms for the protection of biotechnological inventions in Europe – as proposed by the Biotechnology Directive of the EU – should be a cause for concern. Apart from this, there is also a growing trend toward patenting research tools. Thus in light of the developments in biotechnology, the profile of patent regimes is changing quickly in developed countries. Needless to say, a large part of this research is emanating from the private sector.

These changes have severe implications for developing countries. Already struggling with the implementation hurdles of the TRIPs regime, there are many developing countries, including India, which have yet to put in place national legislation to position themselves *vis-à-vis* the international negotiations at the WTO. There have been several reasons for this delay, but now it seems to be clear that national legislation would not only adversely affect the access to technology *per se* but the patenting of research tools would also exclude the latecomers in the technology race from imitation or product development in any form. In this context, the role of public research institutions becomes increasingly relevant. In developing countries productivity levels have yet to move any closer to those achieved in developed countries. This requires the continued maintenance of budgetary support for public research institutions in developing countries, and may even require an increase to meet the demand.

In past years, public plant breeding programmes have evolved with a free exchange of germplasm and cooperative scientific endeavours. The post-Green Revolution agricultural production scenario seems to pose several challenges for food security in developing countries. It is high time that agricultural R&D plans prioritised investment on new technologies so as properly to balance – or, rather, supplement – the traditional techniques with new technologies such as biotechnology. It is important to ensure that public plant breeders/laboratories have access to the best science and germplasm. Similarly, capacity in public plant breeding should be enhanced. This increased capacity should be directed towards those crops which are not likely to attract private investment.

NOTES

1. However, at the outset, it is important to clarify that biotechnology is something beyond the developments of Genetically Modified Organisms (GMOs) (Sahai, 1999).
2. In many developing countries Vitamin A is also distributed in the form of pills. Perhaps these are cheaper than the GM golden rice varieties.
3. Similarly, if an application of PVR is made for a variety that contains a patented gene, is the actual making of the application an infringement of the patent? Obviously, the reply would be in the affirmative and PVR cannot be granted to the plant varieties containing a patented gene if the patent holder does not agree.

REFERENCES

Anonymous 2001. 'Monsanto and Syngenta Monopolise Key Gene Marker Technologies'. May 16.

Blochlinger, Karen. 2000. 'A Variety of Interpretations of Plant Variety'. *CASRIP Newsletter* 7, p. 1.

Brenner, Carliene. 1998. 'Intellectual Property Rights and Technology Transfer in Developing Country Agriculture: Rhetoric and reality'. OECD Technical Paper No. 133. Paris: OECD.

Byerlee, D. and Ken Fischer. 2002. 'Accessing Modern Science: Policy and Institutional Options for Agricultural Biotechnology in Developing Countries'. *World Development* 30: 931.

Chaturvedi, Sachin. 1997. 'Cooperation in the Food Sector in G-15 Countries: A Techno Economic Perspective'. *RIS Digest*, September.

———. 1999. 'Would Bio-technology Help Meet Food Security?' *The Economic Times*. 17 July.

———. 2002. 'Agricultural Biotechnology and New Trends in the IPR Regime: Challenge before Developing Countries'. *Economic and Political Weekly*. March 30.

CIPR (Commission on Intellectual Property Rights). 2002. *Integrating Intellectual Property Rights and Development Policy, Commission on Intellectual Property Rights*. London: CIPR.

Cullet, Philippe. 2003. *Food Security and Intellectual Property Rights in Developing Countries. IELRC Working Paper 2003*. Geneva: International Environmental Law Research Centre.

Dhar, Biswajit and Sachin Chaturvedi. 1998. 'Introducing Plant Breeders' Rights in India – A Critical Evaluation of the Proposed Legislation'. *Journal of World Intellectual Property* 1: 245–62.

Economic Times. 2006. 'India to import wheat'. www.economictimes.indiatimes.com/articleshow/msid-1398676,curpg-1.cms. February 3, 2006.

Erratt, Judy, Konrad Sechley and Strathy Gowling. 2000. 'The European Biotechnology Directive and the Patentability of Higher Life Forms'. *Canadian Biotechnology* 21(5): 22–27.

FAO. 1999. Committee on Agriculture. Fifteenth Session: Biotechnology. 25–29 January. Rome.

Fransman, Martin. 1994. 'Biotechnology: Generation, Diffusion and Policy'. in Charles Cooper (ed), *Technology and Innovation in the International Economy*. Maastricht: Edward Elgar/UNU.

Fugile, Keith, Cassandra Klotz and Mohinder Gill. 1995. 'New Group Varieties'. *AREI Update* 14. US Department of Agriculture and Economic Research Service.

Hirai, Akimitsu. 2001. 'Biotechnology and Legal Protection: Current Issues'. *CASRIP Newsletter*, Winter.

Hiraki, Yusuke. 2000. 'Problems Regarding the Patentability of Genomics and Scope of Protection of ESTs in Japan'. *CASRIP Newsletter*, Winter.

James, Clive and Anatole Krattiger. 1999. 'Biotechnology for Developing-Country Agriculture: Problems and Opportunities: The Role of the Private Sector'. *IFPRI 2020 Focus 2, Brief 4* of 10, October.

Kalaitzandonakes, Nicholas. 1999. 'The Agricultural Knowledge System: Appropriate Roles and Interactions for the Public and Private Sectors'. *AgBio Forum* 2: 1–4.

Klotz, Cassandra, Keith Fuglie and Carl Pray. 1995. *Private-sector agricultural research expenditures in the United States, 1960–92*. Staff paper No. AGES-9525. US Department of Agriculture, Economic Research Service.

Kumar, Nagesh and N.S. Sidharthan. 1997. *Technology, Market Structure and Internationalisation: Issues and Policies for Developing Countries*. UNU/INTECH, London: Routledge.

Linder, R. 1999. 'Prospects for Public Plant Breeding in a Small Country'. Paper presented at the ICABR conference, The shape of the coming agricultural biotechnology transformation: Strategic investment and policy approaches from an economic perspective. 17–19 June 1999. University of Rome 'Tor Vergata.'

Massey, David and David Hemming. 1999. AGBIOTECH 99: Biotechnology and World Agriculture. Conference held 14–16 November, London.

ODI (Overseas Development Institute). 1999. 'The Debate on Genetically Modified Organisms: Relevance for the South'. *Briefing Paper 1999* (1).

Padolina, W.G. 2000. 'Plant Variety Protection for Rice in Developing Countries: Impacts on Research and Development'. *Limited Proceedings of the Workshop on the Impact on Research and Development of* Sui Generis *Approaches to Plant Variety Protection of Rice in Developing Countries*. 16–18 February 2000. IRRI, Los Banos, Laguna, Philippines.

Eversheds National Bioscience Group 2001. 'Plants and IP protection'. *Pharmalicensing*. 2 May.

Pinstrup-Andersen, P., Rajul Pandyal-Lorch and Mark W. Rosegrant. 1999. *World Food Prospects: Critical Issues for the Early Twenty-first Century.* Washington, DC: IFPRI.

Reilly, J. M., and K.O. Fuglie. 1998. Future Yield Growth in Field Crops: What Evidence Exists'. *Soil and Tillage Research* 47: 275–90.

RIS. 1988. Biotechnology revolution and the third world: Challenges and policy options. New Delhi: Research and Information System for Developing Countries.

RIS. 2003. 'Cancun and Beyond'. *World Trade and Development Report.* New Delhi: Research and Information System for Developing Countries.

Ruttan, W. Vernon. 1999. 'Biotechnology and Agriculture: A Skeptical Perspective'. *AgBioForum* 2 (1): 54–60.

Sahai, Suman. 1996. 'Protecting Plant Varieties: UPOV should not be our Model'. *Economic and Political Weekly*. 12–19 October.

————. 1999. What is Bt and what is terminator? *Economic and Political Weekly* 34: 81–4. 16–23 January.

Shiva, Vandana. 1996. 'Agricultural Biodiversity, Intellectual Property Rights and Farmers' Rights'. *Economic and Political Weekly*. 22 June.

Spinney, L. 1998. 'Biotechnology in Crops: Issues for the Developing World'. Research paper compiled for Oxfam GB. Available at www.oxfam.org.uk/policy/papers/gmfoods/gmfoods.htm.

Tabor, R. Steven, Willem Janssen and Hilarion Bruneau. 1998. 'Financing Agricultural Research: A Source Book'. Report, ISNAR, The Hague.

Toenneissen, G. 1991. 'Potentially Useful Genes for Rice Genetic Engineering'. in G. Khush and G. Toenneissen, *Rice Biotechnology*. Wallingford, Oxon: CABI, pp. 258–80.

The Economist. 1997. '*Geens v. Geens*'. 19 July.

US International Trade Commission. 1998. *Industry and Trade Summary: Milled Grains, Malts, and Starches.* USITC 3095 (March).

World Bank. 1991. 'Agricultural Biotechnology: the Next "Green Revolution"?' *World Bank Technical Paper No. 133*. Washington, DC: The World Bank.

PART VI

National, international and historical comparisons

Introduction

Abdallah S. Daar and David Castle

Researchers studying the role of intellectual property rights in innovation systems often come to the issue as if it were a found object that needs to be broken into its constituent parts and reassembled to understand it. One can have sympathy for this situation. The role of IPRs in innovation is so complex that organizing one's thoughts about the subject often means peeling away many layers of distracting information and, one hopes, ultimately getting to the heart of the matter. The problem with any kind of analytical approach to a complex topic is that conceptual rigor sometimes can displace from attention the historical context in which complex systems like intellectual property arise. The downside of analysis is forgetting the past. Of course, it is understandable if the needs of today focus attention on the role of IPRs on their contemporary manifestation. But there is a risk of forgetting that IPRs have country-specific histories that significantly determine contemporary policies and practices. For this reason, the final part in this book is dedicated to three studies in which the historical development and national setting of IPRs are considered.

The chapter by Ian Inkster provides a magisterial overview of historical data examining the nexus between IP and innovation. The chapter by Richard Boadi looks specifically at IP and technology transfer in agricultural biotechnology, asking whether IPRs and technology transfer help to energize innovation in the developing world and, if so, are there any lessons to be learnt for other sectors such as health. The chapter by Tina Piper is even more specific: it considers a particular aspect of IP law, namely the medical exception to patentability, to see what role it has played in innovation, and what effect it has had on developing countries.

Ian Inkster makes the argument that IPR systems have been enormously varied over time and have thus worked in contrasting manners in different countries. More than this, IPR systems, in the form of patents in particular, have served to stimulate innovation in a relatively small number of industrially modernized nations, a group that reached its basic mass and density around 1914 and continued to exist as such to around 1971. In such favoured nations, IPR systems have nested into very elaborate innovation cultures, organizations and institutions, themselves parts of wider financial, legal and scientific infrastructures. Around 1900, in the late-developing nations such as Russia or Japan, heroic efforts towards technology transfer-in were made by centralized political regimes, these involving strategic developments and manipulations of patent systems adopted/adapted from systems generated initially by the USA, UK and France.

In the great majority of less developed nations from after that period, an absence of these wider social elements that situate and support, as well as intrinsic features of patent systems themselves, assisted in the process of relative underdevelopment which dominated the years circa 1918–1971. Importantly, there is another crucial period for development of some patent systems in the early 1970s. At that time, the dominant issues were the OPEC crises, the emergence of new micro-electronic, biotechnical, marine and environmental industries, and the quick changes in the systems of international payments. These global events created a new window of opportunity which allowed another select group of newly industrializing countries (the famous NICs, each characterized by partial-nation or city-nation status) to be the last group of nations that would industrialize at the tail-end of the older machinofacture of steel and chemicals and amidst established institutions governing intellectual property rights, knowledge creation, diffusion and application.

There is little evidence from the generalized statistics of patent registration at a global or comparative level that patent systems post-1850 were ever generally conducive to industrial development outside the favored group of nations which had emerged already by 1914–1918. Inster's chapter presents this argument in the form of the overall statistics of patent application and registration. One important lesson of this chapter is that reformers of the global patent systems must take into account the historical fact that IPR systems have also been information systems and technology transfer systems, and that radical reform must not improve the quality of property rights in poor nations of our world if it at the same time threatens or does not address and improve ancillary information systems for just those nations. Over the whole range of history it may well be that most patent systems have been at their best as information systems and at their worst as IPR systems. This general point is captured effectively in the chapters by Richard Boadi and Tina Piper.

Richard Boadi writes from the perspective of someone who has worked closely with the African Agricultural Technology Foundation. He notes that there is evidence that IPRs have indeed helped to stimulate innovation in the developed world, particularly in the United States, in the past two decades in the area of plant biotechnology. He makes the point that the sketchy statistics that are currently available indicate there has not been as much innovative activity in developing countries in the same period. This trend one could surmise by the three-fold difference in investments of GDP in agricultural biotechnology that persists between developed and developing countries. As a result, where there is some indication of innovative activity in developing countries, there is often also corresponding evidence that most applications for IPR protection in African countries had been filed by entities based in developed countries. Are there obvious pathways for reconciling this issue? Perhaps yes, if increasing awareness that the transfer, adaptation and use of IPR-protected biotechnologies could play a major role in improving agricultural productivity and the overall development of African countries, where agriculture remains the key source of food, incomes, employment and often foreign exchange. But, of course, this is easier to say than to implement. For, as Inkster has reminded us, part of the problem facing developing countries is that they often do not have patent systems at all, or patent systems which are appropriate to their domestic needs rather than their former colonial masters. Without a history of the institutions and practices that would support the socially beneficial use of IPRs, it is unrealistic to suppose that change can happen overnight. Analysis tells us what the problem is; history in the respects just outlined tells us how significant a problem it really is.

The World Bank reports that yields of the major staple crops (maize, sorghum, millet, cassava, cowpea, bananas/plantains) of smallholder farmers in Africa have been stagnant or even declined in the past 40 years due to numerous biotic and abiotic constraints such as diseases, pests and drought. Local research efforts to overcome these stresses are hampered by declining financial support for agricultural research and limited access to elite genetic material and other technologies protected by IPRs. Therefore, a new initiative, the African Agricultural Technology Foundation (AATF), has been established to address this challenge by negotiating access to proprietary technologies and facilitating their delivery to smallholder farmers in Africa. In this respect, the AATF is working to make the IPR system more like an effective information system, particularly from the standpoint of potential end users of the information and technologies. Alas, it is still an open question whether the introduction of IPR regimes and the transfer of IPR-protected technologies have culminated in any increase(s) in innovation in Africa in the agricultural sector. Even more remote is an answer to the question whether the experience in the agriculture sector had any application to other areas of innovation such as health.

The third chapter by Tina Piper examines the relationship between law and medicine at the formation of professional medical associations, and the role that IPRs played. She focuses on the medical exception to patentability, a provision of patent law that governs the patentability of many new biotechnologies. Normally, medical exemptions apply to diagnostics but, as Piper points out, current experience with the medical exemption rule shows that it is notoriously imperfect. The reasons for this are not simply a matter of the vicissitudes of contemporary moral and legal tussles, but have deep historical roots. The medical exemption was initially drafted to account for the needs of industrializing England in the early 1800s, and has remained in the UK's patent law since. It is currently being detrimentally imposed through TRIPs and other harmonization efforts on developing countries, without any evidence (historical or otherwise) that it satisfies its purported desired effect of ensuring that valuable medical technologies are available to medical practitioners. Finally, the medical methods exception has been used to block a range of biotechnologies while it is also used as a normative argument that the patent law does in fact ensure that valuable medical technologies are not patented.

Piper's detailed analysis suggests that the relationship between our current understanding of practices associated with IPRs may have highly specific origins in events which are probably not replicable, and perhaps would not even repeat themselves were one to rewind the tape of time. The corollary is that while historical analysis focusing on the events of a country's evolving patent system may be profoundly illuminating about current practices, the lessons may not be transferable to other contexts. This observation turns out to be quite important, because it implies that there are limitations about what one can expect the history of intellectual property to tell us about avenues for reform. What happened in 19th century England conditions the current situation, but it is mute about the paths other countries ought to take now. Equally, one cannot Whiggishly read back into history to find continuity in the use of IPRs to justify current practices.

Where does this leave us, then with respect to thinking about the topic of this book, which is the role of IPRs in innovation? Historical and country-specific studies show, first, that IPRs are one part of a social system that can be said to have innovative capacity. One tentative conclusion one can draw is that framing the question about IPRs' contribution to innovation may appear to be analytically fruitful, but it may not be historically meaningful. As Inkster's chapter suggests, many factors must come into alignment for there to be innovation, and IPRs may not at any given time figure prominently in the overall evolution of what we presently call innovation systems. Piper's chapter suggests that where there may be a thread to follow about the specific contribution IPRs have made to innovation, it may be rooted in some unique historical events, such as professional competition. Finally, Boadi's chapter illustrates that historical factors carry forward to the present and condition how implementation and reform of IPRs may transpire. In these respects, country-specific histories of innovative activity may have more to do with the extent and kind of use of IPRs than IPRs have on innovation generally.

16. The role of intellectual property rights in biotechnology innovation: national and international comparisons

Richard Y. Boadi

INTRODUCTION

Intellectual property rights (IPRs)[1] refer to a category of rights which confer protection for creations of the human intellect.[2] They are a form of legal entitlement which sometimes attaches to the expressed form of an idea, or to some other intangible subject matter. In general terms this legal entitlement sometimes enables its holder to exercise exclusive control over the use of the IPR. IPRs are designed primarily to provide incentives for innovative behavior; induce investment to develop and commercialize technologies; provide incentive for the disclosure of information; and facilitate the transfer of technology. Yet, IPRs could serve as a barrier to innovation, particularly in developing countries, if technologies needed as research inputs are protected by such IPRs and are not accessible to the research institutions that develop products specifically for these countries where farmers constitute approximately 70 per cent of the general population and 90 per cent of the agri-food workforce.

For smallholder farmers in Sub-Saharan Africa, yields of major staple crops (maize, sorghum, millet, cassava, cowpea, bananas/plantains) have remained stagnant or even declined in the past 40 years. Numerous biotic and abiotic stresses have contributed to this dire scenario. Local research efforts to overcome these stresses are hampered by declining support for agricultural research, limited access to elite genetic material and other technologies protected by IPRs, and the absence of commercial interest in these crops by private owners of agricultural technologies. Despite the availability of agricultural technologies (such as improved seeds and farm inputs), crop productivity in Africa has remained low or stagnant. This is mostly because improved crop varieties that are resistant to biotic and abiotic constraints are not being planted. High costs and the unavailability of technologies in times of need have made drought-tolerant or disease- and pest-resistant seeds inaccessible, particularly to smallholder farmers in developing countries.[3] This situation is compounded by IPRs. Patents and plant breeders' rights[4] attempt to strike a balance between protecting the rights of an inventor/innovator and providing a benefit to the society as a whole, but they also often raise the cost of accessing new and improved products. The decline in agricultural productivity and the rise of IPRs create a challenge as to how best to stimulate innovation while providing mechanisms that support the smallholder farmers' access to these technologies.

This chapter examines the correlation between IPR and innovative behavior with reference to the United States; traces the evolution of IPR regimes related to biotechnology in Sub-Saharan Africa, and explores the impact, if any, thereof on innovation; and describes several initiatives designed to meet the challenge of stimulating innovation while creating the regulatory and policy environment necessary to catalyze biotechnological innovation and utilization in Sub-Saharan Africa.

IPRs AS A MOTIVATOR FOR INNOVATION IN THE USA

There is evidence to suggest that IPRs helped to stimulate innovation in the developed world, particularly in the United States of America (USA), in the past two decades[5] in the areas of health and plant biotechnology. The results of scientific research had historically been treated as public goods. However, there was a gradual paradigm shift with several changes in the legal and policy framework, making it possible to seek statutory IP protection for biotechnological innovations. This shift spurred greater public, as well as private, sector involvement in innovative activities. For instance, the US Council on Governmental Relations reports that 'the Bayh-Dole[6] Act and its subsequent amendments created incentives for the government, universities, and industry to work together in the commercialization of new technologies for the public benefit'.[7] For the first two decades following the enactment of the Act, there was close to a 10-fold increase in institutional involvement in innovative activities, leading to the grant of approximately 8,000 US patents to academic institutions for the inventions of their researchers. During the same period, over 2,200 new companies were formed based on the licensing of inventions from academic institutions, including over 330 companies formed in Financial Year 1997 alone.[8] A national survey conducted by the Association of University Technology Managers (AUTM)[9] in 1997 reported that 70 per

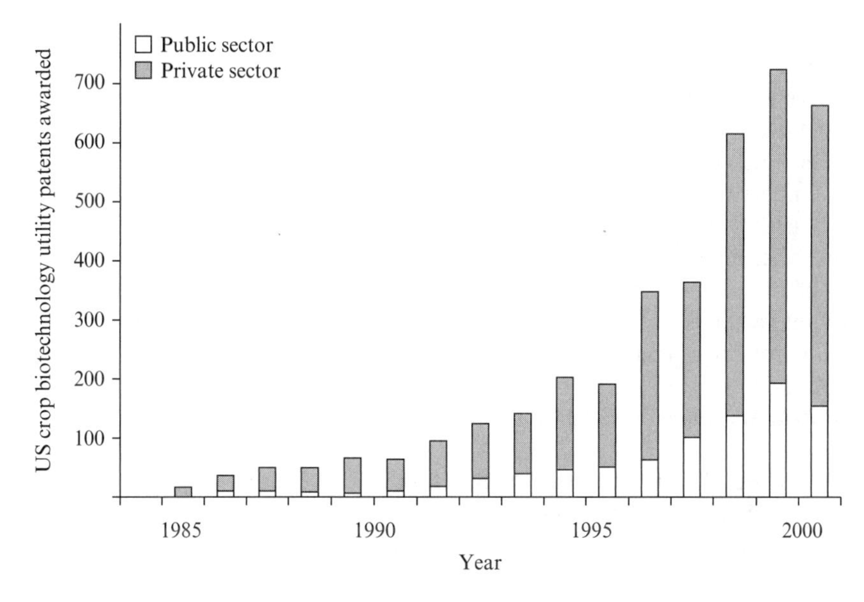

Figure 16.1 *Patents awarded by the US Patent and Trademark Office in the area of crop biotechnology assigned to either private sector or public sector institutions between 1985 and 2000*

cent of the active licenses of responding institutions were in the life sciences, yielding products and processes that diagnosed disease, reduced pain and suffering, and saved lives. Figure 16.1[10] below demonstrates the increase in patent grants over the period under discussion to both public and private sector institutions. It bears ample testimony to the fact that IPRs did indeed induce innovative behavior in the USA.

THE EVOLUTION OF IPR REGIMES RELATED TO BIOTECHNOLOGY IN SUB-SAHARAN AFRICA

Although IPR regimes for biotechnological innovations have existed in industrialized countries for several decades, developing countries, including most countries in Sub-Saharan Africa, are in the early stages of implementing and/or enforcing IPRs related to biotechnology. The past few years have seen increased attention to the strengthening of IPRs in biotechnology-related inventions. The number of countries that grant such rights has grown, the types of inventions that can be protected have expanded, and the scope of protection offered by IPR systems in different countries has broadened as well.[11] The Agreement on Trade Related

Aspects of Intellectual Property Rights (TRIPS 1993) of the World Trade Organization (WTO)[12] significantly changed the importance of IPRs in Sub-Saharan African countries by requiring all WTO members[13] to introduce a minimum level of protection of intellectual property in their national laws. Article 27.3(b) of the TRIPS Agreement requires all members to provide patent protection for both products and processes in all fields of technology, including biotechnology. While this Article allows member governments to exclude some kinds of inventions from patenting – plants, animals and 'essentially' biological processes[14] – it requires new plant varieties to be protected either by patent or a system created specifically for the purpose ('*sui generis*'), or a combination of the two.

Patent and Plant Variety Protection

Nnadozie (2004) notes that the development of IP laws in most countries in Sub-Saharan Africa took the usual form common to colonial legal development.[15] Before independence, and long afterwards in some cases, most of these countries did not have independent IP laws or offices and therefore did not directly grant any IPRs. For instance, for English-speaking countries, IPRs were applied for, granted or registered in the United Kingdom and subsequently only re-registered.[16] Currently, almost all African countries have patent statutes. As regards the protection of new plant varieties, it has been noted that before the TRIPS Agreement, only two countries in Sub-Saharan Africa – South Africa and Zimbabwe – had fully functioning PVP systems.[17] At present, only Kenya, Morocco, South Africa, Tunisia and the OAPI[18] member countries have enacted PVP legislation.[19] We trace below the evolution of the patent and PVP regimes of selected English- and French-speaking African countries.

Ghana

Prior to 1992, the patent laws of the United Kingdom applied in Ghana.[20] Thus, in order to protect in Ghana an invention made in Ghana, it was necessary to have it first registered in the UK and thereafter re-registered in Ghana. Subsequent legislation such as the Patents Registration Ordnance of 1925 and the Patents Registration (Amendment) Decree of 1972, which excluded pharmaceutical products from patentability and cancelled all such prior patents, continued this re-registration system until 1992.[21] With the enactment of the Patent Law of 1992, PNDCL.305A,[22] pharmaceuticals could be patented and it became possible to obtain a patent in Ghana either directly or through the Patent Cooperation Treaty (PCT) administered by the World Intellectual Property Organization (WIPO) as well as through the African Regional Intellectual Property Organization

(ARIPO).[23] The Patent Law of 1992 has since been superseded by the Patent Act of 2003, which is designed to conform Ghana's patent law to the requirements of the TRIPS Agreement.[24]

Ghana does not have a law that protects new varieties of plants but is currently engaged in consultations with UPOV[25] with a view to adopting a PVP law.

Kenya

Like Ghana's the patent law of Kenya had British historical ancestry. The registration of patents was carried out under a British law, the Patents Registration Act 1962, under which only the grantee of a patent in the UK or a person deriving his/her right from the grantee by assignment or any other operation of law could apply to have his/her patent registered in Kenya.[26] An application had to be made within three years from the date of the UK grant and the patent would remain in force only for as long as the patent remained in force in the UK.[27] This limited patent grant to persons with access to registration in the UK. It also made the process expensive and time-consuming. Moreover, the registration process did not address the criteria for obtaining protection or entail examination of applications. This system remained in force until 1989 with the enactment of the Industrial Property Act 1962. The Act was amended a number of times and finally replaced by Kenya's Industrial Property Act, No. 3 of 2001, which came into force in August 2001. The 2001 Act is designed, inter alia, to bring Kenya's patent law into compliance with the TRIPS requirements. It excludes plant varieties from patentability but provides that parts of plant varieties and products of biotechnological processes are patentable.

Kenya has had a Seed and Plant Varieties Act since 1942. This was replaced by the Seed and Plant Varieties Act, cap. 326 of 1977, which has, in turn, been superseded by the Seed and Plant Varieties Act of 1991 and several regulations enacted pursuant thereto.

French-speaking Africa

In the French-speaking countries of Sub-Saharan Africa, patents are issued by OAPI pursuant to the Bangui Agreement, which has the status of national patent law for all member countries. Each of the OAPI member countries has a local office which is capable of receiving filings of registrations, which are then valid for the entire region and ensure the patent rights of an applicant are simultaneously protected in all member countries. The Bangui Agreement provides that patents can be issued for a product or process, provided it is new, involves an inventive step and is capable of industrial application – the usual criteria for patentability – and makes no distinction between biotechnology and other inventions.

The Bangui Agreement was revised in February 1999[28] in order to harmonize its provisions with those of the TRIPS Agreement. The revised agreement provides for the protection of new varieties of plants, based largely on the terms of the UPOV 1991 Convention.[29]

Institutional and human capacity

Almost all Sub-Saharan African countries have offices handling IPR issues. However, very few of these countries have the requisite institutional and human capacity to handle patents related to biotechnology.[30] Human capacity development efforts currently being spearheaded by institutions such as the World Intellectual Property Organization are expected to help address this obvious inadequacy of the IPR regimes of these countries.

THE IMPACT OF IPRs ON BIOTECHNOLOGY INNOVATION IN SUB-SAHARAN AFRICA

Biotechnology Research and Development

There has been some biotechnological innovation in Africa in the last couple of decades, with the majority of research and development efforts focussed on crop improvement.[31] Glover (2007) reports that Ghana, Kenya, Nigeria, South Africa and Zimbabwe all have several research institutes with some capacity to conduct biotechnological research. A number of high-profile biotechnology research projects and initiatives have been undertaken, or are currently being set up, across Sub-Saharan Africa, with Nigeria and South Africa having the largest portfolios of biotechnology research and development projects.[32] The projects include the development of insect-resistant cowpea and banana resistant to banana bacterial wilt, by consortia both led by the African Agricultural Technology Foundation (AATF); and bio-fortified sorghum by a consortium led by the Africa Harvest Biotechnology Foundation International (AHBFI). African scientists have used both genetically-modified (GM) and non-GM approaches in their research.

In the case of GM foods, there are several other projects that can be mentioned. In Kenya, a project supported by Monsanto and the United States Agency for International Development (USAID) created a virus-resistant variety of sweet potato, although the crop failed in the field. Also, scientists at the Donald Danforth Plant Science Center – a not-for-profit organization in St Louis, United States – are working to introduce virus-resistance into a variety of cassava which is a staple crop in much of Africa. While they have succeeded in introducing the trait, it has not yet proved stable over multiple generations. Further, research centers of the Consultative Group on International Agricultural Research (CGIAR) are exploring GM crop traits relevant to African agriculture. For example, the International Crops Research Institute for the Semi-Arid Tropics (ICRISAT), based in Hyderabad, India and with regional hubs in Nigeria and Kenya, is working on transgenic varieties of pigeon pea, groundnut and pearl millet. And the International Center for Maize and Wheat Improvement (CIMMYT) in Mexico coordinates the Insect Resistant Maize for Africa project in collaboration with the Kenya Agricultural Research Institute (KARI) and with the support of the Syngenta Foundation for Sustainable Agriculture. Meanwhile, the first GM crop plant developed entirely in Africa – a virus-resistant variety of maize created by scientists at the University of Cape Town, South Africa – is scheduled to begin field trials in late 2007.

With respect to non-GM projects, Nigeria is host to over 60 projects exploring non-GM biotechnology in the micropropagation of cassava, date palm and ginger research, among others. And one of the most prominent non-GM biotechnology projects in Africa is KARI's ongoing work to produce tissue-cultured banana plantlets and disseminate them to farmers across East Africa.

While a few Sub-Saharan African nations are moving towards officially releasing GM crops, with individual varieties of maize and cotton, for example, at the field trial stage, only South Africa has, to date, fully commercialized these crops.[33]

The robust biotechnology activities notwithstanding, Sub-Saharan African countries may not be able fully to exploit the vast potential of this technology due to several factors, including inadequate infrastructural and human capacity, lack of access to IP-protected research tools, lack of capacity to access and manage IP and the absence of enabling policy and regulatory frameworks. These factors will be examined later in this chapter.

Impact of IPRs

IPR regimes for plant biotechnology have not been in effect in Sub-Saharan African countries long enough effectively to gauge their impact on innovative activities. There appear to have been very few or, in most cases, no patent applications filed or grants made in the various countries despite the reported robust activities in the area of plant biotechnology.[34] Statistics available at the World Intellectual Property Organization (WIPO)[35] indicate that patent applications filed with the patent offices of the various African countries generally tend to be much lower than those in developed countries. It is estimated that not more than five patent applications have been filed for biotechnological inventions in Kenya since the enactment of the country's Industrial Property Act in 2001.[36] There are several reasons for these relatively fewer patent filings in Sub-Saharan African countries. As noted in this chapter, the IP regimes in most of these countries are not yet well developed. Further, some people may not routinely seek to protect their innovations because of lack of awareness of the benefits of IP. And even the well-informed may not seek IP protection due to the high costs relative to the potential economic returns.

The situation is slightly different with regard to applications for PVP in the countries where this regime exists, and this seems to support the notion that the implementation of a functional IPR regime could serve as an incentive and contribute to an increase in innovation. Sikinyi (2003) notes that the introduction of PVP legislation in Kenya has resulted in a general increase in breeding activities and increased collaboration amongst local and foreign institutions as foreign breeders felt confident enough to introduce their materials into and to invest in Kenya.[37] Indeed, a recent report by the World Bank indicates that

> between 1997 and 2003, Kenya received over 600 applications for PVP, but by mid-2004 only 108 certificates had been granted, 70 percent of them for ornamentals, especially roses and alstroemeria. Among field crops, maize has the highest number of applications (accounting for 10 percent of the total); all of these applications are for hybrids from either the public sector or the parastatal Kenya Seed Company (KSC). The rise in domestic applications is partly due to the amnesty granted in 2001 to previously released public varieties. The fact that this ruling is being contested is one of the explanations for the relatively low number of PVP grants issued so far.[38]

FACILITATING ACCESS TO IP-PROTECTED TECHNOLOGIES FOR BIOTECHNOLOGICAL INNOVATION IN SUB-SAHARAN AFRICA

The introduction of IPR regimes in Sub-Saharan African countries had the potential to stimulate innovation in biotechnology innovation. However, the promise of this technology may not be realized if a number of factors are not addressed. These include lack of access to IP-protected research tools, lack of capacity to access and manage IP, inadequate infrastructural and human capacity and the absence of enabling policy and regulatory frameworks.

A major factor that could constrain biotechnological research efforts is lack of access to IP-protected research tools, including genes, DNA sequences, diagnostic techniques, transformation and regeneration methods, as well as the know-how associated with these technologies.

Private sector companies own most of these IP-protected research tools, but the rather small African market does not provide much of a commercial incentive for these companies to invest in specific technologies, varieties and traits adapted to the unique agricultural conditions of the region. On the other hand, public sector organizations have vast experience working on regionally important crops but need access to such IP-protected research tools. However, issues pertaining to the availability, complexity, high transaction costs, licensing, testing, safety and liability tend to serve as barriers to accessing these technologies by African researchers, development specialists and resource poor farmers. Therefore, there is the need for new and innovative approaches requiring the support and collaboration of both the public and private sectors to address these issues. Several initiatives designed to meet this challenge have emerged in recent times.

The African Agricultural Technology Foundation (AATF)[39]

AATF is addressing the challenge of reversing the negative trend in agriculture by negotiating access to proprietary technologies and facilitating their delivery to smallholder farmers in Sub-Saharan Africa (SSA). AATF was designed in 2003 to respond to Africa's food security challenges through a twin approach of negotiating, on a humanitarian basis, for access to proprietary agricultural technologies from anywhere in the world and forming public-private partnerships involving various institutions to ensure safe and sustainable delivery of products made from such technologies. AATF has, to date, negotiated for the right to use IP-protected technologies and has sublicensed these to research institutions for the development of improved crop varieties which could reduce several constraints faced by smallholder farmers.[40] The terms of the license agreements entered into by AATF are very beneficial as they allow for fee-free and royalty-free access to proprietary technologies, thus making it possible to maintain low costs of the products developed using such technologies.

The Public Intellectual Property Resource for Agriculture (PIPRA)[41]

PIPRA is a US-based initiative with global reach which seeks to pool publicly owned and patented technologies for use by research institutions in developing countries and specialty crops in the developed world. PIPRA's core activities include IP policy analysis, IP landscape analysis on particular technologies, development of biotechnology resources,[42] the provision of research consortia support, IP management workshops at public institutions and the provision of regional IP resources, mainly in Latin America and Southeast Asia.

The Centre for the Application of Molecular Biology to International Agriculture (CAMBIA)[43]

CAMBIA is an Australian-based initiative which aims to provide technical solutions that empower local innovators to develop new agricultural innovations. CAMBIA is an international, independent non-profit research institute pioneering Open source biology and informatics to support patent transparency. For more than a decade, CAMBIA has been creating new tools and enabling technologies to foster innovation and a spirit of collaboration in the life sciences and exploring new research and development paradigms, practices and policies to address neglected priorities of disadvantaged communities.[44]

The Public Interest Intellectual Property Advisors, Inc. (PIIPA)[45]

PIIPA is an international non-profit organization which makes intellectual property counsel available for developing countries and public interest organizations which seek to promote health, agriculture, biodiversity, science, culture and the environment. PIIPA has been working on expanding a worldwide network of IP professional volunteers; operating a processing center where assistance seekers can apply to find individual volunteers or teams which can provide advice and representation as a public service (free or *pro bono*); and building a resource center with information for professionals and those seeking assistance. PIIPA has, to date, facilitated the provision of IP management services for organizations based in Sub-Saharan Africa.[46]

It is expected that the collective efforts of AATF, CAMBIA, PIIPA and PIPRA will facilitate access by research institutions to IP-protected research tools and thereby catalyze biotechnological innovation in Sub-Saharan Africa.

Apart from the lack of access to IP-protected research tools, research institutions in most Sub-Saharan African countries may not have the required infrastructural and human capacity fully to exploit the vast potential of biotechnology. This has the potential to militate against effective biotechnological innovation. It is, therefore, heartening to note that efforts are currently underway to improve the infrastructure in most of these institutions. One notable example is an initiative by the UK-based Kirkhouse Trust not only to improve the research laboratories of several countries in East and West Africa but also to help develop the capacity of African scientists to handle marker-assisted selection processes. Other initiatives include the establishment of centers of excellence by the New Partnership for Africa's Development (NEPAD) in the various subregions of Africa designed to help African scientists and institutions to become significant technological innovators as well as users.[47]

The absence of adequate policy and regulatory frameworks in most Sub-Saharan African countries could also hinder biotechnological innovation in the region. Kenya, South Africa and Zimbabwe are commonly regarded as Africa's most advanced countries in terms of combining biotechnology research capacity with the necessary policy frameworks and biosafety regulatory systems. However, a large number of the other Sub-Saharan African countries have acceded to the Cartagena Protocol on Biosafety, an international agreement governing the trans-border movement of GM organisms, and are in the process of developing their policy and regulatory frameworks. Most of these countries have received technical and capacity-building support from the USAID-supported Program for Biosafety Systems (PBS) as well as through the United Nations Environment Programme (UNEP) and the Global Environment Facility (GEF) to help them set up risk assessment and regulatory systems for managing the trade in GM organisms.

CONCLUSION

There is no doubt that there is a relationship between IPRs and innovation. Biotechnology has the potential to promote food security and the alleviation of poverty because it can increase crop yields, enhance nutritional content and improve resistance to pests, viruses, pesticides and environmental stresses that constrain agricultural production. The emergence of strong IPR regimes in Sub-Saharan African countries could, in the long-term, incentivize biotechnological research and development, lead to increased technological inflows and ultimately help achieve food security and alleviate poverty in the region. In the short term, however, there would be the need to create an enabling policy and regulatory environment for the potential of biotechnology to be fully realized in Sub-Saharan Africa. Facilitating access by research institutions to IP-protected research tools, developing the infrastructural and human capacity of African countries to handle biotechnology-related research as well as the capacity to access and manage IP would be crucial.

NOTES

1. IPRs include copyright, patents, plant breeders' rights, trade and service marks and utility models.
2. Gardner B.A., (ed.). 1999. *Black's Law Dictionary*. St. Paul, MN: West Group.

3. See Omanya, G., R. Boadi, F. Nang'ayo, H. Mignouna, M. Bokanga (2005). 'Intellectual Property Rights and Public-Private Partnerships for Agricultural Technology Development and Dissemination', Paper presented at the Kenya National Conference on Revitalizing the Agricultural Sector, 21–24 February 2005, Nairobi, Kenya.
4. The IPRs most relevant to biotechnological innovations are patents and plant breeders' rights (also referred to as plant variety protection – PVP).
5. Delmer, D.P., Greg D.G. Nottenburg and A.B. Bennet. 2003. 'Intellectual Property Resources for International Development', *Plant Physiology* 133, December, pp. 1666–1670.
6. The Bayh-Dole Act, 35 USC § 200–212, allows for the transfer of exclusive control over many government funded inventions to universities and businesses operating with federal contracts for the purpose of further development and commercialization. The contracting universities and businesses are then permitted to license the inventions exclusively to other parties. The federal government, however, retains 'march-in' rights to license the invention to a third party, without the consent of the patent holder or original licensee, where it determines the invention is not being made available to the public on a reasonable basis.
7. See Committee on Governmental Relations Report (1999), http://www.ucop.edu/ott/bayh.html.
8. AUTM press release, December 17, 1998.
9. AUTM Licensing Survey, Fiscal Year 1997.
10. Source: Delmer, et al.
11. The World Bank Agriculture and Development Dept. (2006), 'Intellectual Property Rights: Designing Regimes to Support Plant Breeding in Developing Countries'. Available at http://siteresources.worldbank.org/INTARD/Resources/IPR_ESW.pdf.
12. World Trade Organization, Agreement on Trade Related Aspects of Intellectual Property Rights, Annex 1C to Marrakesh Agreement Establishing the World Trade Organization (TRIPS), Apr. 15, 1994. Available at http://www.wto.org/english/tratop_e/trips_e/art27_3b_background_e.htm.
13. Most African countries are members of the WTO. The few exceptions are Algeria, Cape Verde, Equatorial Guinea, Ethiopia, Libya and Sudan. Source: WTO website. http://www.wto.org/english/thewto_e/whatis_e/tif_e/org6_e.htm.
14. Article 27.3(b) provides an exception by requiring micro-organisms, and non-biological and microbiological processes to be eligible for patents.
15. Nnadozie, K. (2004), 'Intellectual Property Protection in Africa: Status of Laws, Research and Policy Analysis in Ghana, Kenya, Nigeria, South Africa and Uganda'. African Centre for Technology Studies (ACTS), Ecopolicy Series no. 16. Available at http://www/acts.or.ke/pubs/monographs/pubs/ecopolicy16.pdf.
16. Ibid, at p. 6.
17. Kuyek, Devlin (2002), 'Intellectual Property Rights in African Agriculture: Implications for Small Farmers'. Available at http://www.grain.org/briefings/?id=3.
18. The African Intellectual Property Organization (Organisation Africaine de la Propriété Intellectuelle or OAPI in French) is a central registration system for industrial property and was established pursuant to the Bangui Agreement, effective March 2, 1977. The OAPI currently has 16 member states: Benin, Burkina Faso, Cameroon, Central African Republic, Chad, Republic of Congo, Cote d'Ivoire, Equatorial Guinea, Gabon, Guinea, Guinea Bissau, Mali, Mauritania, Niger, Senegal, and Togo.
19. Statistics available at http://www.upov.int/en/about/members/index.htm.
20. The British introduced into the then Gold Coast Colony (Ghana since 1957) Patents Ordnance No. 1 of 1899, making UK patent law applicable to the Colony. Under this system, patents could only be registered in the UK and then re-registered in the country.
21. Supra, note 15 at pp. 8–9.
22. Provisional National Defence Council 1992. Patent Law of 1992. Law No. 137. PNDCL.305A.

23. The current member countries of ARIPO are Botswana, the Gambia, Ghana, Kenya, Lesotho, Malawi, Mozambique, Sierra Leone, Somalia, Sudan, Swaziland, Tanzania, Uganda, Zambia, and Zimbabwe. Patent applicants from any of these countries can file their applications with either their national offices or directly with the ARIPO Office. Under this system, one application is effective in all member states designated in the application. More information on ARIPO is available at www.aripo.wipo.net.
24. One major change effected by the Patent Act of 2003 is the removal of the government's right, under Section 7 of the 1992 Law, to exclude inventions, including pharmaceuticals, from patentability 'in the interest of national security, economy, health or any other national concern'.
25. UPOV is the commonly used acronym for the International Union for the Protection of New Varieties of Plants, an intergovernmental organization with headquarters in Geneva. The acronym is derived from the French name of the organization, which is *Union Internationale pour la Protection des Obtentions Végétales.* It was established by the International Convention for the Protection of New Varieties of Plants. The Convention was adopted in Paris in 1961 and subsequently revised in 1972, 1978 and 1991. The objective of the Convention is the protection of new varieties of plants by an intellectual property right. More information about UPOV is available at http://www.upov.org/.
26. Kameri-Mbote, P. 2004.'Intellectual Property Protection in Africa: An Assessment of the Status of Laws, Research and Policy Analysis on Intellectual Property Rights in Kenya', International Environmental Law Research Centre Working Paper 2005–2. Available at http://www.elrc.org/africa/ipr.php.
27. Kingarui, Joseph 1989.'Towards a National Patent Law for Kenya', in Juma Calestous and J.B. Ojwang (eds), *Innovation & Sovereignty: The Patent Debate in African Development,* Nairobi: ACTS Press.
28. The text of the revised Bangui Agreement is available at http://www.oapi.wipo.net/doc/en/bangui_agreement.pdf.
29. Since April 24, 1999, the 1978 Act of the Convention has been closed to further accessions, as the 1991 Act entered into force after the ratification of Bulgaria and the Russian Federation to this Act (the 1991 Act), bringing to six the number of States which have ratified the Convention. This is in accordance with its Article 37(1) which stipulates that 'the 1991 Act shall enter into force one month after five States have ratified it'. For a comparative analysis of the 1978 Act and the 1991 Act see Kongolo, Tshimanga (2001), 'New Options for African Countries regarding Protection for New Varieties of Plants', *Journal of World Intellectual Property*, Vol. 4, No., 3, May 2001, pp 349–371. Available at http://www.grain.org/docs/kongolo-jwip-en.pdf.
30. The few exceptions include Kenya (where patent applications are handled by the Kenya Industrial Property Institute, which employs about 20 examiners, several of whom have experience in biotechnology) and South Africa.
31. Glover, D., 'Agri-biotech in sub-Saharan Africa: Facts and figures', *Business Day Online*, June 17, 2007. Available at http://www.businessdayonline.com/?c=50&a=13953.
32. Ibid.
33. These include insect-resistant or herbicide-tolerant varieties of cotton, maize and soyabean.
34. The likely reasons are that the various projects may still be at the discovery phase or that IPR may not necessarily be the real motivation for work being carried out, after all. Indeed the World Bank notes that a number of private national seed companies have recently emerged in Uganda in the absence of PVP legislation, 'which may be explained more by the gradual decline of the public seed enterprise (now privatized) during the past decade than by the prospect of protection'. See note 11.
35. http://wipo.int/ipstats/en/statistics/patents/.
36. Personal communication from Fredrick O. Otswong'o, Patent Examiner, Kenya Industrial Property Institute.
37. Sikinyi, E. (2003), 'Experiences in Plant Variety Protection under the UPOV

Convention', paper presented at a WIPO–UPOV Symposium on Intellectual Property Rights in Plant Biotechnology [WIPO-UPOV/SYM/03/09], http://www.upov.int/en/documents/Symposium2003/wipo_upov_sym_09.pdf.

38. Supra, note 11 at pp. 18 and 19.
39. http://www.aatf-africa.org/.
40. For instance, in May 2005, AATF obtained the right to use Monsanto's *cry1Ab* gene for the development and commercialization in all of Sub-Saharan Africa of cowpea varieties resistant to insect pests (*Bt* cowpea). Under license from AATF, the Australian Commonwealth Scientific and Industrial Research Organisation (CSIRO) and the Nigeria-based International Institute for Tropical Agriculture (IITA) are currently engaged in cowpea transformation work. CSIRO has so far successfully transformed cowpea using the Bt gene. Also, in August 2006, AATF obtained a license from the Taiwan-based Academia Sinica to use the plant ferrodoxin-like protein *(pflp)* gene from sweet pepper for banana transformation to produce banana varieties resistant to banana bacterial wilt. AATF has since sublicensed the gene to IITA for its banana transformation work being conducted in collaboration with the National Agricultural Research Organisation (NARO) of Uganda.
41. PIPRA is a coalition of over 40 universities, foundations, and other research institutions which pool together their patents for humanitarian purposes: http://www.pipra.org/.
42. PIPRA has developed a plant transformation vector with a transposon module and is currently engaged in negotiations to license it to AATF for use in the development of nitrogen use efficient and salt tolerant rice traits for smallholder farmers in Sub-Saharan Africa.
43. http://www.cambia.org/daisy/cambia/home.html.
44. CAMBIA's activities include the following: BiOS Initiative™ (Biological Innovation for Open Society), articulating a new means of invention, improvement and sharing of life sciences technologies, and the capabilities to use them; Bios (Biological Open Source) licenses draw inspiration from the open source software movement but are adapted for patented technologies in the life sciences. They create a 'protected technology commons' in which an invention can be improved by the ideas of many, without exclusive capture by any one entity. CAMBIA has seeded this movement with its own technologies, and other technology owners may also provide licenses to their technologies using this framework.
45. http://www.piipa.org/.
46. For instance, at the request of AATF, PIIPA arranged for the Intellectual Property and Business Formation Legal Clinic of the University of Missouri Law School to conduct a comprehensive freedom to operate (FTO) assessment for a project to develop banana resistant to banana bacterial wilt for smallholder farmers in Sub-Saharan Africa.
47. These include the Kenya-based Biosciences eastern and central Africa (BeCA), the South Africa-based Southern African Network for Biosciences (SANBio) and the Senegal-based West African Biosciences Network (WABNet).

17. Intellectual property, information and divergences in economic development – institutional patterns and outcomes circa 1421–2000

Ian Inkster

INTRODUCTION: MAIN THEMES AND CLAIMS

Allowing for small faults with estimation, there are probably about four million patents in the world. Of these, about 1 per cent are owned in the third world, a proportion reached first around 1913 and maintained ever since. In contrast, the third world today produces around 20 per cent of world income and 25 per cent of world trade.[1] So, as illustrated by patent ownership, intellectual property represents a greater bias against late development than other standard measures, and this bias – as this chapter shows – is of long standing. The patent system, yet at the heart of an intellectual property regime that embraces trademarks, utility models, copyright, inventors' certificates and the like, was introduced to colonies and subject peoples by a very small group of industrial nations during the late 19th century. Whilst such dominion undoubtedly insured something of this long-term bias, we argue here that the prime elements at work have been variability combined with complexity. An original variability amongst national systems allowed the successful late entry of such nations as Germany or Japan. The resulting institutional and cognitive complexity inhibited industrialization via technology transfers thereafter. Thus the short 20th century (1918–1971) was a disaster for poor nations. Subsequent changes in technique towards areas such as biotechnology and microelectronics opened new windows of opportunity for very late developers, for new industries did not require the tremendous fixed assets of the earlier metal- and chemical-based machinofactures. The rising pressure for patent reform that has been felt since that time may be seen as a reflection of the needs of poor nations to maximize their benefits during this great climacteric, needs which clash with the commercial and political interests of the rich, industrialized nations, whose powerful agencies are demanding greater universality and homogeneity of systems. The historical record suggests that such demands are inimical to the process of late development.

Intellectual property rights (IPR) systems have been enormously varied over time and have thus worked in contrasting manners in different countries. More than this, IPR systems in the form of patents in particular have served to stimulate innovation in a relatively small number of industrial modernized nations, a group that reached its basic mass and density around 1914 and continued to exist as such to around 1971. In such favored nations IPR systems have nested into very elaborate innovation cultures, organizations and institutions, themselves parts of wider financial, legal and scientific infrastructures. In the late-developing nations such as Russia or Japan around 1900, heroic efforts towards technology transfer-in were made by centralized political regimes, these involving strategic developments and manipulations of patent systems adopted/adapted from systems generated initially by the United States, Britain and France.[2] In the great majority of less developed nations from that point on, an absence of these wider infrastructures as well as intrinsic features of patent systems themselves assisted in the process of underdevelopment which dominated the years circa 1918–1971.[3] Only from around the early 1970s with OPEC crises, the emergence of new micro-electronic, biotechnical, marine and environmental industries, and the quick changes in the systems of international payments, did new windows of opportunity open which allowed another select group of newly industrializing countries (the famous NICs, each characterized by partial-nation or city-nation status) to be the last group of nations that would industrialize at the tail end of the older machinofacture of steel and chemicals and amidst the established institutions governing intellectual property rights, knowledge creation, diffusion and application. There is little evidence from the overall statistics of patent registration at a global or comparative level that patent systems post-1850 were ever generally conducive to industrial development outside the favored group of nations which had emerged already by 1914–1918. One lesson of this chapter is that reformers of the global patent systems must take into account the historical fact that IPR systems have been also information systems and technology transfer systems, and that radical reform must not improve the quality of property rights in poor nations of our world if at the same time it threatens or does not address and improve ancillary information systems for just those nations. Over the whole range of history it might well be that most patent systems have been at their best as information systems and at their worst as IPR systems.

INDUSTRIALIZATION AND ITS HISTORICAL CONFUSIONS

Patents go back a long way, and do seem to have developed uniquely in Western Europe if we define them as some form of identifiable mechanical innovations designed for and registered in order to secure a claim of ownership and a monopoly of time in which to exploit an advance for personal profit.[4] An early identifier is the grant to an inventor in Florence in 1421.[5] Britain may claim a lineage back to the Statute of Monopolies of 1623 which exempted patents for new inventions when declaring all other monopoly grants illegal.[6] German states have similar early claims.[7] But even in these systems, national legislation awaited both regulatory devices and nationality itself – thus we may date national patent legislation in the three systems above mentioned as only from 1859, 1852 and 1877 respectively.[8] For the 18th century and early 19th century period of initial industrialization, IPR systems were generally inefficient in the face of state-instigated and supported technological learning and piracy.[9] But by the mid-19th century Britain, France and the United States dominated the modernized patent system with over 128,000 patents granted for the years 1842–1861, around 82 per cent of world patenting. By the years 1877–1894, Germany had replaced France and the three major nations between them produced a total of some 782,000 patent applications for that period, around 50 per cent of which were granted. Clearly, the acceleration in patenting was considerable, coincident with industrialization, and monopolized by early industrialisers in particular.[10] It is also clear that alterations in regulations had immense impacts on behavior: legislative changes in France during 1844 doubled applications and trebled certificates of addition;[11] the new British legislation of 1883 which lowered the cost of patenting also nearly trebled the number of applications.[12]

By the mid-19th century it was more firmly understood in law that inventors should be liberally rewarded 'somewhat in proportion to the debt which humanity owes them, and as an encouragement to such benefactors'. But how is such a debt to be measured? In his attempts in 1852 to secure a further seven-year protection for his 14-year India Rubber patent of 1839, Nathaniel Hayward claimed that he had secured some $3,000 income from a patent worth $150,000 in itself but $1 million 'in conjunction with other patents'. That is 'the process is worth so much when employed in aid of other patented processes'. In fact, most commercial benefit had arisen from the subsequent Goodyear work on the process (for example, the patent of 1844), which had added considerably and independently to the real commercial worth of the basic process.[13] The judge refused Hayward, not on the social worth figures but on the exact role and recompense of

Hayward and the need for another seven years of patent protection. There was doubt as to Hayward's total returns from this patent. These could be estimated as including payments to him by Goodyear as well as rights and special contracts from Goodyear to manufacture rubber cloth, boots and shoes, in total 'an ample recompense for the ingenuity he displayed, estimating it by any principle whatever'.[14] The example suggests that the patent systems of the advanced nations were being fine-tuned in order to bring new technique into market economies whilst rewarding inventive and entrepreneurial efforts. This process was generally taking place in response to those critics of patents who judged that by 'tying up one idea, we stop the whole course of thoughts in a given direction, and thus interfere generally, and to an indefinite extent, with the intellectual activity of other men'.[15] If patents were to survive this common sense, then systems had to develop in order to provide a *quid pro quo* of information dispersal, in which a monopoly grant did not stop, indeed encouraged, an information search for alternatives and incremental additions or adjustments.

The most powerful of the systems that developed after 1850 was by no means the most representative, and indeed illustrated the varieties of patent systems even at the core of industrialization. By the 1850s in the United States no patents of introduction were allowed, and this may well have reduced foreign patenting there. Furthermore, patents related to acts of ingenuity applied to 'any new and useful art, machine, manufacture, or composition of matter, or any new and useful improvement of any art, machine, manufacture, or composition of matter', and not for a new principle as such, which must be 'made available in some practical form'. This provides a nice distinction between useful and reliable knowledge on the one hand and technique on the other. A hot air blast as an idea in the production of iron was part of knowledge, but it was not yet technology, which required that a machine or process be created that in turn permitted the actual and repeated operation of the hot blast. Nor was the discovery of a 'new natural substance' patent-worthy. A mere 'change in proportions' was not sufficient unless it could be proved that a 'useful result is effected'. Nor was a substitution of mechanical equivalents (for example, cog-wheels for belting) an adequate action for patenting, but new combinations that improved or cheapened processes were legitimate. New applications of existing processes for example, the use of flame to burn off loose fibers of lace, were allowable. The term useful meant only 'that a discovery serves any valuable purpose, though comparatively trivial'. Combinations of machinery could be embraced in one patent except where components were independent and might 'effect objects entirely distinct'. A caveat could be filed to protect an inventor for one year with possible renewals 'who is not a foreigner' at the stage when he is perfecting his invention.

Thus the payment of $20 prevented anyone getting a patent, but a caveat must 'disclose the mode of effecting the object' to such a degree that the patent examiner could see if a later patent by another might infringe the potential patent, and that the potential patent when realized did in fact accord with the caveat. In the American case, differential fees were used to inhibit foreign entry. For example, caveats could be so used, whilst patent applications for United States citizens or resident foreigners (of at least one year and who swore an intention of becoming citizens) cost $30, the identical application from a United Kingdom citizen cost $500, an application from any other foreigner $300. There was clearly a huge negative sanction on foreigners, especially the best-practice British engineering community, hence the small proportion of United States patents lodged by foreigners.[16]

The United States system also departed from others through its examination for novelty. New legislation in the 1830s imposed novelty examinations which demanded a library of scientific works, with each application requiring an exhaustive comparison to caveats already filed, to other pending applications, to patents issued in the United States and overseas, indeed to 'the published inventions of the whole world'.[17] This was to be accomplished within an annual budget of $1,500 in 1836. Such examinations required 'scientific qualifications, an acquaintance with foreign languages, and the attainment of an extensive knowledge of the whole range of invention and discovery in Europe and America. These are possessed by but few.' Given that the Patent Office was also giving information on shaping specifications etc. to avoid trespass, then the provision by law for only two examining clerks was clearly a check on the efficient application of the legislation.[18] Other nations avoided the issue and so the expense, but this might also have meant that on registration United States patents had automatically captured a 'presumptive right to its claims, and immediately a value in the market', whilst in Britain and elsewhere 'the patent is utterly without value until the patentee has, by litigation in the courts, established the fact' of novelty or usefulness, the cost and difficulty of which might well have put a modest inventor 'entirely at the mercy of capitalists'.[19] Only with the 1902 Patent Act did the United Kingdom introduce novelty examinations via the searching by 260 examiners of over 50 years of previous specifications using the 1,022 volumes of abridgements of specifications in 146 classes, going all the way back to number 1 of 1617 for 'engraving and printing maps, plans etc.'.[20] Several important patenting nations, such as Prussia, France, Austria and Belgium, attempted preliminary examinations for usefulness or novelty but abandoned them as too costly.[21]

The 1850–1870 years were those of the great patent debate throughout

Europe.[22] Interesting positions were taken by major figures. Thus Bismark, as Secretary to the Confederation of German States, addressed the North German Federal Parliament in December 1868 to the effect that patents should be abolished throughout the then *Zollverein* in order to 'free industrial pursuits from all artificial restrictions adherent to them'. In this judgement Bismark had been persuaded by general Prussian attitudes, but particularly the opinions of the influential Prussian Technical Committee for Industry in its reports of 1853.[23] Out of such debates came the re-specification of a range of theories justifying patents, from notions of natural law (the inventor has a natural right to full possession of his ideas), monopoly reward (proper remuneration for inventive effort of use to the community[24]), monopoly incentive (protection as an instrument for promotion of technical progress) and ideas of exchange-for-secrets (patents impel disclosure). All these were more or less held as valid for the rest of the machinofacture era. Tensions remained between the commercial need for the protection of one's own information as against one's self-same need of the useful information of competitors or other experts, mostly settled by the 20th century emergence of automatic disclosure after 18 months from first application – a system resisted in the United States but exploited by all commercial sectors in Japan and Germany and elsewhere. The market test has seemed to suggest that patent systems have been frequently viewed by users as much as systems of information (where the focus is directly upon the 'other' in order ultimately to benefit the self) than as systems of protection (where the focus is directly upon the self).

Even the most confused history of intellectual property shows clearly enough that by the 19th century national systems were enormously varied in formal rules, and even more varied in the manner, the sites and the agencies with which they managed or manipulated such rules.[25] Before the Paris International Convention of 1883[26] there was no international system in operation, though the rules, if not the ethos, of most systems by that date followed more or less on the grounds of the major systems as established by Britain, France and the United States. In the case of colonies and dependencies, with few exceptions the institutionalized rules were dictated by the colonial center, an arrangement that was just one small part of a massive global movement whereby the 20th century was to become clearly bifurcated between the rich minority and the poor majority of the nations of the world. Most of the nations that joined the IPR system after 1918 followed rules instigated by the small rich pale of industrializing nations in approximately 1870–1914. No nation connected via trade or investment to leading industrializers could escape the world of IPRs. Even the infamous instance of the Netherlands, which abandoned any patent legislation between 1869 and 1912, was no exception, and fails

Table 17.1 National patent systems and the international convention of 1883

Nation	Year of First Patent Law[27]	Year of Accession to 1883 Paris Treaty
Great Britain	1623	1884
United States	1790	1887
France	1791	1884
Netherlands	1809	1884
Austria	1810	1909
Russia	1812	non-member
Sweden	1819	1885
Spain	1826	1884
Brazil	1830	1884
Mexico	1832	1903
Chile	1840	non-member
Portugal	1852	1884
Belgium	1854	1884
India	1859	non-member
Italy	1859	1884
Argentina	1864	non-member
New Zealand	1865	1891
Canada	1869	1923
Germany	1877	1903
Luxembourg	1880	1922
Turkey	1880	1925
Norway	1885	1885
Japan	1885	1899
Switzerland	1888	1884
Tunis	1888	1884
Denmark	1894	1894
Hungary	1894	1909
Finland	1898	1921
Australia	1903	1907

to make a proper case for successful industrialization in the absence of patenting – technical change there in those years was surely influenced by the 4,535 Netherlands registrations of the period 1817–1869 and by the continuation after that time of patent lodgments by nationals into the patent systems of other industrial nations.[28]

As Table 17.1 above suggests, the introduction of the International Convention had very limited immediate effects on reducing the huge degree of variation between nations.[29]

There is little or no evidence of 'an international IPR system' in this table. The Netherlands, with no national patent regulations between 1869 and 1912, nevertheless was an early signatory to the international convention; nations without specially designated laws until later in the 1880s nevertheless signed in 1883–1884; major systems such as those of Germany and Austria did not sign until the 1900s! Signing of the treaty had no implications for substantial convergence of national systems. Most nations issued patents on the basis of purely formal requirements and without substantive legal or technical investigations or penalties. In some countries notions of novelty were based on global originality; in others on national originality – so that imported patents novel to the nation of importation were legitimate. Searches for novelty varied greatly also, extending from prior domestic patents to prior patents of a select number of or all countries, to the technical literature of other countries. Methods of opposing patents pre- or post- registration were very varied. The institutions of provision, challenge and sanction varied from patent offices to privy councils. So, a mechanism for easing the task of patenting and protecting in more than one country was not in fact in place during the time of the great heyday of the system, when patents undoubtedly acted as prime systems of both information and technology transfer between industries and nations.

Some of the resultant effects were of central importance in the history of industrial development and may only be summarized here. First, the degree to which major systems encouraged applications and registrations from foreign nationals varied greatly. Second, information provisions associated with patent institutions also varied. Third, there was also variation in the manner in which late-developers such as Germany or Japan or Russia attempted to benefit from patents in order to promote their own programs of industrial modernization. We may hazard the interim conclusion that until 1914–1918 patent systems fostered technical innovation and industrialization in a relatively small group of Atlantic-based nations. Japan was the only system to join this group and appears to have done so under the auspices of a very particular usage of intellectual property organization.[30] Whatever the differences between advanced systems, they appear to have shared certain institutional and technical imperatives relating to the role of engineers and tacit knowledge, the circulation and continued re-specification of useful and reliable knowledge ancillary and in proximity to the process of invention and its registration, the systematic incrementalism of both metal- and chemical-based techniques and the experimental and knowledge systems upon which they were based. It was this metal-bedded technique that produced the super-modern culture of patent agency, urban technical associations, professional expert systems and polytechnical training. In the industrializing systems and in the culture of machinofacture resided and developed notions of efficiency, good order, competition and individualism, elegance and simplicity and new combinations of knowledge and skill. Thus was modernity epitomized.[31]

Major technological events occurred within a stream of manufacturing improvement centered on machinofacture processes of measurement, fixing, sliding, turning and re-locating, of true flat and cylindrical surfaces and of screw threads, which together created a total system based on standardization, precision and interchangeability, production requirements which became the very symbols of social and cultural modernity. This was also the center of the great surge of industrializing modernity which occurred in the second half of the 19th century, a first climacteric which embraced much of Europe, North America and Japan, and which demanded and produced new institutional relations between technique, work and legal regimes. By 1900, British annual patenting reached 30,000 registrations, 60 per cent of the United States gross figure, but greater in *per capita* terms, and twice those of the French. This is one reasonable explanation of the dynamic underdevelopment of the 20th century. Despite the labors of many and the theories of a few, the institutions of 'planning' and the panoply of foreign aid, industrialization did nor arise at new points on the globe as long as machinofacture remained the dominant technical mode. The second climacteric after 1971, ushering in the end of the abnormality of modernity, awaited the shift from machinofacture into electronics, biotechnologies, new transport and communication systems and their associated new industries.

Machinofacture gave industrial modernity its professionalism, expertise, scientism and arrogant urbanism, and thus too produced a deep embedment of machine technologies at little cost to government but of profound importance to all systems outside the pale of success. If in the 18th century Euro-American supremacy was based on technological breakthrough, in the 19th it was maintained through the barrier effects of the national cultures of machinofacture. Nations lying outside this machinofacture complex were at a fatal disadvantage, and the last nation to breach this dualism of structures and experiences was Japan, which did so only at tremendous cost and through a unique national effort.[32]

The second industrial revolution period was associated with both a massive growth of patenting and a material and cultural bifurcation between developed industrial/industrializing nations on one hand and the majority group of non-industrializing nations on the other. The existence of national patent regulations in places such as Brazil, Mexico, Chile and Argentina, or India and Korea, does not seem to have aided industrial development significantly, although in dependencies such as Canada and Australia patents do seem to have conferred sustained technical and industrial benefits. Our argument here is that the social, legal and scientific infrastructures in later developers were not such as to support the complex tacit and formal knowledge structures required in order to enter the world of intellectual property rights at a level bearing even a family resemblance to that of the predominant industrial nations. The late industrializers of the pre-1914 years in the main failed to transfer technologies or capture them for their own industrial purposes, and such countries included the majority of the nations entered in Table 17.1 above. Nations that surmounted this problem were either colonized dominions of relatively small population, nations close to the cultures, trade and scientific base of the euro-industrializers (such as Tsarist Russia), together with the singular example of Japan which succeeded through utilizing patents within a massive, planned program of industrial development and technology transfer that incorporated a host of new socio-legal institutions. Thereby Japan substituted for missing legal and technical elements, whilst regions of 'white settlement' incorporated essential features of euro-industrialization within predominantly expatriate micro-cultures and sites.[33]

PATENTS AND DIFFERENTIAL EXPERIENCES IN THE SHORT 20TH CENTURY (1918–1971)

Although many tried and several (for example, Argentina) succeeded for a short time, no nation joined the winners' club between circa 1918 and circa 1971. The reasons for this have been hotly debated and clearly relate to matters of trade, investment and governance. Amongst this wide array of elements resides the prevailing systems of intellectual property rights, which did not much alter in their substance from those established by the small dominant group just prior to 1914. This refusal to alter the substantive character of IPR systems was one of the most salient failures of the short 20th century, as was at that time emphasized by such critics as Patel and Vaitsos. Surandra Patel concluded that over 90 per cent of foreign-owned patents in underdeveloped nations were never used in production processes in those nations.[34] On the basis of these sorts of very strong trends, Vaitsos stated that 'practically the totality of patents granted in developing countries are foreign owned, and their major function is clearly not to encourage inventive activity (domestic or even world wide) but to assist the profit maximisation of large transanational corporations'.[35]

Patents could be used to secure a monopoly position in third world markets and thereby to dampen severely any positive relationship between imported patented technology and indigenous technological capability.

Methods of technological blockage involved owners of patents pursuing incremental R&D to create improvements on the original, allowing its extension or the lodgement of further related patents. Owners of patents commonly erected huge production and marketing facilities during the period of protection, such that the enterprise would have a considerable advantage in terms of market information and contacts even when patent protection nominally ceased. Again, knowledge of the specific content of a foreign patent amongst third world agents did not imply its usage or the replication of the related industrial process. Replication normally required more knowledge than was located in the patent itself and, further, often expensive R&D activity.[36] Finally, the licensing of patented knowledge usually came with highly restrictive conditions or 'ties', and was overpriced. Materials and equipment required to manufacture and priced by the licensor would be purchased from the licensor, and there were frequently limitations on the sale of the product to the domestic market only. 'Grant-back' provisions ensured that any benefits from improvements generated during the operation of the process by the licensee went to the licensor.

Because of such factors, any estimate of the cost of technology transfer by patenting during the short 20th century is bound to be flawed. Host governments certainly attempted to increase the benefits of foreign patenting by insisting that patented knowledge moved into the public domain if the owner enterprise failed to use or license the technology. India and other large nations imposed limits on royalty levels. Nevertheless, such legislation was difficult to implement and could not address all the problems associated with patents as a mechanism of technology transfer. If international patenting was by individuals or small firms most of such problems could be greatly reduced. It was the association of patenting with large transnationals that made it so difficult for host nations to achieve a bargaining position in which they could optimize the net productive effect of patenting activity.

So, international patenting was nominally a system of technology transfer, but for large, poor nations tended to operate so as to block effective transfers. Thus in India during 1947–1961 some 42,000 patents were lodged, around 85 per cent by foreigners. At that time this compared with foreign patenting ratios in the major industrial nations lying at around 47 per cent for the United Kingdom, 32 per cent for West Germany, 24 per cent for Japan, and only 20 per cent for the technologically dominant United States. But in India, this vast surge of foreign activity took place 'with the main object of protecting the export market from competition by rival manufacturers, particularly those in other parts of the world'.[37] Licensing agreements to local Indian firms were dominated by exorbitant royalties and a series of highly restrictive conditions that often demanded that licensors provide raw materials and skilled workers and controlled the final price of outputs.[38] Such restrictions served to reduce the learning process and the employment benefits that were commonly associated with the use of industrial technologies in the home of those technologies, the advanced nations of the world.

With reference to Latin America, Sunkel has written that 'for a techno-scientific infrastructure to influence the process of development there must be close institutional links with the productive structures of society as well as with the potential and administrative structures that control the decisions in the field of development policy. The lack of these links, rather than the actual amount of research, is the main characteristic of technical and scientific underdevelopment.'[39] Yet, despite Sunkel's warning it was principally 'the amount of science', particularly of 'big' or prestige science, that so engaged the western commentators and the policy analysts of the 1960s and 1970s. Estimates of R&D expenditure for the middle of the 20th century showed that underdeveloped nations fared very badly and that population size was nowhere near as important as income levels and economic structure in determining levels of R&D expenditure.[40]

During those years there was no concensus on the identification of those transfer mechanisms most likely to encourage a strong linkage between transferred techniques and indigenous technological systems. The array of transfer mechanisms included the *licensing* of patents and trademarks, the *importation* of machinery and equipment, turnkey plants and other forms of international industrial cooperation, as well as various illegal means of transfer. In *turnkey projects* foreign firms supplied total production systems, which often involved all design facilities, installation works, personnel training and the initiation of production itself. At the other extreme, transfer was effected in such a way that the 'donor' was unaware of the scale or significance of what was happening, as with small legal purchases of machinery for *reverse engineering* (where the analysis of the technical features of a process enables its receiver/user to reproduce the machine/product) or with industrial espionage. Transfer mechanisms also included all institutions or organizations involved in the collection, translation and dissemination of production-specific scientific or technical information, scientific exchange programmes, 'search' *bureaux* (often set up as locally chartered companies in foreign nations and designed to purchase and examine advanced foreign techniques in their country of origin), and the use of *intermediary firms* (for example, firms which would obtain valid export licenses for an apparent end-user in an authorized country of destination but then reship the goods to another, final-user nation).

On the face of it, it would seem that different transfer mechanisms

Table 17.2 United States corporate and individual patenting (patents issued) 1956–1980 (in thousands)

	1956–1960	1961–1965	1966–1970	1971–1975	1975–1980
All Patents Issued	237.5	260.0	325.1	375.5	312.3
Individual Issues	76.6	69.9	73.4	87.0	65.0
Corporate Issues:					
US	133.9	152.5	190.6	192.9	150.4
Foreign	21.0	31.0	52.8	86.0	90.4
US Government	5.9	6.5	8.5	9.6	6.5

would have different but predictable impacts on indigenous technological systems. Licensing allowed the licensee to obtain by purchase the rights to make and market specific products incorporating inventions and processes developed by the licensor. As such, importing tended to transfer products rather than methods. Foreign direct investment through transnational corporations was often seen as the most viable mechanism of transfer. Either through affiliates or as joint ventures, transfers via the corporation could embrace long-term, frequent, repeated and incremental flows of firm-specific as well as general 'know-how' technology, and training and exchange programmes. Corporate activity might be associated with major turnkey projects which themselves could lead to technical cooperation agreements between receivers and users of technology for the further servicing of equipment, for updating of the technology initially purchased or for marketing services.

Corporations soon dominated the world of patenting, and so international patent lodgements. Of 343,000 patents issued to corporations in the United States between 1939 and 1955 just less than one-third went to the 176 largest United States corporations, less than two-thirds to other United States corporations, and some 7 per cent to foreign corporations. By 1955, 59 per cent of all patents lodged in the United States derived from corporate research, 39 per cent stemmed from individuals, 2 per cent from government.[41] At that time the biggest patentees were the research-based firms such as General Electric, ATT, Westinghouse and Du Pont.[42] Table 17.2 provides an outline of the pattern of United States patent issuing since that time and shows the increasing importance of foreign patentees.[43] The process of technology transfer was often entirely internalized by the transnational corporation. In many cases, licensing of patented knowledge could not transfer all information, and there existed uncertainty about the value of yet 'unknown' knowledge amongst receiver or buyer firms. This led to agreements which failed to return the full value to the licensor, especially if the licensor was operating in a high-income market. So it has been argued that such failures in knowledge/technology markets led to the 'internalisation' of the transfer process.[44] Market transactions were by-passed through the development of sophisticated technology packages, which drew not only upon the physical technology of the transnational corporation but 'upon their organisational and managerial abilities, their easy access to capital, and their extensive marketing skills'.[45] Firms engaged in especially lengthy R&D programs and in the 'sunrise' industries of the period would, in this model, fail to use trade and licensing as a transfer mechanism (involving high transaction costs) and would rather develop an overall strategy of 'taking over various market transactions and internalising information and knowledge-related activities within the corporate structure'.[46] Could this sort of phenomenon benefit underdeveloped economies?

Transnational corporations wielded patent rights as a weapon in their expansionist strategies. Ownership of property rights was commonly used as a defensive strategy against potential competitors. These included both other transnational corporations and indigenous enterprises operating within developing regions or newly industrializing countries. Patent registration in an underdeveloped nation could be used to block emulation by indigenous enterprise even where such production technology was not used by the owner. Patent-using firms within the indigenous system could be locked out of admission to ancillary know-how, fail to compete and be taken over by the patent-owning transnational.[47]

However sophisticated science policy regimes may have become in industrialized systems, in the underdeveloped nations of the years approximately 1918–1970 science policy was firmly geared to technology and industrialization under the guidance of a development paradigm which centred on technology transfer, large projects, skilled manpower and high levels of investment, all designed to work in combination to shock low-level-equilibrium systems into becoming dynamic industrializers. The contemporary 'choice of technique' literature presumed that there was an array of techniques for major industries or sectors which existed in the world economy, and that this could be selected from and transferred in order to fulfil the requirements of development strategies in poor nations. The predominate Lewis model really presented only two alternatives – a capital-intensive, fixed technique in industry and a labour-intensive, variable technique in agriculture. In addition, planning for science policy was further constrained by the notion that choice was only of the physical character of technique, not its organizational imperatives, nor in its susceptibility to alteration. The alternative 'appropriate technology' schemas tended to ignore the large amounts of capital-intensive technique already established in third-world manufacturing, and the need to maintain massive existing infrastructures for transport, ports and harbors and so on, and also ignored all dynamic effects which can result from a mix of techniques organized in the form of a bundle of development projects. The poor nations and their nascent science policy programs were often caught amidst contradictory potentials, ranging from craft traditions to the newest of nuclear energy sources. It seemed sensible enough that if technologies were to be learned and absorbed they should be amongst the most modern, that the cost per unit of production of low technology might rise due to lower productivity, that older equipment required more maintenance, that high technologies might boost existing capacities, and that labor-using techniques might employ smaller amounts of capital very inefficiently. It could also be argued that development projects which embraced both small scale indigenous technologies and transferred high technologies, might be the ideal environments in which to nurture intermediate technologies.

SINCE 1971 – PATENT SYSTEMS IN THE SECOND CLIMACTERIC

At the end of the machinofacture era, Jack Baranson could still hold out much hope that 'technology provides a means of transition and adjustment' in an essay that took little account of prevailing institutions.[48] Recognizing fully the problems of inappropriate technique, he did not wrestle with the more trenchant difficulty of inappropriate institutions. Thus attitudes might inhibit, and public policies prohibit, the efficient application of superior technique drawn from the machinofacture world. Although Baranson pointed to something called 'technological elasticity' (the adaptive capacity of any system) he seems not to have considered this in relation to the institutions of transfer and prescription, at the heart of which lay the dominant patent systems and their associated sub-systems of machine shops and sub-contracting. Western patent systems nested safely within such industrial complexes, nurtured them and in turn benefited from them. Because Birminghams did not multiply, outsiders did not access such systems, and transfer of technique required the purchase and assimilation of a mass of inputs if purchased technique was to be rendered usefully elastic.

The world economy underwent significant, climacteric transformation during 1970–1973, the years when Japanese trade and technology began to challenge the position of the United States, when yen revaluation began to cheapen foreign assets in the minds of Japanese investors and initiated the surge of new international investing that peaked during the mid-1980s, and when the group of Organization of Petroleum Exporting Countries (OPEC) nations first exerted their commercial and environmental power at a global level. From then also, and for the first time on such a scale, a number of East Asian nations undertook a series of institutional and political reforms (to be followed in China in the later 1970s and beyond) which acted amongst the necessary conditions for the full sovereign capture of Western technologies and investments, breaking the older paradigm of the Golden Age.

Once more the transition surrounded a group of new technologies as the methods of Henry Ford were replaced by those of Bill Gates. Microelectronic fusions and genetic manipulations have now challenged the automation and office technologies of the earlier Golden Age. Many of the new technologies of the second climacteric which were centred in the older core sites such as IBM, Bell Labs, Texas Instruments, and Western research teams were most closely associated with major core breakthroughs, such as monoclonal antibodies or fiber optics. But applications and commercialization came increasingly from such companies as NEC, Hitachi and Fujitsu. In particular, whereas private companies dominated Western core breakthroughs, public policy in combination with private agency was important to Japanese and Asian transfers, improvements and applications – the Fifth Generation computer program, the combined enterprises project to develop the 64 megabyte D-RAM memory chip, and the establishment of the MITI Laboratory for International Fuzzy Engineering.

This new Climacteric has seen fundamental alterations in the commanding imperatives of technology. That is, the move away from large fixed investments, huge infrastructures, and the raw material and energy dependency of the old machinofacture, and towards highly specific knowledge, expert systems and networks, and low usage of raw materials and labor, meaning that industrial advance has become less site-specific and less dependent on older, Western-based institutional frameworks, including patent systems. This yielded the prospect of a spread of the East Asian edge of advancement on a much wider front, wherein Western-led national frameworks and institutions might become less important than regional and international sites and institutions in dictating the patterns of future technological advancement. In this view the present and continuing efforts on the part of leading Western corporations to embrace every living creature within the maw of intellectual property, and of leading Western nations to force a harmonization of patent systems (since the Uruguay Round of GATT in 1994) are doomed to either immediate failure or ultimate irrelevance.

In reporting the views of a World Intellectual Property Organization (WIPO) congress in Moscow 1974, Kolle states that there was

> no doubt that the most extensive, important and qualified source of information is the technical knowledge contained in the patent system. It includes at present over sixteen million patent specifications and is growing yearly by several hundred thousand. That this literature contains information concerning the technological achievements of the last 150 years is not its only significance. No less important is the fact that it constitutes a source of usually readily available information [that] lends itself especially to a systematic order, the retrieval of information contained therein and multiple technical, economic, and legal analyses as well as trend studies of future technological developments.[49]

That is, the patent system remains a huge locus for the production, assessment and expansion of useful and reliable knowledge. Indeed, Beier and Straus[50] have gone so far as to argue that there is now a prime problem of too much diffused information in patent systems, with Germany from the early 1970s beginning its Programme for the Improvement of Information and Documentation, followed by similar plans in both the United States and the European Community (for example, Euronet).

However, at the present time such hugeness and complexity forces out smaller and less powerful users from the information system, and *en masse* does the same for the poor and developing economies, so there is a strong and coherent call globally for some centralized information agency stemming from patent systems. A cheap information system for third world nations would allow both better technology transfer and more powerful negotiating positions, for through such they could 'ascertain what protected technologies exist, what alternatives are available in particular cases and also where they can be found; the procurement and the development of technologies by adaptation, improvement and enlargement is thereby facilitated'.[51]

Present laws seem quite different from earlier ones.[52] The notion of an optimal patent regime is usually couched in terms of the degree of protection of useful intellectual property, and in this there is usually some failure to distinguish between diffusion of knowledge and diffusion of use of knowledge. There exists often the simple assumption that patent monopoly is detrimental to diffusion of technique because it is detrimental to diffusion of usage within the fairly long period of protection. But this is so only in the sense of recipe-copying – more realistically, a patent lodged is a knowledge asset available to competitors, from which they may launch new patents which modify the originals. Thus the conclusion of public policy economists that property rights in knowledge assets should be minimized is not necessarily sound, for this assumes that monopolized knowledge is not available for such incremental and circumnavigational purposes, and comes up against the economistic argument that weak rights may curtail risks in the leadership of innovation activity. Accepting that patent systems are useful knowledge systems modifies such perspectives and dampens the distinctions between diffusionist/static (weak patents) and exclusionist/dynamic (strong patents) arguments.

We may note that during the 19th century there were quite frequent discussions of the introduction of some equivalent of a utility model in Britain, an instance being the well-known patent agent J. Imray, who advocated a shorter period, cheaper style of patent 'for semi-inventions which could be put into work at once and a profit made'.[53] In Japan the utility model, or *jitsuyo shin-an*, was an early feature of the system of intellectual property rights, and 'any new industrial form relating to the configuration, construction, or combination of articles, may be registered as a UM', so they were always tangible designs, shapes or artefacts relating to products and to their form, not to processes.

CONCLUSIONS – HISTORY AND REFORM

Using a combination of analytical institutional narrative and descriptive historical statistics this chapter has argued for some basics. Until the 1970s IPRs were led by a small group of industrial nations originally established as aggressive leaders in the years circa 1850–1914. This initial system permitted late entry at high cost, success being associated with national, public sector programs designed to increase capabilities and accelerate transfers in. In the subsequent period to the 1970s there was a relative fall of the state and a rise of the corporation in determining the operations of the national and international IPR systems. During these machinofacture years the predominant national patent systems were characterized by great variability, a relative simplicity of definition of the objects and subjects of patenting, a frequency of technology and information transfer between them, a great predominance of incremental innovation associated with the mass employment of technical and legal expertise operating across a wide range of courts, and a common understanding of the social outcomes of limited technical monopolies. Because of the institutional complexity of patent systems they worked so as to encourage technological interactions amongst industrial nations whilst increasingly excluding new entrants, especially those from what were then termed third world countries. If in the years prior to 1914 patent systems were somewhat in advance of the other institutions of industrial capitalism, during the short 20th century (1918–1971) the working of patents increasingly reflected the normal workings of the global economic system, and were thus merely an exaggerated feature of the global bifurcation of material experience which dominated and marred that period. One characteristic within this was related to the exact locations of useful and reliable knowledge (URK). Advanced patent systems brought such URK not only into close spatial and temporal proximity to points of innovation, but also into close *social and cognitive proximity* to such sites. This was not so for poor nations – the problem of the world was not merely the mal-distribution of knowledge creation, nor the spatial biases of new and lucrative URK, for of equal importance was the imbalance in social and cognitive locations. Given that information and knowledge generation, diffusion, re-affirmation and application were all functions of patent systems, then it might be concluded that after 1918 IPR regimes contributed to the institutional and knowledge barriers faced by underdeveloped economies in their efforts to escape relative technological backwardness.

Within this macro-regime of institutions the early late-developers such as Germany and Japan gained a successful entry through the use of deliberate strategies. In the present chapter we have illustrated the seemingly important roles of caveats, automatic disclosure and utility models in permitting knowledge and technology transfers for late-developing economies. More broadly, the historical material may illustrate the value of distinguishing between institutions and organizations when considering the direction of patent reform in the future. If we follow Douglass North[54] and define institutions as the rules of the game, and organizations as the players in and the playing of such games, then we may summarize that late-comers may design strategies which optimize their outcomes within an imperfection of institutions.[55] Institutions are imperfect in so far as they allow organizations to manipulate and at times traduce institutional rules. Within an international IPR system we may think of the global rules as the institution, and the players within particular nations as the wider organization. The more imperfect the institution the better the chances are that peripheral players will find niches of advantage and low-cost entry, such as utility models or leakage of knowledge, or be able to derive some advantage from penalties for non-usage. Historically, the exploitation of the imperfection of institutions is the way in which late-comers into IPR systems succeeded, and this may yet remain a feasible strategy in a world where intellectual property rights are commanded, not by benign agency attempting to optimize benefits to the widest possible extent, but by powerful, competitive players determined to maximize their own immediate commercial outcomes.[56]

The present post-TRIPS international environment is one dominated by the notion of tightening up of property rights, creating a global uniformity, and perfecting the international intellectual property institutions within the optimizing perspectives of the major players in North America, Europe and Japan. Thus the historical 'wiggle room', what this chapter has referred to as a window of opportunity opened by the imperfection of institutions, is to disappear.[57] Any such attempt to close those windows of opportunity that have historically been widened by the confusion, variability and imperfection of national and international patent regimes are likely also to close out potentials for late development in the 21st century. This is especially true given the re-entry of the state into the global decision-making system and the growing importance of trade and foreign investments in determining national technological capabilities amongst developing nations.[58] A historian's conclusion might well be that the best outcome and prospect for late-development would be a global reform which left IPR systems imperfect as systems of rights but efficient as systems of information flow, combined with a liberalization of those protective trade and investment regimes that now make it so very difficult for late-developing systems to increase their own technological capabilities.

NOTES

1. S.B. Brush and D. Stabinsky, 1996. *Valuing Local Knowledge. Indigenous People and Intellectual Property Rights,* Washington, DC: Island Press.
2. For the historical context and detailed examples see Ian Inkster, 1996. 'Technology Transfer and Industrial Transformation: an Interpretation of the Pattern of Economic Development circa 1870–1914', in Robert Fox (ed.), *Technological Change. Methods and Themes in the History of Technology,* London: Harwood Academic Publishers, pp. 177–201: idem., 1999. 'Technology Transfer in the Great Climacteric. Machinofacture and International Patenting in World Development circa 1850–1914', *History of Technology,* 21, 87–106.
3. Here we consider technological capability only in terms of internal knowledge, technical and institutional support, often promoted primarily by government. We do not consider the more contemporary external elements in limiting technological capability, such as trade, foreign investments and debts, often created and maintained by the private sector. See however, K. Maskus 2004. 'Patent Rights and the Form of Technology Transfer', available at spot.colorado.edu/maskus/papers/MSP-paper_6-04.doc. It should be noted that late developers of the 19th century utilized foreign resources such as skills and machinery in developing capacity as part and parcel of public sector industrialization strategies employing private agents: see for example 2001. Inkster, I. 2001. *Japanese Industrial Economy: Late Development and Cultural Causation.* New York: Routledge, chs 1 and 4.
4. Seeming contradictions to this particular euro-centrism weaken on examination. So, W.P. Alford, 1995. *To Steal a Book is an Elegant Offense. Intellectual Property Law in Chinese Civilization,* Stanford, CA: Stanford University Press, finds little functional equivalence between the early Chinese systems of statist recording and registration and the European systems of patenting.
5. Earlier Venice had granted awards (in 1332) for bringing new knowledge from other

places (windmills in this case), but there seems to have been no general patent law until that of 1474.

6. W.H. Price, 1906. *The English Patents of Monopoly*, Boston, MA: Houghton, Mifflin and Co.
7. H. Pohlmann, 1961. 'The Inventor's Right in Early German Law', *Journal of the Patent Office Society*, XLIII, no.2, pp. 121–139.
8. This fails to count in particular the surge of patenting in the individual German states during the years prior to 1877, amounting to 5,193 in the years 1842–1861 alone: see Ian Inkster, 1991. *Science and Technology in History. An Approach to Industrial Development*, London: Macmillan, pp. 162–164.
9. For examples see Ian Inkster, 1990. 'Mental Capital: Transfers of Knowledge and Technique in 18th Century Europe', *Journal of European Economic History*, 19, no. 4, pp. 403–441; John Harris, 1992. *Essays in Industry and Technology in the Eighteenth Century*, London: Variorum, and Doron S. Ben-Atar, 2004. *Trade Secrets: Intellectual Piracy and the Origins of American Industrial Power*, New Haven, CO: Yale University Press. In fact, the last is strong on borrowing but weaker on the notion of piracy. For a more convincing account of the United States history see Catherine Fisk, 2001. 'Working Knowledge: Trade Secrets, Restrictive Covenants in Employment, and the Rise of Corporate Intellectual Property, 1800–1920', *Hastings Law Journal* 52, pp. 441–535.
10. *Report of the Commissioners Relating to Letters Patent for Invention*, London: House of Commons and House of Lords, 1865, table p. 153; *Twelfth Report of the Controller-General of Patents, Designs and Trade Marks, with Appendixes for the Year 1894*, London: HMSO, 1895; *Journal of the Franklin Institute*, vols. 109 p. 309 and 113 p. 61.
11. The French law of 15 July 1844 increased 15-year patenting and brought the total of all French patenting, including foreign and certificates of addition, to some 52,000 in the years 1845–1858. See 'Numerical List of Patents of Invention and Certificates of Addition in France 1845–1859', *Commissioner of Patents Journal* (United Kingdom), VII, 1860, p. 70.
12. *Industries*, 13 August 1886, pp. 157–158. Applications rose from just over 6,000 in 1882 to over 16,000 in 1885, and readership in the improved patent office library almost doubled. The Act reduced provisional protection fees from £5 to £1, and the first payment on registration from £20 to £3.
13. Utilizing the expert services of such British patent agents as Moses Poole and William Newton, between 1844 and 1855 Goodyear lodged a series of patents that were designed both to improve the basic chemical-heat process and apply it to a huge variety of products, ranging from small objects of art to coating of metals and insulation. For examples see *Abridgements of the Specifications Relating to the Preparation of India-Rubber*, Great Britain Patent Office (1875), London: George E. Eyre and William Spottiswoode.
14. For a full account of the case see *Report of the Commissioner of Patents for the year 1852: Agriculture*, Washington, DC: United States Senate, 32nd Congress, Robert Armstrong, 1853, pp. 469–484. The claim was complicated by his having assigned his interest in this patent to Charles Goodyear for the 14-year period at the point of filing the application, hence the latter becoming in law the patentee. In Britain, however, the outcome for Hayward might have been more positive as a judge could exercise discretion in the terms of extension, in the United States this was fixed at a lengthy 7 years.
15. J. Stirling, 'Patent Rights, A Paper to the Glasgow Chamber of Commerce', reported in Macfie, *Recent Discussions on the Abolition of Patents for Inventions. Evidence, Speeches and Papers*, London: Longman, 1869, op. cit. p. 119.
16. *Report of the Commissioner of Patents for the year 1852*. Supra n. 14, pp. 447–448, 466, 469–484. By 1866 patents granted to United States citizens numbered 6,428 annually, to British subjects 82, and to French 40: *Annual Report from Commissioner of Patents for 1865* vol. 1 Washington, DC: Government Printing Office, 1867. p. 4.
17. *Patent Office Report from Commissioner of Patents for 1837*, 25th Congress, Doc. 112, Washington, DC: Thomas Allen, 1838, quotation at p. 3.

18. Ibid, p. 3.
19. *Annual Report of the Commissioner of Patents for 1871*, vol. 1, Washington, DC: United States Patent Office, 1872, quotations at p. 18. Available at http://www.archive.org/details/annualreportcom19offigoog.
20. Anne-Marie Mooney Cotter (ed.). 2003. *Intellectual Property Law*, London: Cavendish Publishing, p. 5.
21. It was also argued that the true examiner was the patentee and the responsibility also his, this especially so in Britain where relevant technical information was abundantly available: see *Report of the Commissioners Relating to Letters Patent for Invention*, Supra n. 10, p. 31.
22. Fritz Machlup and Edith Penrose, 1950. 'The Patent Controversy in the Nineteenth Century', *Journal of Economic History*, 10, pp. 1–29.
23. 'Bismark's Speech to the North German Federal Parliament, Berlin, 10 December, 1868', reported in (W. Armstrong, et al.) *Recent Discussions on the Abolition of Patents for Inventions. Evidence, Speeches and Papers*, London: Longman, 1869, pp. 185–197.
24. This following a classic Enlightenment position, as expressed in Goethe's notion of the inventor as *Lehrer der Nation* (teacher of the country), and where the emphasis is on usefulness.
25. By the early 1860s some 51 countries or dependencies granted patents in some form, as listed in December 1862 evidence of L. Edmunds, *Report of the Commissioners Appointed to Inquire into Laws relating to Letters Patent for Invention, Presented before Both Houses of Parliament*, London: HMSO, 1865, questions 576–577, pp. 28–30.
26. The International Convention for the Protection of Industrial Property concerned all forms of industrial property and was signed in Paris in March 1883 by 11 countries. The main aim was to prevent discrimination in any Convention country against the nationals of any other Convention country, and in patents to provide a 7-month period of grace for applicants to apply for patents in any Convention country, so that the application date would be that of the first application. A revision was provided in an Act of 1901 (Brussels) which extended this period to 12 months.
27. This relates to formal patent legislation at a national level. Thus Belgium registered patents prior to 1854 but under French and Dutch laws; the individual Australian colonies had thriving patent systems from the 1850s; prior to 1869 Lower and Upper Canada had separate patent laws from 1823 and 1826 respectively.
28. G. Doorman, 1947. *Het Nederlandsch Octrooiwezen en de Techniek der 19e Eeuw*, The Hague: Martinus Nijhoff.
29. 'Historical Patent Statistics 1791–1961', *Journal of (United Kingdom) Patent Office Society*, XLVI, February 1964, pp. 89–171. The Paris Convention was amended by protocols of Madrid in 1891 and the Act at Brussels 1900; then the 1911 Washington Convention saw further amendments, and this was effective in all member countries by May 1913. Initial signatories to the amended convention were Austria, Ceylon, Dominican Republic, France, Great Britain, Germany, Hungary, Italy, Japan, Mexico, Norway, Netherlands, New Zealand, Portugal, Spain, Norway, Switzerland, Tunis, Trinidad, Tobago, United States. By a congress of 1910 in Buenos Aires the idea extended to an agreement of 18 North and South American states, but not Canada. Mutual rights were secured between members, with no obligations of domicile. Inventors may secure the same date of award for foreign applications as in their original application. Anyone who has lodged in one member country shall have rights of priority in filing in all other nations for 12 months. The international office at Berne, the Bureau International pour la Protection de la Propriété Industrielle, acted as a clearing and information house for whole system.
30. For details of which see Ian Inkster, 2008. 'Policies, Patents and Reliable Knowledge – An Institutional Approach to the Climacteric 1850–1914', *Economies et Sociétés*: Série 'Histoire économique quantitative', 38, 461–482.
31. See for institutional and statistical detail Ian Inkster, 2004. 'Engineers as Patentees and the Cultures of Invention 1830–1914 and Beyond – The Evidence from the Patent

Data', *Quaderns d'Historia de l'Enginyeria*, VI, 25–50; I. Inkster, 2004: 'Finding Artisans. British and International Patterns of Technological Innovation 1790–1914', *Cahiers d'Histoire et de Philosophie des Sciences*, no. 52, 69–92.

32. Ian Inkster, 2001. *The Japanese Industrial Economy. Late Development and Cultural Causation*, London: Routledge.
33. For the last see Ian Inkster, 1990. 'Intellectual Dependency and the Sources of Invention: Britain and the Australian Technological System in the 19thc', *History of Technology* 12, pp. 40–64.
34. Extracted from table 1, S.J. Patel, 1974. 'The Patent System and the Third World', *World Development* 2, p. 6 and pp. 9–11.
35. C. Vaitsos, 1973. 'Patents Revisited: Their Function in Developing Countries' in C. Cooper (ed.), *Science, Technology and Development*, London: Routledge, pp. 71–98, quotation at p. 77.
36. The whole field of principles and scientific discoveries was generally precluded from patenting.
37. Ajit Roy, 1967. *A Marxist Commentary on Economic Development in India 1951–1965*, Calcutta: National Publishers, quotation at p. 25.
38. R.K. Hazari and H.G. Lakhani, 1967. 'Pharmaceutical Companies in Maharashtra, Financial Structures and Ownership', *Economic and Political Weekly*, II, 26 July, pp. 1169–1182.
39. O. Sunkel, 1971. 'Underdevelopment: The Transfer of Science and the Latin American University', *Human Relations* 24, 1–18, quotation at p. 14.
40. S. Dedijer, 1962. 'Measuring the Growth of Science', *Science* 88, pp. 781–788; S. Dedijer, 'Underdeveloped Sciences in Underdeveloped Countries', in Edward Shils (ed.), *Criteria for Scientific Development: Public Policy and National Goals*, 1968. MIT Press, table 1, p. 62; The World Bank, 1980. *World Development Report 1980*, London: World Bank, OUP, tables in appendix, 1–20.
41. P.J. Frederico, 1957. *Distribution of Patents Issued to Corporations, 1939–55*, Subcommittee on Patents, Trade Marks and Copyright of United States Senate, 84th Congress, Washington, DC.
42. Ibid., pp. 16–18.
43. *Statistical Abstract of the United States, 1982–83*, Bureau of the Census, Washington, 1983, p. 548, table 925. In a less successful industrial nation such as Britain, the presence of foreign technology was more evident. Between 1974 and 1982 over 50% of all applications for United Kingdom patents were of foreign origin. The nations of origin of most importance in deciding order were the United States, Germany and Japan, with Japan outranking Germany after 1980: see J. Phillips, 1984. 'Patents and Incentives to Invent', *Endeavour* 8, table 1 etc., pp. 90–94.
44. For TNC strategies see C.P. Kingleberger, 1969. *American Business Abroad*, Cambridge, MA: The MIT Press; R.E. Caves, 1971. 'International Corporations: The Industrial Economics of Foreign Investment', *Economica* 38, 1–27; T.G. Parry, 1984. 'International Technology Transfer: Emerging Corporate Strategies', *Prometheus*, 2, 220–232.
45. N. Rosenberg, 1962. *Inside the Black Box*, Cambridge: Cambridge University Press, quotation at p. 278.
46. Parry, *supra* n. 45, p. 221.
47. Edith Penrose, 1973. 'International Patenting and the Less-Developed Countries', *Economic Journal* XVI, 83: 768–86; As Richard Caves argued, *supra* n. 43, in most cases of licensing from TNCs 'either the information cannot be transferred independent of the entrepreneurial manpower, or uncertainty about the value of the knowledge in the foreign market will preclude agreement on the terms of a licensing agreement that will capture the full expected value of the surplus available to the licensor' (p. 7).
48. Jack Baranson, 1967. 'The Challenge of Underdevelopment', in Melvin Kranzberg and Carroll W. Pursell (eds), *Technology in Western Civilization* vol 2, London: OUP, 516–531.

49. G. Kolle and J. Straus, 1976. 'Patent Documentation and Information: Its Significance and Actual Development', *International Review of Industrial Property and Copyright Law*, 7, no. 1, 1–26.
50. F.K. Beier and J. Strauss, 1977. 'The Patent System and its Informational Function – Yesterday and Today', *International Review of Industrial Property and Copyright Law*, 8, no. 5, pp. 387–407.
51. Ibid., p. 5.
52. J.A. Ordover, 1991. 'A Patent System for both Diffusion and Exclusion', *Journal of Economic Perspectives*, 5, no. 1, 43–60. This paper does not refer to Kinmouth.
53. J. Imray in discussion of G.G.M. Hardingham's 'On the Relative Cost and Durations of British and Foreign Patents', *Transactions of Institute of Patent Agents* volume 8, 1890, London: IPA, 1889–90, pp. 74–93, quotation at p. 89.
54. Douglass C. North, 2005. *Understanding the Process of Economic Change*, Princeton, NJ: Princeton University Press, see especially pp. 59–64, 120–126.
55. Ian Inkster, 2002. 'An Imperfection of Institutions', *History of Technology*, 24, pp. xvi–xxi.
56. For a recent example discussing such issues as changes in patent regimes, entrepreneurship and software see OECD, 2004. *Patents, Innovation and Economic Performance*, Brussels: OECD.
57. For this expression see Jerome H. Reichman, 1998. 'Securing Compliance with the TRIPS Agreement after *United States v. India*', *Journal of International Economic Law*, vol. 1. no. 4, pp. 585–597.
58. The International Institute for Management Development based at Lausanne has recently attempted to measure the extent to which government policies are conducive to innovation, and their resultant government efficiency rankings place Hong Kong, Singapore, Finland, Denmark and Australia at the top, with the United States, China and the United Kingdom in middling positions, and France, Russia and Italy in low positions. For summary see European Commission 'Governments set the Framework for Innovation', *European Innovation*, July 2005 (European Commission), p. 32.

18. Watch what you export: the history of medical exceptions from patentability

Tina Piper

INTRODUCTION

The analysis and implementation of intellectual property rights (IPRs) are often based on the assumption that IPRs will work similarly in different countries. For example, stated generally, patent protection which incentivizes innovation in the United States will spur innovation in the developing world. The export and implementation of IPRs may further rely on the notion that they are critical to the regulation of a particular industry or technology. Without the particular provision, useless or overbroad patents or no patents at all would be granted, unnecessarily blocking innovation and its commercialization.

Both propositions rely on a common foundation. They presume that it is understood how and why particular IPRs function in the domestic regime. In fact, this may not often be the case. The patent law of the dominant law exporters (United States, United Kingdom, Europe) developed in response to local economic, political and social conditions. In addition, often limited work has been undertaken to understand the effect of particular provisions of the patent law under different political, social and economic conditions.

This chapter presents a brief case-study of IPRs related to medical methods. Medical methods are an important exception to the general patentability of biotechnology subject matter. Medical methods are excluded from patentability by the patent laws of over 80 countries. Their exclusion is a common, almost automatic, feature in the harmonized, globalized intellectual property patent law world.[1] While the medical exception purports to protect valuable, life-saving work from commercial interference through an exclusion from patentability, it fails to meet this goal. The reason for this failure can be traced, historically, to the development of the legal and medical professions – in fact a fight over the 'jurisdiction' to regulate medical technologies between lawyers and doctors. These roots have made it difficult to both justify the exception and implement its promise within the broader context of healthcare regulation. The plural, contextual development of the medical methods exception has serious implications for the export and harmonization of patent law. From this analysis, I impressionistically delineate four indicators which may help determine how an IPR will function when exported to a foreign jurisdiction: the existence and involvement of professional or special interest groups; the role of events unique to a country's history that lead to changes in the law; the type and stage of industrial development and finally the stage of development of complementary regulatory regimes. This chapter will conclude by suggesting that a contextual study of these indicators should be a pre-condition to requiring their import into domestic law.

Protection of Medical Methods

A typical statutory provision for the exclusion of medical and diagnostic methods is found in the European Patenting Convention (EPC), which holds under the heading 'Exceptions to Patentability' that:

> 53. European patents shall not be granted in respect of: . . .
> (c) methods for treatment of the human or animal body by surgery or therapy and diagnostic methods practised on the human or animal body; this provision shall not apply to products, in particular substances or compositions, for use in any of these methods.[2]

At its most basic level, a medical method is a procedure performed by a medical professional on a patient. However, this varies by country and national patent system.

Defining the medical exception: a European example

For the purposes of this analysis I will focus on the European medical diagnostic methods exclusion as an example of how the medical exception operates more generally. The UK[3] and Europe both exclude medical and diagnostic methods from patentability. The diagnostic exception is generally construed narrowly which consequently provides minimal protection.[4] In *Thomson-Csf/Tomodensitometry (No 2)*[5] the Technical Board of Appeal held that a diagnostic method cannot comprise merely a method of analysis, but must also include a concrete diagnostic result.[6] Thus it must encompass *all* the steps involved in making a medical diagnosis.[7] Methods which provide interim results or which are mere data-gathering are not diagnostic methods even if they can be used in making a diagnosis.[8] Thus a mere investigation into the state of the body, for example by measuring blood pressure, would not be enough to qualify as a diagnostic method. The condition measured must itself indicate the problem.[9] A 'diagnostic method' must be performed *on* or *in* the human or animal body, but does not apply to substances removed from the body.[10] Thus these legal rulings leave a very narrow band of medical technologies which may be excluded from the patent law. Some even argue that in effect the exception no longer exists.[11]

The Promise and its Problems

The exclusion of medical and diagnostic methods of treatment is based on a promise which has been articulated by judges and academic commentators: the 'life-saving' work of a doctor should not be hampered by the exercise of patent rights[12] or, as Jacob J noted in relation to medical methods generally, '[t]he purpose of the limitation is . . . merely to keep patent law from interfering directly with what the doctor actually does to the patient'.[13] This exciting promise is very topical in light of current conflicts between intellectual property regimes, human rights and access to health care goods.

However, although courts and legislatures promise to protect doctors' valuable, life-saving work, the promise has remained unfulfilled in three ways. First, while the medical exception exists on paper in case law and statute, in practice the substantive legal protection it provides is emaciated[14] or, as some argue, non-existent.[15] As the discussion of European case law demonstrates, the legal definition of diagnostic methods does not reflect the true nature of a medical diagnosis. Modern diagnoses are rarely final, and few occur without the aid of data and quantitative results from laboratory testing. Moreover, the legitimacy of the patent system is undermined when there is limited substantive protection of what the lay citizen would comprehend as a valuable health care good that ought to be exempt from the patent monopoly. For example, lay observers have brought oppositions challenging the patents over genetic diagnostic tests owned by Myriad Genetics over the BRCA1 and BRCA2 genes. These oppositions ask, among other grounds, why the exception has not stepped in to exclude inventions which clearly impede public access to healthcare.[16] In May 2005 the EPO revoked one of Myriad's BRCA1 patents; however Myriad has appealed this decision.[17]

Second, judges, lawyers and commentators[18] have questioned the legitimacy of the exception since the 1970s;[19] in fact, a survey of the literature on this topic has not found a single commentator who supports the continued existence of the exception in its present form. This questioning has been brought to its logical conclusion in Australia, where the exception has been abolished altogether. Though quelled somewhat in Europe by the passage of legislation, the skepticism may be passively expressed through the progressive hollowing-out of the exception. This question merits serious attention in the harmonized, globalized intellectual property patent law world where including a provision excluding medical methods is almost automatic for countries seeking to legislate IP protection at acceptable levels for international trade (that is, TRIPS).[20]

Third, what protection of medical methods does exist shows little evidence of a principled legal approach with objectively justifiable outcomes. The distinctions drawn in law appears to many to be arbitrary, fruitless and riven with contradiction, as judges since the 1970s have observed.[21] An example is the requirement that diagnosis must be performed *on or in* the body. This means that a diagnostic method of testing for allergies performed on the skin will be excluded from patentability, whereas a blood test to the same end will be patentable. It is unclear why apparatus and drugs are patentable but medical and diagnostic methods are not, particularly when both can be applied in or on the body.[22] Further concerns are raised as to why doctors are given blanket protection from patents in the course of their 'life-saving' work, while other professions and trades the work of which is arguably as important to preserving life are not treated similarly. Examples are fire fighters and emergency workers, paraprofessional health workers, municipal water suppliers, sanitation and other public authorities.

What History Tells Us

This chapter briefly recounts a historical moment that led to the creation of the medical exception to patentability. The historical approach enjoys broad support from intellectual property theorists as a key tool for understanding the present scope and future application of intellectual property rights.[23] It is explained best by Sherman and Bently who state that

> working from the basis that the past and the present are intimately linked, we believe that many aspects of modern intellectual property law can only be understood through the lens of the past . . . Paradoxically, the more the past is neglected, the more control it is able to wield over the future.[24]

The exclusion of medical methods is no exception. This chapter will outline and analyse the modern aspects of intellectual property law through the lens of the past to understand how the exception entered into the law eventually to become a global norm of patent law. In so doing it will explain the disjunction between the supposed 'promise' of the medical exception and the true reason for its existence. This incongruity has led to some of the current dissatisfaction with the exception and its inconsistent results. The historical analysis focuses on the UK as the birthplace of the medical methods exception.

HISTORY OF PATENTS AND MEDICINE

Medicine in the Patent System

Patent law is not promulgated in a vacuum. Its terms reflect a political, social and economic compromise. This section will first consider the early interactions of medicine with the patent system, primarily through patent medicines, and then explore how industrialization spurred medical professionalization which led doctors to create their own parallel 'patent' regime that excluded regulation by lawyers.

Some of the first medical remedies in England were patented in the Tudor and Stuart periods. In 1620 John Dickson obtained a patent for a 'commodious apparatus' used to relieve sick people who were 'distempered or troubled with heate of theire backes through continual keeping or lying on theire beddes'.[25] By the 18th century patents had grown increasingly popular due to the perception that the patent was no longer a royal favour but an instrument of protection.[26]

Patenting activity grew over the next century as an instrument of protection and accreditation in the market; a key example of this is patent medicines. Patent medicines also demonstrate the growing importance of lawyers in the regulation and constitution of medical economic activity. Patent medicines, which were also known as 'proprietary remedies', 'secret remedies' and 'quack medicines', were exploited and vindicated through a range of means in addition to the patent law, such as trademark, breach of confidence, passing off and advertising.[27]

Patent medicines accounted for 22 per cent of the patents granted between 1740 and 1760. 'No other industry ever dominated the system to such an extent.'[28] As MacLeod shows in her study, patentees in the professions (which were 'almost entirely apothecaries, surgeons and "doctors" patenting proprietary medicines') accounted for 13.5 per cent of patents from 1660 to 1699, 16.4 per cent from 1700 to 1749 and 11.5 per cent from 1750 to 1799.[29] The number of patent medicines and 'surgical aids' was well out of proportion to their economic significance.[30] Many patent medicines were dangerous and some proved fatal.

Patents played an expansive role in the early internal regulation of medical treatment as opposed to objectively rewarding the protection of invention. The patent functioned as a means of gaining a competitive edge and publicity, by establishing the brand as original, genuine and government approved,[31] much in the same way that trademarks function in the present market.[32] Importantly, the growth of patent medicines placed the locus of control of publicity, reputation and commercialization of medical practitioners in the hands of lawyers. Doctors or apothecaries often built up their practices through attaching their name to popular patent medicines, but their 'reputation with [their] peers frequently suffered in inverse correlation with . . . popular success'.[33] Not surprisingly, the Royal College of Physicians (RCP) indicated a general hostility to patenting.[34]

The excessive patenting of dubious proprietary medicines amplified demands for stricter scrutiny of patent applications and legislative reform of the patent system in the 1800s.[35] Importantly, the unregulated growth of dangerous patent medicines adversely affected the reputation of medical practitioners and increased the profile and income of 'quacks'.[36] In part this resulted from a patents registration system managed by lawyers with no input from relevant practitioner associations.[37]

The effect of the patent medicines crisis was long term: it chilled the demand for seemingly 'tainted' patents from pharmaceutical companies and other inventors, until the mid-1930s by some accounts.[38] Members of the medical profession repeatedly confirmed that patent medicines played a large role in the strident 'ethical' objections of the medical profession to medical patents.[39]

The patent system, functioning as an accreditation or 'trade mark' of quality for patent medicines, allowed the legal profession and the Crown to judge the stature of a medical practitioner. However, this stamp of approval often had no bearing on the safety of the medical product to which it was attached. Thus the patent system gave control of medical practitioners to lawyers and government. The patent system also created public distrust due to its role in the marketing of unsafe products with the names of doctors attached to the remedy. For good reason the patent system was not well-regarded by credible medical practitioners who wished to institutionalize a reputable profession.

The Patent Law: Too Small for Two Professions

The previous discussion described how patents supported proprietary medicines, giving lawyers an important role in medicine in the 18th and 19th centuries. The next section will consider how excluding those same lawyers was important to the professionalization of medicine. Thus the medical exception reflected this jurisdictional struggle between doctors and lawyers over regulating new medical technologies. The medical methods exception was first judicially articulated in 1911 in *Re C&W's Application.*[40] A formal medical methods exception was not introduced into UK statutory patent law until the Patents Act 1977.[41] This historical study focuses on the moment between the late 1800s and early 1900s when the medical exception first came to be recognized by law and medicine alike. A broader study of the evolution and recent history of the exception is outside the scope of this study.

The reasons to professionalize medicine

As the patent law developed to regulate new inventors, including doctor-inventors, the practice of medicine faced unique problems which limited its income generating potential and market power. These problems included a rigidly hierarchical workforce, a large body of low wage practitioners, some with very doubtful training and practices, no formal body of knowledge, practice, training or educational standards and, by the 18th century, no regulation to exclude unqualified practitioners. Further, the law and the state were not supportive of doctors' initial efforts to self-regulate their trade.

The medical profession in the past was traditionally divided into physicians, surgeons and apothecaries, who belonged to separate colleges (or guilds). Physicians stood at the top of the social and medical hierarchy, and apothecaries at the bottom, with surgeons in the middle.[42] The law regarded medicine practiced by some as an art and not as a trade. It prohibited physicians from suing for their fees (just as low-ranking attorneys could sue, but barristers could not).[43] It regarded the services of physicians, but not surgeons or apothecaries, as purely philanthropic.[44] As Starr relates, 'to be placed outside the market was an honour in an aristocratic culture but a penalty in a democratic and commercial one'.[45] As surgeons and apothecaries attempted to improve their position in the hierarchy, they distanced themselves from their 'guild' status, which included patents.

There is evidence that these attempts at improving the credibility of medicine were not supported by all, particularly by those in power and by members of the other key profession, law. For example, in *Dr Bonham's Case,* Coke CJ held that the court could not convict someone of malpractice merely because they were unlicensed.[46] Judges held that the principle of *caveat emptor* should apply to common medical practice as much as to any other field of the provision of services.[47] The medical profession did not gain state support until it had proved it was of some benefit to the public interest.

During the 18th and early 19th century the number of doctors kept increasing, as did the demand for affordable medical services.[48] By the end of the 19th century only a small proportion could become full-fledged physicians, a position which was either purchased or acquired through nepotism.[49] Thus, the profession was divided amongst the few who earned a high salary, a reasonable number of GPs with a moderate salary and a large number of very poorly paid doctors who 'engaged in competitive undercutting, fee-splitting, canvassing for patients and other unethical practices'.[50] As documented by many commentators of the period, there was no standardized and broadly accepted body of knowledge or practice called medicine. Training was largely voluntary and asystematic.[51] As well, as Porter has noted, practitioners in the 1830s 'were up to the chin in quackish practices'.[52] There is no evidence that this nascent medical profession had any interest or engaged in any debates over a medical exclusion from patentability.

The growth of the medical profession

As we have seen, the medical market prior to proper professionalization was extremely competitive with unlimited demand for healthcare. A monopoly was easily organizable since consumers were disaggregated by the private nature of the transaction. Medicine would be attractive to state regulation if it proved itself effective, since it provided a community service in the best interests of citizens.[53] Doctors and their representative bodies saw this opportunity and realized that a professional monopoly would best secure the benefits of this market, instead of allowing doctors to be controlled by the unregulated market.[54] Professionalization, with its promise of increased security and income to bourgeois practitioners without patrons, was supported in an occupation where the vast majority of practitioners were poorly paid and regarded.[55]

Key to professionalization was establishing control of both medical practitioners and medical sub-trades. This was achieved through the development of codes of ethics, professional associations, standardization of education, knowledge, practice and training, and licensing. Control increased the quality of the medical commodity, which then justified state support and protection. However professionalization was to be achieved independently of the IP system. In fact, IP law and its guardianship by lawyers was often antithetical to the needs of the emerging profession, which helps explain the exclusion of medical methods of treatment from the patenting regime.

Controlling the medical commodity. To secure the benefits of this market it was necessary to present it with a distinctive and consistent product, a strategy recognized centuries earlier by the craft guilds.[56] The distinctive commodity here was the doctor whose 'production' had to be controlled, in contrast to the existing chaotic market for medical services.[57] Standardizing the product stimulated the market (by increasing demand) but also limited supply (by increasing the qualifications required of doctors to enter the market), which increased the profit potential for doctors.[58] Standardization and increased effectiveness justified state support in the creation and protection of the professional monopoly. As with the earlier guilds the medical profession would be supplying a public good that the state did not provide. The medical profession went about standardizing the product, and increasing the authority and power of the medical doctor[59] by first of all standardizing the content of knowledge (through a scientific education) and the requisite practical training to become a doctor.[60] Best medical practice was promoted through the creation of peer-reviewed journals[61] which allowed doctors to evaluate, approve or reject new medical developments. The growth of scientific medical knowledge and its 'cognitive exclusiveness' to medicine allowed boundaries to be built around newly developed technologies.[62] Technologies also helped standardize the services doctors provided.[63]

Formal medical education was standardized by GMC guidelines after 1858 and unqualified apprenticeship eliminated in 1892.[64] By 1858 there were 20 licensing bodies and 61 occupational qualifications which could be registered under the Medical Act; two of the most common and prestigious licences could be obtained upon examination from the RCS (after 1800) and the Society of Apothecaries (after 1815).[65] In this way the medical profession controlled and limited membership much as earlier guilds had done.

Key to standardizing the product, asserting medical authority and conveying legitimacy was the need to create consensus, loyalty and collegiality within the profession in the early and middle 19th century.[66] Loyalty and consensus were critical to controlling colleagues and mobilizing the power of collective action.[67] The unwillingness of practitioners to criticize one another reinforced the correctness of medical judgement.[68] A key tool was the widespread adoption of ethical codes, such as Percival's *Medical Ethics*[69] and the Hippocratic oath. Codified prohibitions on price competition, advertising and bargaining reinforced internal consensus.[70] Further, the introduction of fee schedules in the 1870s limited competition between doctors,[71] and in 1911 the National Health Insurance (NHI) scheme provided a guaranteed fee per patient.[72]

Codes and etiquette also marginalized competitors and built public trust in the early to mid-19th century by foregrounding the profession's reliability.[73] As a consulting profession, medicine had to ensure its external legitimacy to the public to justify its privilege.[74] Business morality was recast as specialized 'professional ethics',[75] distancing medicine from the lowest standards of market morality, for example, *caveat emptor*.[76] Codes and etiquette frowned on connections with pharmaceutical firms, money-making medical museums and the sale or endorsement of patent medicines.[77] The GMC was able to remove practitioners 'for infamous conduct or after conviction for a felony'.[78] Codes also controlled information available to the public, ensuring a sense of mystery and respect for the profession.[79] The medical profession played up the fact that medicine served the most important 'sacred' commodity of life.[80] Finally, improved results, primarily those wrought by improved sanitation and antibiotics enhanced public support.

Controlling the sub-trades. In addition to manufacturing internal consensus and broader societal legitimacy, doctors also had to control medical sub-trades to validate their claim of sole superior medical expertise in the early to mid-19th century.[81] This they did first through the state and second through technology.

Critical to gaining state support was persuading the public and government that the medical profession had a collective service orientation and thus trust it to be autonomous and self-regulating.[82] Freidson, in particular, argues that there is no evidence that the medical profession has a greater service orientation than other professions, merely that it has been the most effective at persuading people that it does.[83]

State regulation was used to exclude unqualified practitioners from the medical practice monopoly.[84] The state grant of a patent monopoly did not play a role in the medical profession's search for professional control, largely because it had been overwhelmed by disreputable patent medicines and quacks and would have undermined, not strengthened, professional authority.[85] Practically, this professionalization project was implemented through the British Medical Association (BMA) founded in 1832,[86] an organization with a range of members, from surgeons to hospital physicians and GPs and with a goal of engaging in scientific discourse, in notable contrast to the RCP.[87] The BMA became very involved in government,[88] for example, lobbying unsuccessfully for dedicated seats for medical doctors in Parliament.[89]

In 1858 the BMA supported the passage of the Medical Act,[90] which established the General Medical Council (GMC) and the Medical Register, which listed qualified practitioners. The Medical Act 1858 made it an offence for the unqualified to claim to be qualified. Further, the GMC was given the power to supervise education, sue for fees and provide an authorized pharmacopoeia (giving doctors control over pharmacy).[91] It also made the divisions between the RCP, RCS and Society of Apothecaries irrelevant.[92]

In addition to licensing and even assimilating important sub-trades, the medical profession attained control over its sub-trades by making the physician the only person with ultimate control over privileges, benefits and tools that others needed to operate, for example, access to hospitals and the power of prescription denied to midwives, chiropractors and osteopaths.[93] Paramedics were controlled by requiring that equipment or treatments could only be performed under the request and supervision of a physician, for example, X-rays, CAT-scans, laboratory tests.[94] Thus only physicians had simultaneous access to technology and the market.[95] In addition, much of the training and work conducted by the sub-trades had to be conceived of and approved by doctors. Thus the sub-trades depended on doctors,[96] who retained the power to direct their activities.[97]

As the profession eliminated competitors, in the face of a universal and disorganized clientèle, the institutionalization and standardization of the production of producers, and an affinity with dominant institutions (such as the government), the market for medical services became a monopoly.[98] Improved technologies and the growth of scientific medicine, the products of which could easily be patented, played a key role in the transformation of medicine and the entrenchment of the medical profession's status.[99]

Damaged by association: the medical profession and the patent law

Patents for invention were now under the effective guardianship of a growing cadre of lawyers and patent agents. Key to the professionalization of medicine just described was the exclusion of intellectual property law and lawyers. The consolidating medical profession had six major reasons to exclude the patent law.[100]

First, control over practitioners could not be achieved while lawyers and patent agents were still regulating practitioners by 'accrediting' their products through patents, as the example of patent medicines demonstrates. Patentees were using the Crown grant of a patent monopoly as evidence of royal endorsement of their product; patent lawyers and agents did nothing to dispel this misunderstanding. Further, rightly or wrongly, the patent system was perceived to play an important role in regulating the safety of medicines and setting standards for new technologies. Patents, however, were not granted based on efficacy, reliability or safety. Lawyers thus unintentionally undermined the medical profession's ability to control and standardize the medical commodity, which eroded trust and public support. A leading article in the *Lancet* on the well-known case of *Morison v Moat*[101] confirms this perspective, highlighting the dangers of unsafe patent medicines as compared to mechanical inventions and the patent lawyers' failure to consult with the relevant medical societies about the quality of the medicines in issue.[102]

Further evidence that the legal system undermined the medical profession's ability to control and standardize the medical commodity is given by the holding in *Clark v Freeman*.[103] Sir James Clark tried to sue the defendant apothecary Freeman in passing-off for selling an anti-consumption patent medicine, which contained both mercury and antimony, using both his name[104] and his purported endorsement. Sir James claimed that the 'injurious' pills undermined public and professional confidence in his professional ability and thus his professional income.[105] The court found that it could not grant an injunction to prevent the use of Sir James's name, in particular because he himself was not in the business of manufacturing and selling pills.

Second, doctors had framed medicine as a non-economic profession in the public interest subject to professional ethics separate from the regulation of the trades. The patent system required an industrial application, useful purpose or vendibility of an invention and, as discussed above, it often helped to establish a doctor's business. Contemporary documents were preoccupied with ensuring that patents were not used to publicize a doctor's practice and that patenting medical technologies would not lead to conflicts of interest. Patenting undermined public confidence and the autonomy granted to the medical profession by questioning its service orientation.[106] The most influential indication of this is the resolution passed by the BMAs Central Ethical Committee in 1903 (and 1920), holding:

> [t]hat it is contrary to the ethics of the medical profession to attempt to secure a monopoly in the sale of any article used in the treatment of disease, and especially by patenting any such article in the name of a medical practitioner whose name would necessarily be used in its advertisement.[107]

The BMJ frequently elaborated upon this injunction in response to specific questions by readers.[108] However, the injunction against advertising did not apply 'where the sale would be only to professional men or other practitioners'.[109]

Third, professional medical knowledge was constructed through public lectures, books, scientific journals, conferences and discussions between colleagues. Incentives to advance medical knowledge were adequately provided for by colleague approval. The medical profession rejected secrecy among professionals,[110] and even went to court to enforce these norms.[111] A general professional obligation existed to publicize and disseminate discoveries to the profession through communications fora.[112]

While attempting to create a monopoly on medical treatment, doctors vehemently opposed the monopolization of knowledge and techniques within the profession.[113] Monopolies perpetuated internal stratification (that is, between physicians and other practitioners) which the profession was trying to eliminate in favour of collegiality.[114] The medical profession's aversion to monopoly (other than its own beneficent one), as well as the emphasis on collegiality and group learning, provided further justification for rejecting patents. Most inventions were the product of a collaborative effort, with no one inventor.[115] Granting a patent to one inventor undermined collegiality and teamwork.[116]

Allowing examiners in the PO to usurp or even play a role in defining what constituted medical knowledge surrendered professional control over 'knowledge' and thus the power to self-regulate. During the early 20th century the BMA even proposed schemes to give doctors power of patent grant, including a review by doctors of medical patent applications[117] and a Medical Trustee to oversee a system of 'dedicated patents'.[118]

Fourth, surrendering control over technologies and methods to the PO undermined the profession's control of its sub-trades and para-professions.[119] Exempting medical inventions from the patent regime emphasized the control of doctors (not lawyers) in determining who should use and exploit inventions.

Fifth, the patent system encouraged a public contest in the courts which undermined collegiality and sharing critical to the creation of the profession, public trust and state support. The medical profession was ill-disposed to the public resolution of disputes between doctors in the courts of law, commenting that '[t]here is nothing more lamentable and nothing more reprehensible than unseemly quarrels amongst medical practitioners'.[120] An earlier issue of the *Lancet* pleads that 'there is one class of disputes which should always be kept from public observation: we allude to the personal differences of medical practitioners on professional matters'.[121] Advertising (particularly through patents) was similarly enjoined for undermining collegiality and lowering public esteem of the profession. Advertising was trade, and trade meant profiting and putting one's interests ahead of others in his or her trade.

Finally, the medical monopoly supplanted a need for the patent monopoly. If its terms (fraternity, trust, lengthy training, etc.) were accepted then the rewards of a comfortable, stable income could be realized. The medical profession itself acknowledged this later in the early 20th century when it addressed limited patent reforms which would allow patenting of some medical substances by medical workers not involved in clinical practice. The BMA acknowledged that 'there is no sound ethical distinction between the direct remuneration of the medical research worker by way of royalties under the Patent Laws and that of the clinician by patients' fees'.[122]

CONCLUSIONS

Patents which granted proprietary control over new technologies grew to encroach on medicine in the 18th and 19th centuries by effectively publicizing and accrediting practitioners. When medicine began aggressively to professionalize in the mid-19th century, it had to extract itself from the rule of lawyers in order to assert the control over its work which would identify it as a profession. Control was achieved by creating a parallel system to the patent regime which led to a medical exception in the common law.

Medicine became a state supported monopoly; inventions were evaluated as to merit in collegial fora such as conferences, journals, hospital teaching and public lectures. Incentives to innovate were provided by peer approval and the importance of reputation, reinforced by the ethical codes and collegiality. Disclosure and dissemination were achieved by teaching, conferences, journals and ethical prohibitions of secrecy within the medical profession. The lucrative reward of a professional income was guaranteed.

The judiciary and then the state explicitly exempted medical methods from the patent regime. Thus excluding regulation by the patent regime was critical to medical professionalization. First, situating medicine outside the law helped convince the public and government that medicine sat apart from lapsed trade morality. Second, excluding patents removed lawyers, which facilitated consolidation of control over professionalization. Third, excluding IP allowed *internal* quality control of the medical commodity, critical to standardizing medicine and generating public and state trust. Finally, excluding the patent regime discouraged public intellectual property fights in the courts and encouraged cohesiveness and collegiality.

This historical analysis of the medical exception demonstrates that its promise might accurately be restated from 'patents will not hamper the valuable, life-saving work of doctors' to 'patents will not hamper doctors' pursuit of a valuable income'. Reframing the purpose of the exception thus may explain some of the present dissatisfaction and inconsistent results that judicial elaboration of the exception has engendered. As the product of a regulatory contest, the medical exception lacks a convincing and enduring guiding rationale that would ensure its healthy growth over time. The exception is unsuccessful at regulating healthcare as a whole and balancing competing healthcare priorities. It is based on a late 19th century vision of medical care provided solely by qualified medical practitioners, outside the market, with a limited role for technology. This does not accurately convey the modern reality of medicine, which is ever more reliant on medical para-professionals, privatized schemes, and highly evolved biotech, instrumentation and medical research industries.

The problems created by the medical exception illustrate four major perils inherent in exporting the medical methods exception to other jurisdictions. The medical exception did not arise from a Law Commission inquiry into how best to regulate healthcare. It evolved as, first, professional groups and, second, industries responded to the pressures and opportunities of industrialization in the UK. Different stages of development may require different levels of patent protection that should flexibly adjust to current conditions.[123] Third, the patent medicines crisis played a unique role in the creation of the medical exception suggesting that the law may be created to respond to unique local events which might not have occurred in countries to which law is exported. Finally, the creation of the medical exception responded to a lack of appropriate health and safety legislation over medical products. Where appropriate regulations currently exist, a medical exception may be unnecessary or unjustifiable. A contextual study of these indicators should be a pre-condition to import provisions into domestic law.

NOTES

1. Eg Article 27(3)(a) of the Agreement on Trade-Related Aspects of Intellectual Property Rights (TRIPS) (adopted 15 April 1994), (1994), 33 ILM 81; North American Free Trade Agreement (NAFTA), Article 1709(3)(a) (opened for signature 17 December 1992) (1994) 32 ILM 289.
2. Art 53(c).
3. Patents Act 1977 (UK) c 37 s 4(2).
4. *Eisai/Second Medical Indication* [1979–85] EPOR B241 (Enlarged Bd App); *Bruker* [1988] EPOR 357 (Tech Bd App). This conclusion was recently confirmed by an opinion of the Enlarged Board of Appeal G0001/04 ('Opinion').
5. [1979–85] EPOR C917 (Tech Bd App). The patent application was for a method of tomodensitometry.
6. This was confirmed in *Philips/Diagnostic Method* [1979–85] EPOR C937 (Tech Bd App); *Siemens* [1988] EPOR 365 (Tech Bd App); *Thomson-Csf/Tomodensitometry (No 1)* [1979–85] EPOR C763 (Tech Bd App). Note, however, the contrary holding in *Baxter* where the Board found that a method of blood extraction could be unpatentable if it constituted one step of a diagnostic method (note not the final result), eg, blood analysis for determining the cause of disease: *Baxter* [1998] EPOR 363 (Tech Bd App) 367.
7. E.g. *Ultrafem* [2002] EPOR 35 (Tech Bd App) where a female vaginal discharge collector was not considered a method of diagnosis since it did not produce results that made it directly possible to decide on a particular course of treatment; Opinion, n. 4 above.
8. *Bruker*, n. 4 above.
9. Ibid.
10. EPO Guidelines ch IV 4.3.
11. R. Moufang, 'Methods of Medical Treatment Under Patent Law' (1993) 24(1) IIC 18, 47. The requirements of the case law have been fully incorporated into the EPO Guidelines ch IV 4.3.
12. *Georgetown University* [2001] EPOR 21 (Tech Bd App); *Wellcome/Pigs I* [1988] EPOR 1 (Tech Bd App) [3.7]; *Cygnus* (11) [13]; *Bruker*, n. 4 above.

13. *Bristol-Myers Squibb Co v. Baker Norton Pharmaceuticals Inc* [1999] RPC 253 (Pat Ct), aff'd in part [2001] RPC 1 (CA).
14. This is further elaborated by Moufang, n. 11 above 47.
15. This is further elaborated ibid., 47; W. Cornish, *Intellectual Property: Omnipresent, Distracting, Irrelevant?* (Oxford: OUP, 2004) 11.
16. Institut Curie, 'European-Wide Opposition Against the Breast Cancer Patents', Press Release (26 September 2002).
17. Wellcome Trust available at http://www.wellcome.ac.uk/en/genome/geneticsandsociety/hg14n020.html (5 June 2004).
18. O. Mitnovetski and D. Nicol 'Are patents for methods of medical treatment contrary to public order and morality or "generally inconvenient"', *Journal of Medical Ethics* 30, 470–75. 2; D. Thums 'Patent Protection for Medical Treatment – A Distinction Between Patent and Medical Law' (1996) 27(4) *International Review of Industrial Property and Copyright Law* 423.
19. *Eli Lilly & Co's Application* [1975] RPC 438, 445; *Re Schering AG's Application* [1971] 3 All ER 177 (PAT) 181; *Re Dow Corning Corporation (Bennett's) Applications* [1974] RPC 235 (Pat App Trib).
20. Eg Article 27(3)(a) of TRIPS, n. 1 above; NAFTA, n. 1 above, Article 1709 (3)(a).
21. In *Eli Lilly & Co's Application* [1975] RPC 438, 445 the Patents Appeal Tribunal held that it has been long established that the medical exception is based on ethics, not logic; this was most recently reinforced in a UK by Jacob J in *Bristol-Myers Squibb Co v Baker Norton Pharmaceuticals Inc* [1999] RPC 253 (Pat Ct), aff'd in part [2001] RPC 1 (CA): '[t]he thinking behind the exception is not particularly rational: if one accepts that a patent monopoly is a fair price to pay for the extra research incentive, then there is no reason to suppose that that would not apply also to methods of treatment'. See also 'Grounds given with regard to the ratification of international patent conventions' R/1181/74, p 18.
22. *Anaesthetic Supplies Pty Ltd v Rescare Ltd* (1994) 50 FCR 1 (NSW Gen Div) 6; *Bristol-Myers Squibb Co v F H Faulding & Co Ltd* (2000) 97 FCR 524.
23. E.g. C. Macleod, *Inventing the Industrial Revolution: the English Patent System, 1660–1800* (Cambridge: CUP, 1988); B. Sherman and L. Bently, *The Making of Modern Intellectual Property Law: The British Experience, 1760–1911* (Cambridge: CUP, 1999).
24. Sherman and Bently, n. 23 above, 2.
25. British Medical Association. 1873. 'Early Medical Patents', *British Medical Journal* 2, 440.
26. Macleod, n. 23 above, 52.
27. E.g. the classic contract case *Carlill v Carbolic Smoke Ball Co* [1893] 1 QB 256 (CA) which enforced an offer made in an advertisement for 'The Carbolic Smoke Ball', a medical preparation which purported to prevent influenza.
28. MacLeod n. 23 above, 154.
29. Ibid., 134–135. As MacLeod highlights, 'Their apogee came in the 1750's, when they obtained 25 per cent of all patents, before settling down to 8 per cent for the remainder of the eighteenth century': at 134. H.I. Dutton, *The Patent System and Inventive Activity During the Industrial Revolution 1750–1852* (Manchester: Manchester University Press, 1984) 206 highlights that the process of 'medical surgery' constituted 1.68% of patent activity between 1750 and 1851. However no indication is provided as to what type of activity is encompassed by the term 'medical activity'.
30. Ibid., 97.
31. British Medical Association. 1886. 'The Patent Medicine Delusion', *British Medical Journal* 1, 31–2; R. Porter, *Health for Sale: Quackery in England 1660–1850* (Manchester: Manchester University Press 1989) 28.
32. MacLeod, n. 23 above, 85. Other means of enhancing the respectability of a product were to gain the sanction of a member of the Royal Family or, failing that, the sanction of the RCP: ibid., 86.

33. Ibid., 86.
34. Sir G. Clark, *A History of the Royal College of Physicians* (Oxford: OUP, 1966), 1, 235, 335–336.
35. MacLeod, n. 23 above, 190.
36. Porter, n. 31 above.
37. MacLeod, n. 23 above, 45.
38. J. Liebenau, *Medical Science and Medical Industry* (London: MacMillan Press, 1987) 35.
39. BMA Central Ethical Committee Minutes 1929–30, index 40, Subcommittee re the Ethics of Remuneration and Reward for Research and Inventions, 'Memorandum by Dr Myer Coplans' 1; see also Central Ethical and Science Committee Minutes 1930–31, index 22, Research and Invention Subcommittee, Tuesday, 24 February 1931.
40. (1914) 31 RPC 235.
41. (UK).
42. R. Porter, *Health for Sale: Quackery in England 1660–1850* (Manchester: Manchester University Press 1989) 26.
43. P. Starr, *The Social Transformation of American Medicine* (New York: Basic Books, 1982) 61.
44. E.g. *Chorley v Bolcot* (1791) 4 TR 317, 100 ER 1040; *Lipscombe v Holmes* (1810) 2 Camp 441, 170 ER 1211.
45. Starr, n. 43 above, 62.
46. (1610) 8 Co Rep 107a, 118a; 77 ER 638, 652.
47. Porter, n. 42 above.
48. M. Sarfatti-Larson, *The Rise of Professionalism: A Sociological Analysis* (Berkeley, CA: University of California Press, 1977) 14.
49. J. Jacob, *Doctors and Rules: A Sociology of Professional Values* (2nd edn New Brunswick, NJ: Transaction Publishers, 1999) 102; Larson, n. 48 above, 12, 88.
50. N. Parry and J. Parry *The Rise of the Medical Profession: A Study of Collective Social Mobility* (London: Croom Helm, 1976) 133.
51. Larson, n. 48 above, 90.
52. Porter, n. 42 above, 226.
53. Larson, n. 48 above, 21–23.
54. Ibid., 9 who draws a comparison between the medical profession and the craft guilds established in 11th century Europe which wished to take advantage of new methods of social production.
55. Ibid., 222.
56. Ibid., 13.
57. Ibid.
58. Starr, n. 43 above, 25.
59. E. Freidson, *Profession of Medicine: A Study of the Sociology of Applied Knowledge* (Chicago, IL: University of Chicago Press, 1988).
60. Ibid.; Larson, n. 48 above, 36.
61. Such as the *Lancet* established in 1823 and the *Provincial Medical and Surgical Journal,* which later became the *British Medical Journal* (BMJ).
62. Larson, n. 48 above, 17.
63. Ibid., 37.
64. A. Digby, *The Evolution of British General Practice, 1850–1948* (Oxford: OUP 1999) 54.
65. Ibid., 50.
66. Starr, n. 43 above, 80; Jacob, n. 49 above.
67. Larson, n. 48 above, 68.
68. Starr, n. 43 above, 94.
69. T. Percival *Medical Ethics* (Manchester: S. Russell, 1803).
70. J. Securis 'A Detection and Querisome of the Daily Enormities and Abuses Committed in Physic' in A. Wear 'Medical Ethics in Early Modern England' in A. Wear, J. Geyer-

Kordesch and R. French (eds), *Doctors and Ethics: The Earlier Historical Setting of Professional Ethics* (Amsterdam: Rodopi, 1993) 100; J. Gregory, *On the Duties and Qualifications of a Physician* (Amsterdam: Rodopi, 2nd edn 1820) originally published in 1770; Starr, n. 43 above, 94; Jacob n. 49 above.

71. Ibid., 100.
72. Ibid., 307–308.
73. Jacob, n. 49 above.
74. Freidson, n. 59 above.
75. Larson, n. 48 above, 63.
76. Starr, n. 43 above, 23.
77. A. Digby, *Making a Medical Living* (Cambridge: CUP 1994) 61; P. Bartrip, 'Secret Remedies, Medical Ethics, and the Finances of the *British Medical Journal*' in R. Baker (ed.), *The Codification of Medical Morality* (Dordrecht: Kluwer 1995) 194.
78. Digby, n. 64 above, 38.
79. Larson, n. 48 above, 41.
80. Ibid., 39.
81. Ibid., 13.
82. Freidson, n. 59 above.
83. Ibid.
84. Larson, n. 48 above, 68.
85. Bartrip, n. 77 above, 192.
86. Though it was called the Provincial Medical and Surgical Association (PMSA) until 1856.
87. British Medical Association 'About the BMA: an Outline History of the British Medical Association', available at http://www.bma.org.uk/ap.nsf/Content/About+the+BMA+-+An+outline+history+of+the+BMA (12 September 2003).
88. *The Lancet* 'BMA, Exeter Hall, January 18, 1842' [1841] 1 Lancet 593. See also: *The Lancet* 'Leading Article' [1852] 2 Lancet 157; *The Lancet* 'Leading Article' [1868] 2 Lancet 117; British Medical Association 'Select Committee on "Patent Medicines"' [1911] 2 BMJ 1672.
89. *The Lancet* 'Leading Article' [1852] 2 Lancet 157.
90. (UK) 21 and 22 Vict c 90.
91. First published in 1864. Digby, n. 64 above, 38.
92. R. Porter *Bodies Politic: Disease, Death and Doctors in Britain, 1650–1900* (Ithaca, NY: Cornell University Press, 2001) 255.
93. Starr, n. 43 above, 223.
94. E. Freidson, *Medical Work in America: Essays on Health Care* (New Haven, CT: Yale University Press 1989) 67.
95. Starr, n. 43 above, 223.
96. Freidson, n. 94 above, 67.
97. Larson, n. 48 above, 38.
98. Ibid.; Freidson, n. 94 above.
99. Starr, n. 43 above, 16.
100. Evidence is found in the documents that express the views of the medical profession: the *Lancet*, the BMJ and confidential committee proceedings from the BMA, RCP and RCS.
101. (1851) 9 Hare 241, (1851) 68 ER 492.
102. *The Lancet* 'Leading Article' [1851] 2 Lancet 586, 586.
103. (1848) 11 Beavan 112, 50 ER 759.
104. They were called 'Sir James Clarke's Consumption Pills'.
105. *Clark*, n. 103 above, 116, 759.
106. British Medical Association 'Patents by Medical Men' [1903] BMJ 1242, 1242.
107. BMA Central Ethical Committee Minutes 1931–32, index 16, Conference on Medical Patents Wednesday April 20th 1932, 'Professional Opinion and Practice in Relation to Medical Research and the Patent Law' 2.

108. British Medical Association 'Medico-Legal and Medico-Ethical: Patent Appliances' [1888] 2 BMJ 158. See also British Medical Association 'Patents by Medical Men' [1903] 1 BMJ 1242.
109. BMA Central Ethical Committee Minutes 1929–30, index 41, Subcommittee re The Ethics of Remuneration and Reward for Research and Inventions, 2 June 1930, 'The Ethics of Remuneration and Reward for Research and Invention, Memorandum' 2.
110. Note that the specification had to disclose the invention where the remedy was 'patented' as opposed to trade marked.
111. *Abernethy v Hutchinson* (1824) 1 H & Tw 28, 47 ER 1313. See commentary in *The Lancet* 'Abernethy v the Lancet in the Court of Chancery' [1824] V Lancet 378.
112. *The Lancet* 'Leading Article' [1852] 2 Lancet 61.
113. *The Lancet* 'Abernethy v the Lancet in the Court of Chancery' [1824] V Lancet 378; *The Lancet* 'London College of Surgeons, Court of Examiners' [1823–24] 2 Lancet 198. See also *The Lancet* 'BMA, Exeter Hall, January 18, 1842' [1841] 1 Lancet 593.
114. This is evident in *The Lancet* 'Bye-Law of the College of Surgeons' [1825] 8 Lancet 344.
115. British Medical Association 'The Patenting of Diptheria Antitoxic Serum' [1898] 2 BMJ 643. See also Central Ethical Committee Minutes 1929–30 indices 5, 10, 12, 22 and 39.
116. BMA Central Ethical Committee Minutes 1929–30, index 4, Research and Inventions Subcommittee Meeting, Wednesday, September 18th 1929 at 1. See also indices 5, 10, 12, 13 and 39.
117. BMA Central Ethical Committee Minutes 1930–31, index 28, Central Ethical and Science Committees: Research and Invention Subcommittee, March 27 1931 'Minutes' 1.
118. BMA Central Ethical Committee Minutes 1929–30, index 21, Report of Research and Inventions Sub-committee, November, 19 1929, 'The Ethics of Remuneration and Reward for Research and Inventions' 4. Neither initiative was ultimately adopted.
119. Starr, n. 43 above, 223.
120. *The Lancet* 'Medical Annotations' [1863] 2 Lancet 72. The case in issue was a complaint brought by a physician against a surgeon in Durham Police Court because the surgeon had failed to fill the physician's prescription.
121. *The Lancet* 'Leading Article' [1863] 1 Lancet 610.
122. BMA Central Ethical Committee Minutes 1931–32, index 12, Conference on Medical Patent, Wednesday, April 20th 1932 'Professional Opinion and Practice in Relation to Medical Research and the Patent Law – Historical Review' 3. See also Central Ethical Committee Minutes 1929–30, index 41, 1.
123. H.J. Chang masterfully elaborates this thesis in *Kicking Away the Ladder* (London: Anthem, 2001).

Index